Advances in STEM Education

Advances in STEM Education is a new book series with a focus on cutting-edge research and knowledge development in science, technology, engineering and mathematics (STEM) education from pre-college through continuing education around the world. It is open to all topics in STEM education, both in and outside of classrooms, including innovative approaches and perspectives in promoting and improving STEM education, and the processes of STEM instruction and teacher education. This series values original contributions that view STEM education either in terms of traditionally defined subject-based education or as an educational undertaking involving inter-connected STEM fields. The series is open to new topics identified and proposed by researchers internationally, and also features volumes from invited contributors and editors. It works closely with the *International Journal of STEM Education* and *Journal for STEM Education Research* to publish volumes on topics of interest identified from the journal publications that call for extensive and in-depth scholarly pursuit.

Researchers interested in submitting a book proposal should contact the Series Editor: Yeping Li (yepingli@tamu.edu) or the Publishing Editor: Melissa James (Melissa.James@springer.com) for further information.

More information about this series at http://www.springer.com/series/13546

Pratim Sengupta • Marie-Claire Shanahan
Beaumie Kim
Editors

Critical, Transdisciplinary and Embodied Approaches in STEM Education

Springer

Editors
Pratim Sengupta
Werklund School of Education
University of Calgary
Calgary, AB, Canada

Marie-Claire Shanahan
Werklund School of Education
University of Calgary
Calgary, AB, Canada

Beaumie Kim
Werklund School of Education
University of Calgary
Calgary, AB, Canada

ISSN 2520-8616 ISSN 2520-8624 (electronic)
Advances in STEM Education
ISBN 978-3-030-29488-5 ISBN 978-3-030-29489-2 (eBook)
https://doi.org/10.1007/978-3-030-29489-2

This Springer imprint is published by the registered company Springer Nature Switzerland AG
The registered company address is: Gewerbestrasse 11, 6330 Cham, Switzerland

Acknowledgments

We would like to thank the International Society for STEM in Education for its support and the University of Calgary's Werklund School of Education for graciously hosting the "STEM as Critical Literacies" symposium, September 27–31, 2017, at the Banff Center for Arts and Creativity in Banff, AB, Canada. Earlier drafts of these chapters were presented at the symposium, which sought to bring together STEM Education researchers from universities across different countries and continents, as well as practitioners (artists, K-12 teachers, and museum designers) who have been collaborating with some of these researchers.

The countries represented include Canada, Australia, USA, and China (in no particular order), while the authors represent the following universities: Université de Montréal, Canada; University of California, Berkeley, USA; York University, Canada; Beijing Normal University, China; Shanghai Normal University, China; University of Sydney, Australia; Simon Fraser University, Canada; Queensland University of Technology, Australia; Brooklyn College, USA; University of Maryland, College Park, USA; The Pennsylvania State University, USA; Harvard University, USA; Stanford University, USA; Vanderbilt University, USA; University of Ontario Institute of Technology, Canada; University of British Columbia, Canada; and University of Calgary, Canada.

In particular, we would like to thank Dr. Dennis Sumara, Dean, and Dr. Sharon Friesen, Vice Dean, Werklund School of Education, University of Calgary, for their leadership and financial support, without which the symposium and the book would not be possible. A special note of thanks also to the Learning Sciences PhD students at the University of Calgary's Werklund School of Education, who provided invaluable logistical support for the symposium: Marilu Lam-Herrera, Zulay Diaz Caceres, Sandra Becker, Chris Ostrowski, and Stefan Rothschuh. Their reflections from this symposium were published as a short paper at the International Conference of the

Learning Sciences, 2018,[1] and they asked some of the most insightful questions during the symposium, which only deepened our collective engagement with issues of power and equity in STEM education.

Finally, the symposium would not have been possible without the invaluable support of the administrative staff in the Dean's office at Werklund School of Education. We express our sincere thanks to Michelle Downey, Cam Smith, Melanie Girard, and Violet Baron, whose silent efforts made everything possible.

The Editors

[1] Ostrowski, C. P., Becker, S., Caceres, Z. D., Lam-Herrera, M., & Rothschuh, S. (2018). Exploring the Margins of the Field: Rethinking STEM in Education. In Kay, J. and Luckin, R. (Eds.) Rethinking Learning in the Digital Age: Making the Learning Sciences Count, 13th International Conference of the Learning Sciences (ICLS) 2018, Volume 2. London, UK: International Society of the Learning Sciences.

Contents

About the Authors

Jennifer D. Adams is a Tier 2 Canada Research Chair of Creativity and STEM and Associate Professor at the University of Calgary where she holds a dual appointment in the Department of Chemistry and Werklund School of Education. Her research focuses on the intersection between creativity and STEM teaching and learning in postsecondary contexts. She has scholarly expertise in STEM teaching and learning in informal science contexts including museums, national parks, and everyday learning. She was awarded a National Science Foundation Early CAREER award to study informal learning contexts and formal/informal collaborations for STEM teacher education. Her research portfolio also includes youth learning and identity in informal science contexts, with a focus on underrepresented youth and place/identity in transnational communities and environmental education. Her work emphasizes critical and sociocultural frameworks and participatory, qualitative, and poststructural approaches. She has also worked as an educator and researcher in NYC public schools and the American Museum of Natural History.

Gabriela Alonso Yanez is an Assistant Professor in the Werklund School of Education at the University of Calgary, Canada. Her research interests include learning dynamics in the context of sustainability research, transdisciplinary collaborations, environmental sociology, and global environmental change.

Laleh Behjat is a professor in the Department of Electrical and Computer Engineering at the University of Calgary, Canada. Her main research focuses on developing mathematical techniques and software tools for automating the design of digital integrated circuits. She is an Associate Editor of the *IEEE Transactions on Computer-Aided Design of Integrated Circuits and Systems* and *Optimization in Engineering* from Springer. Dr. Behjat is passionate about increasing the status of women in science, technology, engineering, and mathematics (STEM). Her team has been collaborating with researchers in arts and education on developing to teach innovation and creativity in post-secondary education. She acted as an academic advisor of Google Technical Development Guide and was a member of the Google's Council on Computer Science Education. Dr. Behjat led the Schulich Engineering

Outreach Team, which received the ASTech Leadership Excellence in Science and Technology Public Awareness Award in 2017.

Alberto Bellocchi is an Associate Professor and is a science education researcher and micro-sociologist who leads the Science Education Emotions Lab at Queensland University of Technology. He studies the complexities of science classroom life in schools and university preservice teacher education settings. His recent work focuses on emotions, social bonds, and science learning experiences in middle school classes. He is the lead editor of the collection entitled *Exploring Emotions, Aesthetics and Wellbeing in Science Education Research* (Springer, 2017) and co-editor of the sociology of emotions collection *Emotions in Late Modernity* (Routledge, 2019). He serves as Lead Editor for *Cultural Studies of Science Education* and as an editorial board member for *Journal of Research in Science Teaching*.

Janette Bobis is Professor of Mathematics Education in the Sydney School of Education and Social Work at the University of Sydney and a member of the STEM Teacher Enrichment Academy. Her research and teaching focus on *teacher learning* in mathematics education, particularly the development of teachers' knowledge, beliefs, and practices; and *student learning*, with a current focus on the development of young students' motivation and engagement in mathematics while undertaking challenging tasks and the impact of teaching interventions on each of these aspects. Janette is a Senior Fellow of the UK Higher Education Academy and has been awarded two Vice Chancellor's awards (2016 and 2009) for her teaching of preservice teachers.

Suzanne de Castell is Professor Emerita, Faculty of Education, Simon Fraser University, where she has held professorial, administrative, and decanal roles. She has published widely on educational theory, literacies both traditional and digital, research technologies, gender studies, and multimodal analysis of communicative interaction and is currently a Visiting Professor at the University of British Columbia.

Vinesh Chandra is a Senior Lecturer in Technology, STEM, and Global Education. He teaches on a range of undergraduate and postgraduate courses that focus on these areas. He also supervises postgraduate students whose research interests lie in these fields. Dr. Chandra has also worked on a number of research projects (including projects overseas). He leads the SEE Project (http://theseeproject.org/), which is actively involved in advancing the print and digital literacy agenda in rural and remote schools in developing countries. His book *Share Engage Educate: SEEding change for a better world* (Brill Publishers) highlights some of the work he has undertaken with his team in eight developing countries.

Feng-Kuang Chiang is a distinguished professor and vice-director in the Department of Educational Technology, Shanghai Normal University, China. Now, he also is a visiting scientist at RELATE Lab of Massachusetts Institute of

Technology, USA. His research focuses on STEM education, learning spaces, and the integration of information and communications technology into instruction. E-mail: fkchiang@shnu.edu.cn.

Douglas Clark is the Werklund Research Chair of Design-Based Learning at the University of Calgary's Werklund School of Education. His research focuses on students' conceptual change processes, computational thinking, argumentation, modeling, and use of representations in digital and non-digital inquiry environments. Doug has been awarded multiple grants from the US National Science Foundation, Department of Education, and the US National Academy of Education/Spencer Foundation.

John Craig is an application developer and associate of the Mind, Matter and Media lab (M3Lab) at the Werklund School of Education, University of Calgary. His research interests include physics, chemistry, computer science, education, and feminism. John works with the M3Lab to design virtual reality and 2D computer simulations that support public exploration of complex systems.

Shima Dadkhahfard is a PhD student in the Learning Sciences at the Werklund School of Education, University of Calgary. Dadkhahfard's main areas of expertise are user-centered design, environmental design, and media design for young learners. Dadkhahfard's research interests include spatial design, body movements, and STEM learning.

Atasi Das is an educator-activist, adjunct lecturer, podcaster, and doctoral candidate in Urban Education at the Graduate Center, City University of New York. Her doctoral research focuses on enacting critical numeracy across the curriculum—a framework to situate numbers and enumeration as a social and political activity. She is also the co-creator and co-host of Abolition Science Radio, a biweekly podcast series exploring science and math from an anti-colonial stance. She has worked as an educator across rural and urban settings in New York, Massachusetts, Vermont, and Bengaluru, India. As an elementary educator, she studied and worked with the Spark Teacher Education Institute to advance a liberatory praxis—learning and doing to create an equitable and just world.

James P. Davis is a Lecturer in STEM and entrepreneurial education, at the Queensland University of Technology. He has taught across science curriculum and pedagogy, integrated STEM, emotions in teaching and learning, and teacher leadership and entrepreneurial thinking. His research interests focus on enacted pedagogy and the emotive-cognitive interplay in student engagement and learning, via micro-sociological and phenomenological perspectives. He serves as an Assistant Editor for *Research in Science Education*.

Amanda C. Dickes is a postdoctoral research fellow in the Department of Technology, Innovation and Education at Harvard University. Her work emphasizes expanding learner and practitioner agency in the production of STEM knowledge and investigates the processes through which computational practices are taken up and adapted to become the *language* of practice in the elementary science and math classroom. Dr. Dickes received her PhD in Learning, Teaching, and Diversity at Vanderbilt University where she was a member of the Mind, Matter and Media Lab.

Andrew Elby is a Professor in the Department of Teaching and Learning, Policy and Leadership at the University of Maryland, College Park. His previous research has focused on the learning and teaching of science, especially physics, with a focus on students' and teachers' epistemological views. More recently, he has been exploring (with Ayush) the entanglement of conceptual, epistemological, emotional, and identity/identification dynamics, including in the context of engineering ethics.

Amy Voss Farris Assistant Professor of Education at Penn State University, investigates the intersections of scientific modeling and computing in elementary and middle school classrooms. She seeks to understand how learners' experiences in scientific computing can support their development of ideas and practices across STEM disciplines. Her recent work argues that understanding computational thinking requires accounting for the perspectival, material, and embodied experiences in which children's computing work is grounded. Farris received her PhD in Learning, Teaching, and Diversity at Vanderbilt University where she was a member of the Mind, Matter and Media Lab.

Ayush Gupta is Associate Research Professor in Physics and Keystone Instructor in the A. J. Clark School of Engineering at the University of Maryland, College Park, USA. He is interested in modeling learning and reasoning processes. In particular, he is attracted to micro-genetic and sociocultural models of learning and how these models can be integrated. His research is aimed at creating dynamic models of learners' cognition, including how different aspects such as emotions, identities, epistemologies, and conceptions are entangled in moment-to-moment interactions. Lately, he has been interested in (1) engineering design thinking, how engineering students come to understand and practice design, and (2) the role of ideologies in learning, especially in how engineering students think about technology and society.

Diali Gupta is a PhD Candidate in Learning Sciences at the Werklund School of Education. Her research focuses on aesthetic experiences in Minecraft as emergent designs for learning. She has presented her research at a number of international conferences and published a journal article on critical thinking through Minecraft. She has simultaneously been involved in research on gameful learning and has published and presented in this regard.

Stephanie Hladik is a PhD student in Learning Sciences (Educational Research) at the Werklund School of Education, University of Calgary. Her research focuses on the intersections between critical feminist theory, educational design, and computing education. In particular, Ms. Hladik is passionate about increasing diversity in science, technology, engineering, and mathematics. She believes that careful design of learning environments and curriculum can privilege different ways of knowing and being within computing and challenge notions of power and privilege which currently exist in the field. Prior to joining the field of Learning Sciences, Ms. Hladik earned bachelor's and master's degrees in Electrical Engineering, both from the University of Calgary.

Jennifer Jenson is Professor of Digital Languages, Literacies, and Cultures in the Language and Literacy Department, Faculty of Education, the University of British Columbia. She has published on gender, technology, digital games, and making video games as a pathway to STEAM teaching and learning. She is currently the Principal Investigator of a large, International Partnership Grant funded by the Social Sciences and Humanities Research Council of Canada, Re-Figuring Innovation in Games (www.refig.ca).

Shakhnoza Kayumova is an Assistant Professor in the Department of STEM Education and Teacher Development as well as a Research Scientist at the Kaput Center for STEM Education Research and Innovation. Shakhnoza is the founder of the STEAM Language, Learning, and Identity Research lab. Shakhnoza also directs Community Research and Partnership Initiative (CoRPI) at the University of Massachusetts Dartmouth. Shakhnoza is the NSF Early CAREER awardee and recognized for her work that bridges university research and community engagement with schools, teachers, families, and children/youth from culturally and linguistically diverse backgrounds.

Beaumie Kim is an Associate Professor of Learning Sciences in the Werklund School of Education at the University of Calgary. Her research is focused on engaging learners in playing and designing games that model ideas, concepts, and systems, and also express something about themselves. Her research work is carried out in collaboration with teachers and students as design partners and by observing their interactions, discourse, and artifacts.

Marilú Lam-Herrera holds an MA in Industrial Design from Hanyang University (South Korea) and a BSc in Industrial Design from Universidad del Istmo (Guatemala). Currently a Learning Sciences PhD student at the University of Calgary, she is interested in Decolonizing STEM and Complex Systems Education.

Victor R. Lee is Associate Professor in the Graduate School of Education at Stanford University. He conducts research on wearable technologies and self-tracking, Maker education in out-of-school settings, and unplugged and screen-free computing. Lee is past recipient of the Jan Hawkins Early Career Award, a National Academy of Education/Spencer Foundation Postdoctoral Fellowship, and the NSF CAREER award. He completed his PhD in Learning Sciences from Northwestern University.

Wenjing Li is a Math teacher in Wanquan Primary School in Haidian district, Beijing, China. She received a master's degree in Educational Technology from Beijing Normal University. Her research focused on STEM education and teacher development. E-mail: wendywk1027@163.com.

Anders Nygren is a Professor in the Department of Electrical and Computer Engineering and Vice-Dean of the Schulich School of Engineering at the University of Calgary, Canada. Dr. Nygren's research interests are in the area of cardiac electrophysiology and mechanisms underlying cardiac arrhythmias. His research involves a combination of experimental and computational approaches to electrophysiology. He is committed to increasing access to engineering education and has been instrumental in the development and implementation of new entry pathways to the Schulich School of Engineering, as well as a combined business and engineering degree program at the University of Calgary. Dr. Nygren holds master's degrees from the Royal Institute of Technology, Stockholm, Sweden, and the University of Houston, Texas, USA, and a PhD from Rice University, Houston, Texas, USA. Prior to joining the Schulich School of Engineering in 2004, Dr. Nygren completed several years of postdoctoral training in physiology and biophysics at the University of Calgary.

D. Kevin O'Neill is an Associate Professor of Education and Technology at Simon Fraser University in British Columbia, Canada, where he co-founded the graduate programs in Educational Technology and Learning Design. The main foci of his research are the design and use of digital technologies for history learning, design-based research methodology, and the examination of analogies for educational research and practice. He earned his bachelor's degree in Great Books and Computing Science at Brock University and his PhD in Learning Sciences at Northwestern University.

Dylan Paré is a PhD student in the Learning Sciences at the Werklund School of Education, University of Calgary. Dylan researches and designs for public engagement with complex systems, with a primary focus on gender and sexuality, using the affordances of digital media technologies to enable individual meaning-making and socioscientific exploration. Dylan's design work has been exhibited at Beakerhead Calgary science and technology festival, TEDx Calgary, the Banff Centre for Arts and Creativity, and Spark Science Centre.

Thomas Philip is an Associate Professor and the Director of Teacher Education in the Graduate School of Education at UC Berkeley. Professor Philip studies processes of ideological change that are central to how teachers perceive and act on their sense of agency as they navigate and ultimately transform classrooms and institutions toward more equitable, just, and democratic practices and outcomes. He examines the larger ideological contexts that influence teaching and learning and explores how teachers might engage in reshaping the ideologies that enable and constrain their work. His most recent scholarship explores the possibilities and tensions that emerge with the use of digital learning technologies in the classroom, particularly the relationship between data literacy and racial literacy.

Ashlyn Pierson is a PhD candidate in the Teaching and Learning Department of Vanderbilt University's Peabody College of Education and Human Development. Broadly, her research merges theory and practice by using design studies to explore the interplay between learning environments and scientific modeling practices. Ashlyn has received the National Academy of Education/Spencer Dissertation Fellowship and the Bonsal Applied Education Research Award.

Jrène Rahm is Full Professor in Educational Psychology at the Université de Montréal in Canada. She studies diverse youths' learning and identity work in science through a space-time lens, documenting their educational pathways but also agency and transformation in and through a vast array of educational opportunities. Her current work implies partnerships with community organizations to support teacher education, an interest in youth voice and empowerment through co-creation, as well as the documentation of environmental stewardship in Nunavut and Nunavik in collaboration with Inuit. She is an editorial member of the journal *Mind, Activity and Culture* and has published widely in English and French, including two books with Sense.

Stefan Rasporich is an educator at Calgary Arts Academy. Stefan has taught a variety of subjects through a gamification and Arts Immersion lens. He is an award-winning writer, actor, game designer, and CEO of Infinite Mind Pictures Inc. Stefan advocates for fair trade, basic income, alternative currencies, and the truth and reconciliation process, which translates into his teaching and learning efforts.

Pratim Sengupta is a Professor of Learning Sciences at the University of Calgary's Werklund School of Education, where he also held the Research Chair of STEM Education. His work focuses on designing open source, multi-agent-based, computational simulations and modeling platforms for K-12 students, and public engagement. He is a Fellow of the International Society for Design and Development in Education and recipient of the CAREER Award (2012) from the US National Science Foundation and the Paul D. Fleck Award (2019) from the Banff Centre for Arts and Creativity. He directs the Mind, Matter & Media Lab (m3lab.org) along with Dr. Marie-Claire Shanahan.

Marie-Claire Shanahan is an Associate Professor of Learning Sciences at the Werklund School of Education at the University of Calgary, where she also held the Research Chair. She studies diverse contexts including science classrooms, mentoring programs, teacher–scientist collaborations, online science writing, and comment spaces as well as public experiences of scientific art and creativity. She is also the recipient of Paul D. Fleck Award (2019) from the Banff Centre for Arts and Creativity.

Miwa Aoki Takeuchi is an Associate Professor in the Learning Sciences at the Werklund School of Education, University of Calgary. Takeuchi's works are situated in an emerging and transdisciplinary field to address a complex educational issue—designing the learning environments to enrich mathematics learning, with a focus on linguistic diversity, and cultural and historical contexts. Takeuchi approaches this issue through longitudinal and horizontal analyses of learning trajectories and learner identity development.

Ixkoj Ajkem Council The mission of Ixkoj Ajkem Community Council is to protect the Mayan Textile Art; to contribute to the conservation, valorization, and respect toward the Maya attire; as well as to contribute to sustainable economic growth through the elaboration of Mayan textiles. The group of women that comprise the Ixkoj Ajkem Community Committee (of Mujeres Tejedoras, in Spanish) began to work together in 1988 with the help of different loans. Since then the name of the organization has evolved according to the projects in which they participate. First, they worked as Aj Kemola' Group, then Committee Santo Domingo, Group Xenacoj, then Grupo Gestor. Today, the group is organized and registered in the municipality as the Ixkoj Ajkem Community Council. This Committee is subdivided into four subgroups. The subgroups are focused on attending to different projects (Mayan textiles' protection, the weaving school, community's sustainability, defense of weavers' rights, creating opportunities for young people, among others). These subgroups are Guatemaltecas de Corazón (Guatemalan Women's Heart), Grupo Nacoj (Nacoj Group), Arcoiris Grupo (Rainbow Group), and the Aj Ajkemonela weaving school of Ixkoj Ajkem.

Matthew Thompson, a queer, settler Canadian, is an experienced game designer who lives with his partners and canine companions in the Treaty 7 region of Southern Alberta. He is currently developing VR games and prototypes to explore the new experiences these technologies afford us, and the impact they can have on the ways we socialize, relax, learn, play, and dream.

Kurt Thumlert is an Assistant Professor in the Faculty of Education at York University, Toronto, Canada. His research and teaching interests include technology studies and new media literacies, as well as curriculum theory and pedagogical innovation in twenty-first century learning contexts. He is an Executive Member of the Institute for Research on Digital Learning (IRDL) at York University.

Chandra Turpen is Assistant Research Professor in Physics at the University of Maryland, College Park. Chandra's work involves designing and researching contexts for learning within higher education. She focuses on studying faculty as learners within professional development and workplace settings and undergraduate students as learners within interdisciplinary physics and engineering classrooms. Her research draws from the perspectives of anthropology, cultural psychology, and the learning sciences to build and link explanatory theories to the design of learning environments. She focuses on the role of culture and ideology in science learning and educational change. Chandra pursues projects that have high potential for leveraging sustainable change in undergraduate STEM programs and makes these struggles for change a direct focus of her research efforts.

Scout Windsor is a multidisciplinary artist whose digital sculpting work focuses on character and creature design and is informed by previous work in costume construction, traditional clay sculpting, and resin casting. Scout completed a BA in Film Studies at the University of Calgary in 2009 and co-directed *Creative Futures & Creative Futures 360°* with Dylan Paré, which was an official selection of the NSI Canada Online Short Film Festival in 2019. She was invited to share her virtual reality sculpting expertise with the public via Facebook's Oculus Medium official live stream and DesignerCon's 2018 VR art and 3D printing panel.

Part I
Introduction

Chapter 1
Reimagining STEM Education: Critical, Transdisciplinary, and Embodied Approaches

Pratim Sengupta, Marie-Claire Shanahan, and Beaumie Kim

Abstract This chapter presents a critical overview of the contributions in this book, that re-imagines STEM education along two themes. The first theme offers heterogeneous illustrations of transdisciplinarity, and the second theme illustrates the complex relationship between the body, hegemony and decolonization. Overall, we argue that this book advances critical, transdisciplinary, and embodied approaches in STEM education by illustrating the following: (a) how pedagogical design of transdisciplinary activities can lead to the creation of new representational genres where multiple disciplines meet and coexist, and (b) when learners' bodies and our collective and colonial histories become part of such investigations, we can challenge cultural and disciplinary hegemonies in K-12 and teacher education.

Keywords STEM education · Critical theory · Transdisciplinarity · Embodied cognition · Decolonization

1.1 Motivation and Overview

Over the past decade, integrated STEM education research has been identified internationally as an essential area for growth at the K-16 levels. The ever-increasing centrality and ubiquity of technological innovation and computing within and across the STEM disciplines, and a global resurgence of interest in teaching and learning to code at the K-16 levels are visible forces that are significantly shaping this emerging field of research. This, in turn, is driven both by policy guidelines that highlight workforce needs and by a greater emphasis on innovation in public education and

---

P. Sengupta (✉) · M.-C. Shanahan · B. Kim
Werklund School of Education, University of Calgary, Calgary, AB, Canada
e-mail: Pratim.sengupta@ucalgary.ca

P. Sengupta et al. (eds.), *Critical, Transdisciplinary and Embodied Approaches in STEM Education*, Advances in STEM Education,
https://doi.org/10.1007/978-3-030-29489-2_1

professional practice. At the same time, issues of democratization, equity, power, and access—including recent decolonizing efforts in public education—are also beginning to be acknowledged as legitimate issues in STEM education. This book presents a collection of chapters that make theoretical advances at the intersection of these two dimensions.

The images of STEM education presented in the following chapters can be grouped under two broad themes: transdisciplinary approaches (Theme 1) and bodies, hegemony, and decolonization (Theme 2). Chapters in each theme offer heterogeneous (Rosebery et al., 2010) imaginations of disciplinary boundaries and phenomena that can take us beyond disciplinary myopia. They present illustrations of *reflexive* formations and practices, where learning in one discipline can be deepened through experiencing that discipline through *other* disciplinary lenses. Chapters in Theme 2 also offer insights into how we can design such reflexive formations through centering critical and historical perspectives, revealing a complex relationship between bodies, hegemony and decolonization in the context of STEM education.

In the following sections, first, we elaborate on each theme by illustrating the images of STEM education research and praxis in each chapter. We then highlight the design guidelines for STEM integration that emerge from these contributions. Finally, in light of these contributions, we argue for an emancipatory imagination of STEM education, that can challenge technocentrism (Papert, 1987) the problematic discourse of efficacy and employability.

1.2 Theme 1: Transdisciplinary Approaches in STEM Education

The emphasis on "integration" of STEM disciplines implies an emergence of big ideas and representational practices that unify or transcend specific disciplines (Berland, 2013; Nathan et al., 2013; Sengupta & Shanahan, 2017a). But what is the nature of this *emergent* field of knowledge and practice, particularly in K-12 and college classrooms? To this end, the chapters along this theme offer new imaginations of technological infrastructures, multidisciplinary contexts of inquiry, classroom activities, and researcher–practitioner partnerships where multiple disciplines come together in a seamless fashion. In this section, we briefly describe the contributions.

Davis, Chandra, and Bellochi (Chap. 2) begin with an ontological stance where, rather than positioning STEM as "knowledge," STEM is positioned as a confluence of different "ways of knowing" (Markauskaite & Goodyear, 2016). They argue that learning integrated STEM requires epistemic fluency (Markauskaite & Goodyear, 2016), further developing the notion by situating it in the literature on innovation and entrepreneurship (Von Hippel, 2017). Specifically, they argue that introducing learners (in their case, pre-service teachers) to "wicked" problems (Jordan,

Kleinsasser, & Roe, 2014; Rittel & Webber, 1973), i.e., problems involving a social dimension, where the problem definitions and their solution are context dependent and subject to re-evaluation over time, can serve as a productive pedagogical approach for fostering epistemic fluency. The authors then ground their arguments in praxis by presenting hermeneutic, reflective analyses (Van Manen, 1977) of their own experiences of teaching integrated STEM education courses to pre-service teachers at the Queensland University of Technology. What emerges from their analysis is the importance of pre-service teachers struggling with problem definition—i.e., answering the question "what kind of a problem is this"—both in the context of working with structured and ill-structured problems—for developing their epistemic fluency.

Alonzo-Yanez Thumlert, de Castell, and Jenson (Chap. 3) argue that the narratives of STEM "crisis and salvation" can be productively challenged and disrupted by pedagogical approaches that do not ignore pressing ecological, ethical, and social justice exigencies. Instead, they position these issues as sites of multidisciplinary inquiry in the classroom. In this pedagogical approach, learners and teachers jointly conduct inquiry on *critical sustainability*. Here the goal is to develop a deep understanding of the deterioration of natural physical and biological systems and their relationship to technological, social, and economic states of affairs, and to design potential solutions to such problems. Critical sustainability acknowledges that current socio-environmental problems are dependent on the uneven distribution of environmental costs, inequitable access to environmental resources by various groups and nations, and vulnerable groups' inadequate access to decision-making power and governance positions. Such a problem framing and associated curricular redesign has two advantages: (a) they can provide students and teachers opportunities for conducting inquiry that will necessarily integrate STEM disciplines and (b) by centering *social action*, they can help students and teachers challenge neoliberal economies and ideologies of innovation and performativity that have co-opted public education.

Kim, Rasporich, and Gupta (Chap. 4) provide an example of what critical sustainability might look like in a classroom. They report a classroom study grounded in a research–practice partnership. In this semester-long study, ninth grade students worked on designing sustainable villages in Minecraft. This study explores transdisciplinarity between social sciences, sciences, and language construction in a Canadian school, whose curricula integrate arts in all aspects of student learning. The authors position this as a self-study in which the teacher (Rasporich) reflects on his journey through the scholarly conversation with the university researchers (Kim and Gupta). They discuss how Rasporich supported his students in reflecting on their own assumptions about who and what we look like, how we communicate, and how we live and interact with others and the environment, through designing and creating new virtual worlds. Interesting findings include how the students' emerging consciousness and appreciation of indigenous perspectives and relationships with their land shaped their creative work, in terms of both designing their virtual worlds for sustainability and designing the languages that would be used for communication in these worlds.

Li and Chiang (Chap. 5) report Chinese pre-service teachers' perception of STEAM education. This chapter is particularly poignant given the current national educational policy in China, which mandates STEAM education as a key direction in public education. The central agenda is to foster innovation as a way of knowing and learning at the K-12 levels across disciplines. In response to such policy mandates, leading universities in China have adopted problem-based learning as a pedagogical approach in their pre-service teacher education courses. However, Li and Chiang report that pre-service teachers, despite taking these courses, experienced difficulty in identifying deep, unifying contexts and problems that could potentially bring together different disciplines, even when they were able to make superficial claims about the interrelatedness of STEAM disciplines. The findings of this study show that the adoption of problem-based learning as a pedagogical approach for STEAM integration in pre-service teacher education may require more explicit and careful attention to how Art can be integrated with STEM education.

Clark and Pearson (Chap. 6) argue that transdisciplinarity in STEM education can be fostered through designing and using educational games where computational modeling, as part of the central game play, can integrate second language learning with computational thinking and science learning for emergent bilinguals. Their work is premised on the genre of games known as disciplinarily-integrated games (DIGs) proposed by Clark, Sengupta, and colleagues (Clark, Sengupta, Brady, Martinez-Garza, & Killingsworth, 2015; Krinks, Sengupta, & Clark, 2019; Sengupta & Clark, 2016). DIGs are grounded in the "science as practice" perspective (Lehrer, 2009; Pickering, 1995; NRC, 2007). In this perspective, modeling is positioned as central to scientific expertise and as a "language" of science (Lehrer, 2009), where manipulating, interpreting, and translating across formal representations are seen as a key form of expertise. Taking advantage of this synergy between science and language learning, the authors propose collaborative DIGs as a rich context in which emerging bilinguals can leverage a broad range of resources and representations, grounded in Goodwin's (2017) cooperative action framework.

O'Neil (Chap. 7) situates transdisciplinarity as a problem of transfer, particularly in the context of emphasizing coding and computational thinking as a cross-curricular competency. O'Neil presents a critique of the emphasis on coding and computational thinking especially for young children—popularly known as the "kid coding" movement—as being grounded in a "functional rationality" and lacking in focus on "substantive rationality" (Mannheim, 1940). The author then presents the issue of transfer as central to the development of a deep and substantial expertise in coding and computational thinking and reminds us of the lessons that can be learnt from the studies of LOGO programming language in elementary classrooms during the 1980s and 1990s. O'Neil situates his arguments in light of popular reports of children's coding, as well as research studies on teaching children's computational thinking. This is an important chapter because it presents arguments for why disciplinary depth does not stand in opposition to transdisciplinarity, and in fact, can support the latter in the context of integrated STEM education.

Dickes and Farris (Chap. 8) further the position presented by O'Neil by illustrating how computational thinking and modeling can become meaningfully situated within the K-12 science and mathematics curricula through deepening engagement with disciplinary practices and cultures. This work arises from the concern that investigations into children's computing have largely focused on coding and programming as isolated competencies. Adopting a phenomenological approach to supporting and integrating computational thinking in K-12 STEM (Sengupta, Dickes, & Farris, 2018), the authors demonstrate how teachers, in long-term partnerships with researchers, can position coding and computational thinking as epistemic practices that become essential tools in the work of their students for modeling scientific phenomena. They discuss the importance of designing mathematical measures as a key form of activity that can bring together multiple STEM disciplines, while at the same time, deepening scientific engagement.

Hladik, Behjat, and Anders (Chap. 9) provide a new illustration of a phenomenological approach to computational thinking (Sengupta et al., 2018) by adapting the Conceive-Design-Implement-Operate (CDIO) pedagogical framework from the field of engineering education (Hilton & Bracy, 2015). Using this approach, they designed a set of activities that involve computational thinking and modeling with the goal to integrate STEAM disciplines. Hladik and colleagues discuss how the CDIO framework guided their design of four creative computing activities that leverage representational practices from disciplines such as mathematics, physical education, language, and art. These activities were implemented in two settings: in elementary classrooms and in a professional development workshop for pre-service teachers. They argue that the use of the CDIO design framework makes space for the design of activities that can engage even elementary students in computing practices that challenge a technocentric approach to computing in the classroom (Papert, 1987; Sengupta et al., 2018).

In the final chapter along this theme, Sengupta, Kim, and Shanahan (Chap. 10) further the phenomenological stance on educational computing in the context of pre-service teacher education. The authors investigate how pre-service science teachers can be introduced to computational thinking and modeling through playfully designing computer simulations and games for modeling kinematics and ecological interdependence. The notion of play challenges the expectations of disciplinary rigidity that often keep newcomers from participating in scientific inquiry (Kim & Ho, 2018; Sengupta & Shanahan, 2017a). This is particularly important given that many pre-service teachers may not have prior experience with programming (e.g., Yadav et al., 2017). This chapter presents a pedagogical approach that builds upon (instead of ignoring) pre-service science teachers' interpretive dilemmas as they engage with computational science. They illustrate three epistemological and pedagogical stances that the pre-service teachers adopted through their playful engagement with computational science. Science educators have now started paying attention to supporting teachers' productive uncertainty in the context of modeling in the science classroom (Duschl, 2008; Manz & Suárez, 2018; Farris, Dickes, & Sengupta, 2019), and to this end, this chapter shows how playful engagement with computational models and simulations can support such pedagogical approaches in the pre-service classroom.

1.3 Theme 2: Bodies, Hegemony, and Decolonization

Takeuchi and Dadkhahfard (Chap. 11) further deepen O'Neil's critique of functional rationality as the paradigm of STEM education from the perspectives of critical race theory and queer theory. They ask the following question: what images of STEM education can we visualize if we place human *capabilities* (Sen, 1997)—not capital—at the center? Their work challenges the dominant paradigm of reducing learners as disembodied vehicles for knowledge consumption and learning for the production of human capital in our classrooms and instead centralizes learners' bodies as socio-historical sites of learning in STEM disciplines. By integrating perspectives of sociocultural theory with queer theory and critical race theory, they conceptualize learners' bodies as the locus of negotiating disciplinary norms, emotions, and desires and view them as fundamentally cultural and historical. Utilizing counter-storytelling practices framed by critical race theory and queer theory, Takeuchi and Dadkhahfard introduce the stories of two learners, both recent immigrants to Canada learning mathematics in context of a second language classroom. The cases highlight (a) how learners' bodies are subjugated through formal school curriculum and pedagogy and (b) how personal and unsanctioned embodied enactments may queer normative disciplinary representations and can facilitate shifts in learners' participation and positional identities in mathematics learning.

Lee (Chap. 12) also positions learners' bodies as sites and tools for developing scientific and mathematical expertise. Lee presents a case study of sixth-grade students in a low-income, rural school in the western United States who used step data obtained from commercial wearable activity trackers to substantively interpret and discuss distributional shapes produced by their physical activities. The students investigated how their walking activities shaped their data and then designed new forms of mathematical representations for modeling statistical distributions in their embodied data. Lee's analysis demonstrates how leveraging, as opposed to ignoring, the learner's body as integral to the curriculum can re-position the learner to a place of power in relationship to complex, disciplinary knowledge.

Bobis (Chap. 13) argues that while the visions and practices of past reform-based approaches have consistently proven difficult for beginning STEM teachers to incorporate in their classroom practice, embodied enactments and re-enactments can be helpful a pedagogical approach to address this issue. Taking enactivist and embodied perspectives on mathematics learning, Bobis presents findings of a study that explored prospective primary teachers' enactment of targeted practices for teaching mathematics in innovative ways. The opportunities to approximate such practices were provided across university and school settings. The participants in this study were pre-service elementary teachers in their first mathematics methods course and at the end of their first professional experience. Results indicate that pedagogies employed in the course, namely rehearsals, videos, and teacher educator modeling of practices and coaching during co-planning and co-teaching opportunities in a range of designed settings, were considered by novice teachers to be effective in supporting the enactment of targeted practices during their field placements.

Gupta, Turpen, Philip, and Elby (Chap. 14) expand and deepen criticality in STEM education beyond learners' bodies and position social constructivism as a missing element in theorizing K-12 and college-level engineering education. They present a pedagogical framework grounded in *social constructivism* (Bijker, 2015), where issues of power, privilege, policy, and politics in a globalized world shape and are shaped by technology. They contrast this approach with common narratives and pedagogical approaches in engineering education, which assume that engineers are ethically "walled off" from the complex, long-term effects of their work. The authors present a case study that illustrates how focusing on the *macro*-ethics of engineering and technology, grounded in social constructivism, can facilitate transdisciplinarity in engineering education. Such focus encourages and facilitates the convergence of technical elements of engineering and technology design on one hand, and critical theoretical and historical perspectives on militarization and colonization on the other. Their work challenges common notions of design thinking that are grounded in uncritical, ahistorical, and simplistic definitions of empathy.

Rahm (Chap. 15) situates STEM education socio-historically in terms of the current geo-political situation, central to which are issues such as movement, uncertainty, and globalization. Rahm reminds us of neo-liberal ideologies that undergird common approaches in STEM education, positions the youth as agents of their learning and identity in science, and reports ethnographic case studies of joint video production projects in two different contexts: a Saturday Art-Science club in a school with immigrant student participants and a longitudinal study of six young women who were high school students conducted over a span of 7 years. Building on the work of Furman and Barton (2006), Rahm sought to understand in what ways videos can become a means for youth to tell stories about science and reconfigure their relationship with science. Based on the findings, Rahm argues for repositioning informal learning programs as sites of critique and transformations, and researchers as co-constructors of change-making over time.

Urban Environments and Education Research Coven (Featuring Das and Adams, Chap. 16) also presents an investigation at the intersection of critical theory and transdisciplinarity in the context of *critical numeracy* (Strong et al., 2016). Critical numeracy emerges from experiences, reflections, politicization, and research into public schooling as a site of tracking (Oakes, 2005), exploration (Dewey & Small, 1897), death (DeJesus, 2016), and possibilities (Giroux, 2000), indicating an urgency to organize, teach, and practice toward human emancipation particularly as state violence continues to intensify forms of occupation in a settler colonial and anti-Black society (Hudson & McKittrick, 2014). Similar to Rahm (Chap. 15), they argue that STEM education should be centered on the lives and geographies of learners, their local knowledge, and place experiences and accentuate critical, decolonizing, and desettling frameworks to provide learners and educators tools necessary for critical civic participation in STEM.

Paré, Sengupta, Windsor, Craig, and Thompson (Chap. 17) investigate the relationship between gender and sexual identities and the design of immersive, virtual reality environments for STEM education. Their work arises from the concern that the assumed "naturalness" of male/female binary categories in biology is often at the center of the queer, trans, and intersex panics in public education. They present a reimagination of how gender and sexuality can be experienced as critical literacies using 3D sculpting in VR environments. They argue that when engaging in such computational modeling activities in collaboration with their close friends, a safe space can be created for deep and critical engagement with complex narratives and marginalized experiences of gender and sexuality. Epistemologically, this chapter brings together scholarship on critical and queer theory, computational thinking, and the design of VR learning environments in the context of STEM education and highlights the roles that *desire* and *intimacy* can play in shaping and deepening transdisciplinary engagement in STEM.

Finally, Lam-Herrera, Ixkoj Ajkem Council, and Sengupta (Chap. 18) argue for integrating STEM disciplines in the context of K-12 education by bridging scholarship in complex system education, computational thinking, design of learning environments, indigenous studies, and decolonization. They present the design of an immersive learning environment—*Grafemos*—for learning about complex systems in a manner that integrates indigenous perspectives on design and complexity and Western scientific and technological perspectives on computational modeling and complexity. The authors present an example of a partnership with the Ixkoj Ajkem Council in Xenacoj, Guatemala, that illustrates how working with Grafemos draws upon both indigenous practices of storytelling in the context of Mayan women's fabric design and Western scholarship on computational modeling and thinking in order to represent learners' personal, relational narratives, as well as ecological interdependence.

1.4 Design Guidelines for STEM Integration: Some Emergent Coherences

The chapters in this volume, collectively, make two forms of contributions in terms of design guidelines for STEM education. First, they suggest domains of knowledge and practice that can create opportunities for transdisciplinarity in STEM education. These domains are not typically parts of K-16 education in the STEM disciplines, and the proposal here is that focusing on these domains can bring disciplines together as well as support a critical perspective on STEM education. Second, the contributions of the chapters can also be organized thematically in terms of a few but important epistemic and representational anchors for designing such learning environments. We discuss both these forms of guidelines in this section.

1.4.1 Toward New Transdisciplinary Domains for STEM Integration

1.4.1.1 Critical Sustainability

One of the key coherences that emerged from the chapters is how powerful sustainability can be as a context and domain of investigation for a critical and transdisciplinary approach for STEM integration. Alonzo-Yanez and colleagues (Chap. 3) argue that focusing on urban and environmental sustainability offers teachers a pedagogical opportunity to foster students' deep understanding of the complexity of integration across disciplines and divergent perspectives by taking a problem-oriented approach rather than a discipline-driven one. Kim, Rasporich, and Gupta (Chap. 4) illustrate how creating and designing sustainable villages in virtual worlds can be interwoven with learning about critical histories of physical environments, languages, people, and the land. Both these chapters argue how the deeply intertwined relationships between individuals and environmental, ecological, and urban sustainability can be modeled from a social justice perspective. Davis et al. (Chap. 2) posit that it is at such intersectional worlds that risk-taking should be encouraged in STEM classrooms. Dickes and Farris (Chap. 8) further demonstrate that even for much younger students (third- and fourth-grade students), ecological sustainability can also serve as a reflexive disciplinary context for supporting and deepening computational thinking. And finally, Lam-Herrera, Ixkoj Ajkem Council, and Sengupta (Chap. 18) illustrate how issues of critical sustainability are also important to attend to from the perspective of decolonizing complexity education.

1.4.1.2 Macro-Ethics

In their chapter, Gupta and colleagues (Chap. 10) introduce macro-ethics as a potentially new domain for deepening disciplinary understandings in the field of engineering education. Using the lens of social constructivism, they situate macro-ethics at the intersection of the militarization of technologies in the USA, ideological and gendered stances in professional and educational contexts in STEM and user-centered design. The emergent learning experience is one where professional expectations—both epistemological and performative—are re-examined as more nuanced, highly complex for an increasingly diverse world. For example, the notion of the "hypothetical" user is no longer viewed as colorless or ahistorical; students begin to situate users and contexts of use as shaped not only by personal expectations and histories but as sociopolitical and historical constructions. This is, of course, an introductory investigation, and the authors argue that the unavoidably emotional nature of some of these conversations in the classroom calls for more research on the pedagogy and analysis of student learning in such contexts.

1.4.2 Epistemic and Representational Anchors for Design

1.4.2.1 Modeling and Play

Modeling is a theme that is common across all the empirical chapters in this book. For example, the chapters by Takeuchi and Dadkhahfard (Chap. 11), Lee (Chap. 12), Paré, Sengupta, and colleagues (Chap. 17), and Hladik and colleagues (Chap. 9) highlight the role that embodied modeling can play in bringing together STEM disciplines. Sengupta, Kim, and Shanahan (Chap. 10) illustrate how agent-based computational models of scientific phenomena can become rich sites for computational, scientific, and pedagogical inquiry for pre-service science teachers, even without any prior experience in coding, through playful conversations and interactions. Dickes and Farris (Chap. 8) and Hladik and colleagues (Chap. 9) present arguments about how framing computational thinking and programming as modeling can meaningfully integrate computing within the K-12 science curricula. It is then perhaps not too far a stretch to say that integrated STEM education could be conceptualized as a rich imagination for modeling disciplinary practices, making it possible for multiple forms of knowledge and knowing—both within and across disciplines—to come together.

This coming together of different forms of knowing and knowledge should essentially be playful in nature, in terms of the experiences of the learners. Bobis (Chap. 13) illustrates this through a different type of embodiment and modeling, where the teacher positions herself as the learner and thus *plays* the role of the learner to enact possible disciplinary figured worlds. "Playing" these roles involves sharing and enacting moments of vulnerability creating room for making mistakes and building an inviting environment where beginning teachers are more likely to share their concerns and their own vulnerabilities. We found this to be an important commonality, especially because, as Kim and colleagues have argued elsewhere, playful learning environments encourage risk-taking, persistence, and problem-solving not only for children but also for learners of all ages (Kim, Tan, & Bielaczyc, 2015). Shanahan, Burke, and Francis (2016) and Sengupta and Shanahan (2017a) have similarly argued that STEM education should be imagined as boundary objects and boundary play, where multiple disciplinary worlds come in contact with each other as well as with personal histories and interpretations of the learners and public. As Alonzo-Yanez and colleagues (Chap. 3) and Gupta and colleagues (Chap. 14) argue, as we move away from a pedagogical approach that emphasize reproduction of disciplinary forms, grounding transdisciplinary inquiry in complex, real-world problems calls for creative work on the part of learners where risks, vulnerabilities, and the ability to take myriad perspectives on the same objects and issues constitute central elements of the experience.

Play challenges ontology and thus encourages boundary-making, breaking, and re-making. For example, Kim and colleagues (Chap. 4) illustrate how playfully creating virtual societies led learners to challenge what should count as "language" and invent new, hybrid forms of communication where language is not merely embodied in speech and gesture, but also in the form of deeper physiological properties, designed with reducing social inequalities in mind. Hladik and colleagues (Chap. 9), Lee (Chap. 12), and Farris and Dickes (Chap. 8) present different forms of playful engagement with computational modeling, where "code" and "data" are reimagined as physical objects as well as embodied actions and interactions. Playful engagement is therefore an invitation to reimagine disciplinary work and practices as transdisciplinary, and playful modeling therefore emerges as a powerful mode of engagement for learners in STEM education.

1.4.2.2 Making Hegemony Visible: Unearthing Histories, Desire, and Bodies

Many of the chapters, especially those along Theme 2, have explicitly attempted to make visible the hegemonic structures and histories that underlie disciplinary practices. Several of these chapters also positioned this historical reimagination of disciplinary work as a place for transdisciplinarity. For example, Gupta and colleagues (Chap. 14), Rahm (Chap. 15), and Das and Adams (Chap. 16) illustrate that delving deeper into our individual and collective histories has implications for reimagining what should constitute scientific work (Rahm, Chap. 15), engineering practice (Gupta et al., Chap. 14), and mathematical literacies (Das and Adams, Chap. 16).

Arguments to reimagine scientific literacy by connecting science to lives of protest and resistance are not new (e.g., Weinstein, 2009, 2016). Several chapters in this book echo this sentiment. However, Paré and colleagues (Chap. 17) further argue that desire and intimacy can also be a form of pedagogical resistance against hegemony. They illustrate how the hegemony of cisheteronormativity, which has predominantly shaped gender and sexuality education in K-12 biology and social studies curricula, can be countered by creating spaces where learners' desires and narratives about their gender and sexuality can become contexts of both interpersonal and disciplinary inquiry through a form of 3D computational modeling (sculpting) in VR worlds.

Other authors also argue for recognizing the role that bodies can play as contexts and tools for making hegemony explicit. As mentioned earlier, Takeuchi and Dadkhahfard (Chap. 11), Lee (Chap. 12), Paré and colleagues (Chap. 17), Dickes and Farris (Chap. 8), Hladik and colleagues (Chap. 9), and Bobis (Chap. 13) illustrate the role that embodied modeling can play in bringing together STEM disciplines, in both computational and non-computational contexts. However, these chapters also highlight how intricately learners' agencies (and their personal histories, as reported by Takeuchi and Dadkhahfard, Chap. 11) are connected to their bodies. Rahm (Chap. 15) further argues that it is not only the learners' bodies but also their mobilities of their everyday practices in relation to professional disciplines that have implications for the development of their disciplinary identities in professional STEM worlds. Rahm's argument echoes similar calls issued by Leander, Phillips, and Taylor (2010) in the context of new media and literacy.

1.4.2.3 Art and Aesthetic Experience

Several authors in this volume highlight the importance of art in bringing together STEM disciplines. It is interesting to note that some of these chapters highlight how learners and participants created art within virtual environments, either through coding (Hladik et al., Chap. 9) or through designing and modeling using computational platforms for creating virtual worlds, such as Minecraft (Kim et al., Chap. 4) and 3D sculpting in VR using Occulus Medium (Paré et al., Chap. 17). At the same time, Lam Herrera, Ixkoj Ajkem Council, and Sengupta (Chap. 18) suggest bringing together Western and indigenous perspectives on complexity and computation and identify how indigenous fabric design can become an inspiration for reimagining computational thinking and modeling in ways that can challenge the hegemony of Western perspectives on STEM and STEM education.

In these studies, art is not simply a context for STEM integration. These chapters show how art can fundamentally shape the nature of experience of STEM for students, in a manner akin to Dewey's notion of *aesthetic experience* (Dewey, 1934; Farris & Sengupta, 2016; Higgins, 2008). The experience of creating art, particularly in contexts that involve modeling science computationally, can be fundamentally synthetic and transformational (Farris & Sengupta, 2016). It is synthetic in the sense that such experiences invite the learner to bring together experiences from their lifeworlds outside the classroom alongside computational thinking and disciplinary practices. For example, 3D sculpting provided the participants in Paré et al.'s study opportunities to reflect deeply on their experiences of gender- and sexuality-based marginalization. Such experiences are transformational in the sense learners transform "materials" (e.g., modeling tools in Oculus Medium) into expressive media (e.g., 3D sculptures in VR that symbolize participants' gender and sexual identities). But, such experiences are also transformational in a more critical theoretical sense, in that they can create opportunities for challenging disciplinary hegemonies. This can happen by centering the activities around experiences of oppression and marginalization that have been historically been left out of curricular contexts (e.g., learning about plastic contamination due to urbanization in local communities in Lam Herrera et al.'s study), as well as by challenging hegemonic norms as a way to learn about issues such as gender and sexuality (e.g., youth redesigning gender norms in their virtual worlds in Kim et al.'s study). Thinking computationally is then no longer limited to thinking about computational abstractions, and as Lam Herrera et al. (Chap. 18) illustrate, neither does computing need to involve a computer. When viewed in this light, aesthetic experience as part of the STEM curriculum can challenge disciplinary and technological hegemony, as well as hegemonic norms in society.

We believe that this has significant entailments for both the design of learning environments as well as for pre-service teacher education. The survey reported by Li and Chiang (Chap. 5) shows that even when pre-service teachers intuitively understand that "art" can be a synergistic domain of practice for STEM education, in the absence of meaningful experiences of integration, such intuitive understanding remains removed from their classroom practices. Li and Chiang's work therefore suggests that rather than asking the question "Can Art be integrated with STEM education?" we need to think more deeply about *how* such integration can be brought about. The chapters in this volume provide several examples of learning environments and experiences where such integration can be brought about by urging us to think beyond superficial definitional issues of Art and STEAM education. Instead, they illustrate the importance of considering more deeply the underlying epistemological orientations and implications—i.e., the nature of *aesthetic experience*—that can be supported as learners engage in art in the context of STEM education.

1.5 Epilogue: What Form of *Literacies* Can STEM Become?

Simply put, professional practice takes the form of *literacies* in public education. Currently, most of the national and international standards and guidelines situate literacies in each of the STEM disciplines in terms of professional *practice* in these disciplines, which are at once representational and epistemic in nature. For example, modeling is well-accepted as the "language" of science (Lehrer, 2009; National Research Council, 2007), and unsurprisingly, it is also a key element of US Next Generation Science Standards (NRC, 2013). Even more recently, Sengupta and Shanahan (2017b) argued for reimagining scientific literacy as a public experience by centralizing the notion of practice. They argued that rather than making public the results of science, researchers and research institutions should also work toward designing opportunities for the public to engage with directly in scientific modeling practices.

The chapters presented here, collectively, argue for a different imagination of practice, in which practice ceases to belong to *a* discipline. They also illustrate that multidisciplinarity should no longer be considered a trade-off for disciplinary depth, and neither should transdisciplinarity imply that practices must necessarily span across disciplines. Rather, the images of transdisciplinarity presented here show that different practices from different disciplines can work together in a *reflexive* manner. This position is similar to the notion of reflexivity as put forward by Harel and Papert (1991) and Sengupta, Kinnebrew, Basu, Biswas, and Clark (2013), who have argued that learning programming (in specific forms and with appropriate scaffolding) can further deepen the experience of learning mathematics and science. The emergent learning experience then cannot belong to a single discipline, even though different elements of these experiences, at times, might belong to different disciplines. Epistemologically, this is similar to the images of *influx of senses* (Vygotsky, 1980) and *agglutination* (Bakhtin, 1983), where the meaning of a word emerges from a multitude of meanings that surround the word.

The authors of this edited volume illustrate examples of such agglutinated forms of transdisciplinarity in and as STEM education, and further suggest the following: (a) the pedagogical design of such transdisciplinary activities can lead to the creation of new representational genres where multiple disciplines meet and co-exist;

and (b) when learners' bodies and our collective and colonial histories become part of such investigations, we can challenge cultural and disciplinary hegemonies. For example, learning mathematics can become learning critical numeracy, where learners begin to see the history of slavery in the USA as relevant to learning in the mathematics classroom (Das and Adams, Chap. 15). Similarly, learning biology from a critical theoretical perspective can happen through computational sculpting and gender play with close friends in VR (Paré et al., Chap. 17); and jumping around with friends in the school yard can become data for statistical, scientific, and mathematical models (Lee, Chap. 12). This book therefore calls for viewing STEM as *emergent* literacies where multiple disciplines meet and co-exist in reflexive ways, and social and historical injustices and hegemonies become *legitimate* elements of and integral to STEM education.

This does not mean that such curricula should be *free form* or *free choice*, and neither should the commitment to disciplinary depth be at stake here. To this end, we have identified some disciplinary, epistemic, and representational anchors for designing such learning environments in the previous section. In viewing integrated STEM as an influx of senses where meaning arises from heterogeneity, we believe that we can begin to realize Dewey's vision of learning as aesthetic experience (Dewey, 1934). As Higgins (2008) argued, the metaphor here is that of learning as "seeing more," rather than barely "seeing as," The distinction here is between a rudimentary form of recognition, for example, when we simply see something as an instantiation of a category or in service of a specific purpose (seeing as), as opposed to experiencing the richness and complexity of that object (seeing more). Seeing as, Higgins (2008) argues, is a case of arrested perception, which can "insulate us all too fully from the thorny, the surprising, the complex" (p. 14). Several chapters in this book illustrate that "seeing more"—i.e., experiencing the thorny, the surprising, and the complex—should involve seeing more *critically* and *historically*, echoing similar calls issued by Philip, Bang, and Jackson (2018).

This expansive reimagination of STEM also has implications for the design of learning technologies such as programming languages, digital games, virtual worlds, and virtual reality environments. A common theme across the chapters is the importance of reimagining these technologies as experiences, especially in terms of the *heterogeneity* of the learners' experiences: emphasizing materiality, embodied interactions, and playful engagement with complex and historically situated, transdisciplinary phenomena. Furthermore, when these experiences are grounded in the learners' voices and communities that have been historically oppressed and marginalized, they can fundamentally alter both the disciplinary discourse and technological infrastructure. These forms of heterogeneity are particularly important for breaking the trope of technocentrism (Papert, 1987) that largely colors and shapes our experiences with technologies in educational settings (Sengupta et al., 2018), where too often, narrowly defined forms of technological work become the centerpiece of STEM education.

In its most emancipatory imagination, going beyond and even countering the problematic discourse of efficacy and employability, STEM education can help us reimagine the work of disciplining our bodies and minds as practices that offer learners and teachers opportunities to challenge hegemonic discourses. Hegemony, like marginalization, is also heterogeneous in form, but it has a central and defining characteristic in all of its instantiations: hegemony always results in the silencing of voices that can disrupt the disciplinary core. The chapters in this book offer heterogeneous visions that can challenge disciplinary hegemonies in different ways, thus offering new imaginations of STEM education as critical and transdisciplinary practice, with an open invitation to the field to continue this work.

References

Bakhtin, M. M. (1983). *The dialogic imagination: Four essays*. The University of Texas Press.

Berland, L. K. (2013). Designing for STEM integration. *Journal of Pre-College Engineering Education Research (J-PEER), 3*(1), 3.

Bijker, W. (2015). Social Construction of Technology. In J. D. Wright (Ed.), *International Encyclopedia of the Social & Behavioral Sciences, 2nd edition* (pp. 135-140). Oxford, UK: Elsevier Science.

Clark, D. B., Sengupta, P., Brady, C. E., Martinez-Garza, M. M., & Killingsworth, S. S. (2015). Disciplinary integration of digital games for science learning. *International Journal of STEM Education, 2*(1), 2.

DeJesus, I. (2016, May 6). After more than 100 years, American Indian children buried in Carlisle begin a journey home. PennLive.com. Retrieved from http://www.pennlive.com/news/2016/05/carlisle_indian_school_repatri.html

Dewey, J. (1934/2005). *Art as experience*. Penguin.

Dewey, J., & Small, A. (1897). *My pedagogic creed* (No. 25). EL Kellogg & Company.

Duschl, R. (2008). Science education in three-part harmony: Balancing conceptual, epistemic, and social learning goals. *Review of research in education, 32*(1), 268–291.

Farris, A. V., Dickes, A. C., & Sengupta, P. (2019). Learning to interpret measurement and motion in fourth grade computational modeling. *Science & Education*. https://doi.org/10.1007/s11191-019-00069-7

Farris, A. V., & Sengupta, P. (2016). Democratizing children's computation: Learning computational science as aesthetic experience. *Educational Theory, 66*(1–2), 279–296.

Giroux, H. A. (2000). Public pedagogy as cultural politics: Stuart hall and the crisis of culture. *Cultural Studies, 14*(2), 341–360.

Goodwin, C. (2017). Co-operative action. Cambridge University Press.

Harel, I. E., & Papert, S. E. (1991). *Constructionism*. Ablex Publishing.

Higgins, C. (2008). Instrumentalism and the clichés of aesthetic education: A Deweyan corrective. *Education and Culture, 24*(1), 6–19.

Hilton, A., & Bracy, A. (2015). All of Programming. Bracy and Hilton.

Hudson, P. J., & McKittrick, K. (2014). The geographies of blackness and anti-blackness: An interview with Katherine McKittrick. *The CLR James Journal, 20*(1/2), 233–240.

Jordan, M. E., Kleinsasser, R. C., & Roe, M. F. (2014). Wicked problems: Inescapable wickedity. *Journal of Education for Teaching, 40*(4), 415–430.

Kim, B., Tan, L., & Bielaczyc, K. (2015). Learner-generated designs in participatory culture: What they are and how they are shaping learning. *Interactive Learning Environments, 23*(5), 545–555.

Kim, B., & Ho, W. (2018). Emergent social practices of Singapore students: The role of laughter and humour in educational gameplay. *International Journal of Child-Computer Interaction, 16*, 85–99.

Krinks, K. D., Sengupta, P., & Clark, D. B. (2019). Modeling games in the K-12 science classroom. *International Journal of Gaming and Computer-Mediated Simulations (IJGCMS), 11*(1), 31–50.

Leander, K. M., Phillips, N. C., & Taylor, K. H. (2010). The changing social spaces of learning: Mapping new mobilities. *Review of Research in Education, 34*(1), 329–394.

Lehrer, R. (2009). Designing to develop disciplinary dispositions: Modeling natural systems. *American Psychologist, 64*(8), 759.

Markauskaite, L., & Goodyear, P. (2016). *Epistemic fluency and professional education: Innovation, knowledgeable action and actionable knowledge* (Vol. 14). Springer.

MANNHEIM, K. (1940), Man and Society in an Age of Reconstruction. New York: Harcourt, Brace and World.

Manz, E., & Suárez, E. (2018). Supporting teachers to negotiate uncertainty for science, students, and teaching. *Science Education, 102*(4), 771–795.

Nathan, M. J., Srisurichan, R., Walkington, C., Wolfgram, M., Williams, C., & Alibali, M. W. (2013). Building cohesion across representations: A mechanism for STEM integration. *Journal of Engineering Education, 102*(1), 77–116.

National Research Council. (2007). *Taking science to school: Learning and teaching science in grades K-8*. National Academies Press.

National Research Council. (2013). *Next generation science standards: For states, by states*.

Oakes, J. (2005). *Keeping track*. Yale University Press.

Papert, S. (1987). Information technology and education: Computer criticism vs. technocentric thinking. *Educational Researcher, 16*(1), 22–30.

Philip, T. M., Bang, M., & Jackson, K. (2018). Articulating the "how," the "for what," the "for whom," and the "with whom" in concert: A call to broaden the benchmarks of our scholarship. *Cognition and Instruction, 36*(2), 83–88.

Pickering, A. (1995). The mangle of practice: Time, agency, and science. University of Chicago Press.

Rittel, H. W., & Webber, M. M. (1973). Dilemmas in a general theory of planning. *Policy Sciences, 4*(2), 155–169.

Rosebery, A. S., Ogonowski, M., DiSchino, M., & Warren, B. (2010). The Coat Traps All Your Body Heat: Heterogeneity as Fundamental to Learning. *The Journal of the Learning Sciences*, 322–357.

Sen, A. (1997). Human capital and human capability. *World development, 25*(12), 1959–1961.

Sengupta, P., & Clark, D. (2016). Playing modeling games in the science classroom: The case for disciplinary integration. *Educational Technology, 56*(3), 16–22.

Sengupta, P., Dickes, A., & Farris, A. (2018). Toward a phenomenology of computational thinking in STEM education. In *Computational thinking in the STEM disciplines* (pp. 49–72). Cham: Springer.

Sengupta, P., Kinnebrew, J. S., Basu, S., Biswas, G., & Clark, D. (2013). Integrating computational thinking with K-12 science education using agent-based computation: A theoretical framework. *Education and Information Technologies, 18*(2), 351–380.

Sengupta, P., & Shanahan, M. C. (2017a). Boundary play and pivots in public computation: New directions in STEM education. *International Journal of Engineering Education, 33*(3), 1124–1134.

Sengupta, P., & Shanahan, M. C. (2017b, May). Open Science, Public Engagement and the University. Paper presented at the NSF/NIH conference on Open Science and the University. *arXiv preprint arXiv:1702.04855*.

Shanahan, M. C., Burke, C. A., Lydia, E., & Francis, K. (2016). Using a Boundary Object Perspective to Reconsider the Meaning of STEM in a Canadian Context. *Canadian Journal of Science, Mathematics and Technology Education, 16*(2), 129–139.

Strong, L., Adams, J. D., Bellino, M. E., Pieroni, P., Stoops, J., & Das, A. (2016). Against neoliberal enclosure: Using a critical transdisciplinary approach in science teaching and learning. *Mind, Culture, and Activity, 23*(3), 225–236.

Van Manen, M. (1977). Linking ways of knowing with ways of being practical. *Curriculum Inquiry, 6*(3), 205–228.

von Hippel, E. (2017). *Free innovation*. Cambridge, MA: MIT Press.

Vygotsky, L. S. (1980). *Mind in society: The development of higher psychological processes*. Harvard University Press.

Weinstein, M. (2009). Critical science literacy: Identifying inscription in lives of resistance. *Journal for Activist Science and Technology Education, 1*, 2.

Weinstein, M. (2016). Imagining science education through ethnographies of neoliberal resistance. *Mind, Culture, and Activity, 23*(3), 237–246.

Yadav, A., Stephenson, C., & Hong, H. (2017). Computational thinking for teacher education. *Communications of the ACM, 60*(4), 55–62.

Part II
Transdisciplinary Approaches in STEM Education (Theme 1)

Chapter 2
Integrated STEM in Initial Teacher Education: Tackling Diverse Epistemologies

James P. Davis, Vinesh Chandra, and Alberto Bellocchi

Abstract Science, Technology, Engineering, and Mathematics (STEM) each have distinct epistemological foundations for the production of knowledge, yet a recent international trend in education is to integrate these fields as an approach to teaching and learning. According to the literature, *integrated STEM education* involves concurrent teaching of two or more knowledge domains from the collection of traditional knowledge silos that constitute STEM. The rationale for integrated STEM education is grounded in a perceived need to simulate the complexity of real-world situations, where examples of integrated STEM tend to evolve over time, through the need to solve problems in naturalistic contexts by teams of researchers with different disciplinary expertise. In educational settings, each school STEM discipline has evolved with pedagogical responses to simulate real-world contexts such as science inquiry or mathematical problem solving, however, the notion of integrated STEM adds layers of complexity to pedagogical responses. Our aim in this chapter is to address this complexity from the perspective of integrated STEM in *initial teacher education* programs, based on critical reflections of our recent teaching experiences and learning experiences of our students. We explore the demands on initial teacher education STEM students in terms of the diversity of analytical epistemological orientations, and we consider possible strategies for understanding synthetic epistemological orientations that may inform better our understanding of learning through integrated STEM.

Keywords Cognition · Emotion · Epistemic fluency · Integrated STEM · Teacher education

J. P. Davis (✉) · V. Chandra · A. Bellocchi
Queensland University of Technology, Kelvin Grove, QLD, Australia
e-mail: jp.davis@qut.edu.au

P. Sengupta et al. (eds.), *Critical, Transdisciplinary and Embodied Approaches in STEM Education*, Advances in STEM Education,
https://doi.org/10.1007/978-3-030-29489-2_2

STEM, as an acronym, gives recognition to a way of emphasizing the importance of each disciplinary element of Science, Technology, Engineering, and Mathematics within educational contexts (English, 2016). Lyn English (2016) highlights the challenge that STEM introduces for researchers and educators, which is evident in diverse interpretations of what is meant by STEM education or the integration of STEM. Defining *integrated STEM* opens a new level of complexity from educational perspectives due to its conceptualization across a diverse range of international contexts, and the different ways that individual disciplines are represented through integration (English, 2016). For example, the phrase *integrated STEM* in education may be defined as the concurrent teaching of two or more knowledge domains from the collection of traditional knowledge silos that constitute STEM school subjects (Kelly & Knowles, 2016). That phrase may also apply to the integration of STEM elements with real-world contexts (STEM Task Force Report, 2014). The common catchphrase *real-world* is an educational reaction to past school instruction that had students engaging in problems (e.g., solving an equation) with no connection between these activities and the situations in which STEM discipline experts apply such practices. Effectively what real-world refers to is situations and scenarios where STEM experts apply their disciplinary knowledge to explain phenomena or solve problems (Bellocchi, King, & Ritchie, 2016). Integration of STEM with real-world contexts introduces new complexities such as ill-structured problems and innovation (Nadelson & Seifert, 2017), and notions of interdisciplinary, multidisciplinary, or transdisciplinary collaboration (English, 2016). This complexity within the field of STEM education is evident through *inconsistencies* in definitions of integrated *STEM*, especially in the use of terminology such as inter-, multi-, or transdisciplinary (cf. Shernoff, Sinha, Bressler, & Ginsburg, 2017).

The complexity that we, James, Vinesh, and Alberto, recognize in integrated STEM education prompts us to describe this field as *transdisciplinary,* and to acknowledge integrated STEM investigations as a form of complexity research (cf. Davis & Sumara, 2006). In this sense, we recognize the possibility for diverse methods, methodologies, and ways of knowing (i.e., epistemologies) through adaptive, integrated STEM learning as distinct from discrete STEM learning. Such epistemological diversity may be influenced by the particularities of specific objects of inquiry (Davis & Sumara, 2006) *and* the instruments applied by particular investigators as learners (cf. Allchin, 1999a). Our aim in this chapter is to explore the epistemological complexity of integrated STEM in an initial teacher education program.

2.1 Theoretical Framework

This exploration is based on critical reflections of our recent teaching experiences and the learning experiences of some of our preservice teachers. We have structured our exploration by recognizing the notion of *analytical* and *synthetic* orientations toward thinking and ways of knowing (Wickman, 2017). Analytical orientations, also called formal analysis (Liberman, 2007), suggest that we reach a position of knowing based on an inherent nature of things derived through clear processes of rationality, logic, or dichotomous categorizations such as "mind-world, inner-outer, value-fact" (Wickman, 2017, p. 10). In contrast, a synthetic orientation produces knowledge that is contingent upon localized performances of practice and lived experiences of those practices. These practices unfold over time, and are used by people to produce reference points for justifying localized knowledge and their sense of social order. For example, in a school science inquiry, the act of students mutually aligning their bodies, faces, and eye gaze toward a common object, and then making utterances in conversation to describe the object at the focus of their shared gaze would typically become a formal reference point in scientific practice called *observation*. Observation would be the formal analytical way of knowing the focal object that students and science teachers may refer to when explaining the production of scientific knowledge. In contrast the detailed description of how observation unfolded over time through bodily actions and verbal utterances would be a synthetic way of knowing *how* knowledge was produced on a particular occasion of scientific observation (cf. Davis & Bellocchi, 2018).

2.1.1 Analytical Ways of Knowing in STEM

We preface this section with the notion of epistemic fluency to describe a capability of being flexible and adept in problem-solving situations that require diversity in thinking across different ways of knowing (Markauskaite & Goodyear, 2016). Epistemic fluency is described by Markauskaite and Goodyear (2016) as a skill to be learned by professionals, such as teachers, so that they may become effective in working across different ways of knowing in contemporary workplaces. This is an important notion when considering the analytical orientations of the discrete fields of STEM that are currently being integrated in educational contexts.

Definitions of integrated STEM in the education literature tend to address integration by referring to *connectivity* between knowledge domains via the recognition of conceptual links (cf. Nadelson & Seifert, 2017). That approach to defining integrated STEM education may be interpreted as treating knowledge as an object (Markauskaite & Goodyear, 2016) possessed by particular disciplines. What we have found in the integrated STEM education literature is a lack of focus on different *ways of knowing* or *epistemologies* to explain how knowledge is produced differently across the sciences, technology, engineering, and mathematics, and how these are interpreted in educational contexts. As a start point in understanding integrated STEM education, we suggest that researchers should consider the different epistemologies contributing to STEM. This is an alternative to focusing on the products of epistemology in terms of *disciplinary* concepts and connectivity across knowledge domains (cf. Shernoff et al., 2017). We point to a need for understanding better, how STEM concepts may be generated in educational contexts through various epistemological orientations or ways of knowing.

To illustrate these different ways of knowing briefly, we start by considering the mathematics element of STEM. The analytical orientations of mathematical epistemologies may be collectively referred to as an implicit assumption about the "existence of a 'mind-free' mathematics" (Núñez, Edwards, & Matos, 1999, p. 61). That is, an assumption that rational and logical pathways for mathematical reasoning exist independent of thought, waiting to be mapped, discovered, unlocked, explored, or understood by our human minds. As an example, we may consider the use of proof, where the logic of the analytical method is justified as a form of deductive reasoning about mathematics. In mathematics this may be understood as a primary means for producing what people commonly regard as objective mathematical knowledge (Buldt, Lowe, & Muller, 2008). In that context, *objectivity* is referred to in its analytical form suggesting a Baconian independence of mathematics, distanced from the subjectivity of the mind (cf. Davis & Bellocchi, 2018). This is one way of thinking about mathematics as a way of knowing, and it is the way mathematics is typically portrayed via textbook approaches where mathematical problem solving is taught in an instrumental or procedural manner (Fan & Zhu, 2007). The application of this formal, analytical, mathematical reasoning extends into science where modeling may be used to understand phenomena. The use of science models often involves mathematical models as well as physical models, and both forms of modeling may be classified as deductive reasoning (Valiela, 2001). For example, in the earth sciences students may study the impact of waves on sand formations using a wave tank in a laboratory. Conclusions made from a study of the wave tank model may then be translated by students to real-world contexts such as oceans and beaches, which would be a form of deductive reasoning. Such deductive modeling may also involve mathematics where an algebraic relationship between force applied and wave size may be represented mathematically, and applied to understand wave actions in an actual ocean. In this sense, both mathematics and deductive science produce knowledge by developing a priori models as a formal analytical method for understanding real-world phenomena, beyond the classroom laboratory.

In contrast to mathematics and deductive science in education contexts, empirical science is less dependent on a priori models and theory, as it places greater emphasis on collecting and analyzing data direct from observable phenomena (Valiela, 2001). Through observation of phenomena, the empirical sciences rely on scientists' perceptual experiences of reality to be a primary source of data (Chalmers, 2013). However, to transform data from perceptual experience into knowledge, empirical scientists engage with each other, socially, by describing perception through language and preexisting and/or unfolding conceptual lenses (Allchin, 1999b). Empirical science is therefore not value free, or theory free, but it is less explicitly dependent on preexisting theory as a formal analytical way of knowing (Chalmers, 2013) compared with the deductive sciences.

It is also important to recognize the diverse ways of knowing within the empirical sciences. For example, Valiela (2001) identifies empirical science inquiry methods such as descriptive, correlational, comparative, perturbation, and controlled experiments. While highlighting the epistemological implication of these inquiry methods, Valiela (2001) then explains how knowledge about cause–effect relationships is possible only with the use of controlled experiments. Controlled experiments involve scientific methods (i.e., practices) where the majority of variables unlikely to be involved in a causal relationship with a phenomenon can be controlled. This enables scientists to manipulate just one variable (the *independent* variable) and observe any impact on one other variable (the *dependent* variable). From an empirical sciences perspective, this is the only way of producing knowledge about the categories of cause and effect between two variables.

Other scientific practices, such as exploration or observation, do not involve the establishment of causal relationships, and for this reason the practices involved in those areas of empirical science produce different forms of empirical scientific knowledge. These different forms of empirical scientific knowledge are often described as producing knowledge with different degrees of certainty or uncertainty, evidencing complexity (Valiela, 2001). From the perspective of integrated STEM in education, we can therefore identify a high degree of epistemological diversity, or diverse ways of knowing, by simply understanding how knowledge may be produced through mathematics, deductive science, and empirical science. If we now consider technology and engineering, the diversity of epistemological approaches broadens even further.

Engineering is typically associated with *design process* in the STEM education literature (King & English, 2016). In a K-12 educational engineering context, the process of design starts with a problem, and it engages investigators in an iterative process of construction, testing, reconstruction, and re-testing until a technological solution to the problem is produced (King & English, 2016). As a way of knowing, epistemological outcomes of engineering design processes experienced by students and teachers are problem specific, the outcome of which may be represented in the form of a concrete technological artifact, as distinct from the abstract concepts and ideas of scientific and mathematical knowledge.

As experienced by students and teachers, the fields of technology tend to be broader in their approaches to design process, compared with engineering, and typically embrace the notion of *design thinking* (Hong & Choi, 2011). Examples of further epistemological diversity through the technology element of STEM include technologies produced through creative processes such as the Arts (Watson, 2015). For example, Andrew Watson (2015) claims to adopt a STEM design process for an arts technology project and modifies this as *design thinking*, by including creativity through *imagining* as part of the practices of technology production. A further example is the application of design thinking to produce technological innovation in fields such as entrepreneurship (Huq & Gilbert, 2017). As a formal, analytical way of knowing in STEM, *design thinking* provides a pathway for integrating STEM education with entrepreneurship and enterprise education where creativity and problem definition in real-world contexts are foregrounded (Quality Assurance Agency, 2018).

In contrast to looking for conceptual connections (Nadelson & Seifert, 2017) across S.T.E.M. silos (Kelly & Knowles, 2016), our description of different analytical orientations attributable to integrated STEM offers an innovative way to think about this field. Contemporary thinking tends to be focused on the content knowledge of S.T.E.M. domains, rather than the foundations upon which different knowledge are produced. We believe that understanding a variety of epistemologies contributing to integrated STEM is an important way to understand the complexity of this field, particularly if preservice teachers are to become proficient in epistemic fluency. However, what we have touched upon so far, points to the need for epistemic fluency as an analytical skill for teachers to learn. If we are to consider synthetic ways of knowing through integrated STEM practices, we may also need to reconsider epistemic fluency as part of the lived experiences of teachers. What this may look like is considered next.

2.1.2 Synthetic Ways of Knowing in STEM

Synthetic ways of knowing in STEM may enable us to understand epistemic fluency and analytical epistemologies *as* and *through* the contingencies of situated practices and lived experiences. In this section, we will highlight some examples of what we are categorizing as synthetic orientations that have been previously applied in mathematics and science education. In a study of mathematics education (Núñez et al., 1999), synthetic orientations are considered through the field of embodied cognition, where cognition is viewed as being grounded biologically in individuals. Because individuals interact with others, it is claimed that embodied cognition produces a foundation for a socially situated dimension to understanding how mathematical knowledge may be generated. From this perspective, understanding the formation of mathematical knowledge may be achieved by studying bodily and social practices as constituents of a conceptual system as the basis for embodied cognition. In the context of learning through integrated STEM this means understanding observable situated practices as evidence of the embodiment of mathematical reasoning. In this way embodied cognition adopts a constructivist approach to understanding how conceptual schemata are formed as knowable objects. This is different from some of the other synthetic orientations in this paper that we will now consider.

Another synthetic approach for understanding the formation of mathematical knowledge is evident in the work of Noble, DiMattia, Nemirovsky, and Barros (2006) who study the use of a mathematics tool they call a *drawing machine*. That study is informed by phenomenology and evidences how knowledge is generated about the tool, and related mathematics, *through* the lived experiences of *using* the tool. This way of thinking is important as a form of epistemology. For example, if we broaden our notion of tools from the physical instrument applied in that study we may find it possible to regard mathematical, scientific, or design methods as knowable *through* their *use*, rather than in an a priori sense. That is, we may understand the performance of the *analytical* methods of STEM reasoning where the *in-the-moment performance* of analytical epistemology, such as a scientific method, is itself a lived epistemological phenomenon. Like the notion of embodied cognition, this phenomenological approach has implications for the way knowledge is understood in an integrated STEM learning context that is different from analytical traditions.

A further, different approach to the study of ways of knowing through practices is evident in the work of Miwa Aoki Takeuchi (2016) who explored friendship and the influence of social relationships on the formation of knowledge about the self *and* mathematics in a linguistically diverse classroom. That study examines agency and the circulation of power through relationships that influenced student actions by acting through opportunities to learn. Unlike the earlier two approaches for knowing in STEM, Takeuchi's (2016) approach evidences the interplay between the formation of self-knowledge as student identity and the formation of localized and personalized knowledge *of* mathematics. This has implications in the present study where we expect preservice science and mathematics teachers to embrace epistemic fluency and integrated STEM which may be viewed as something that challenges disciplinary professional identity.

Finally, in this overview of synthetic approaches to epistemology we consider ethnomethodological (Liberman, 2013) and micro-social studies (Collins, 2004) of learning in science classrooms. We refer here to studies that adopt a social ontological position (Davis, 2017) and a social epistemology (Bellocchi, 2017) informed by the work of Émile Durkheim (1912/2008) and the phenomenon of emotional energy (Collins, 2004). One example of this approach by Davis and Bellocchi (2018) foregrounds students' performing *objectivity* in a school science inquiry as an occasioned and particular collection of situated cultural practices. In that study, the analysis of objectivity as a social phenomenon enables an understanding of its formation, giving order to situations *as* meaning, which unfolds through the emotionality of subjective experience (Davis & Bellocchi, 2018). As a synthetic approach to possible understandings of integrated STEM, Davis and Bellocchi's (2018) study of objectivity highlights the interplay between emotion and cognition *as* and *through* contexts of learning at sites of localized knowledge production.

The collection of synthetic orientations, described so far, as possible ways of knowing in learning contexts of integrated STEM education are important for teaching and learning. They provide a means for working across the analytical epistemologies of integrated STEM. In learning contexts where preservice teachers do not just learn *about* STEM but they actually enact practices to learn *through* STEM, it may be possible to observe a dialectical relationship between analytical traditions and synthetic orientations, as demonstrated in the study of objectivity in school science inquiry (Davis & Bellocchi, 2018). This is important because the synthetic epistemologies we have pointed to in this chapter draw on social and cognitive resources that are important for understanding in-the-moment epistemological performances and experiences. We now consider these thoughts in the context of some reflective, empirical data from the STEM classes of Vinesh and James.

2.2 Method of Data Production and Analysis

In this section, James (author 1) and Vinesh (author 2) adopt a reflective protocol (van Manen, 1997) to describe briefly the STEM courses they teach, and to document their most prominent experiences. The documented experiences relate to teaching and learning issues that we later analyze from the perspective of our conceptual framework. James has also included some data from a recent study of preservice teacher learning experiences during their participation in his integrated STEM investigation course. Both James and Vinesh teach STEM to preservice teachers across Bachelor of Education programs at the Queensland University of Technology, a large city university located in South East Queensland, Australia.

2.3 Findings

2.3.1 Teaching Real-World STEM in Education: Vinesh's Story

I teach a 9-week elective course within the Bachelor of Education program designed for preservice teachers across early childhood, elementary, and high school settings. *Some* of these preservice teachers undertake several courses in a specialist discipline such as physics, chemistry, biology, earth sciences, or mathematics, but *many* have not formally studied science or mathematics since elementary school or high school. The purpose of the course I teach is to develop preservice teachers' knowledge and skills on teaching *integrative STEM* in their future classrooms (Sanders & Wells, 2010). *Integrative STEM* is a variation to integrated STEM, used to describe further integration of STEM with other school subjects such as languages and the arts, often referred to as STEAM. Students are presented with real-world problems informed by technological/engineering design-based learning approaches. These approaches intentionally integrate the content and processes of science, technology, engineering, and mathematics—either in whole or in parts. In this course, project-based learning is adopted as one of the key pedagogical strategies (Darling-Hammond et al., 2015). I teach by modeling STEM pedagogy that facilitates practical engagement, promotes the 4C's (critical and creative thinking, collaboration, and communication) and leads to the development of concrete (i.e., physical) artifacts in response to real-world problems. The sequence of teaching and a summary of my experiences are shown in Table 2.1.

In the first half of the course, the preservice teachers were presented with weekly design challenges that led to the creation of products (e.g., paper planes, model houses). As part of their portfolio tasks, preservice teachers had to demonstrate their understanding of design and engineering processes and their conceptual connections with STEM. These weekly projects also enabled preservice teachers to engage with a real-world problem and student entrepreneurs who visited the class. These

Table 2.1 Real-world STEM in education

Timeframe	Course sequence	Teaching issues
Weeks 1–5	Engaging in real-world activities through the application of project-based learning strategies	Developing a culture of integrative STEM in preservice teachers who were accustomed to thinking in silos
	Integrating STEM knowledge and skills	Activities were well-defined and lacked complexity in these early weeks
	Exploring entrepreneurship in STEM through experienced entrepreneurs	
	Disrupting education: Multidisciplinary innovation workshop with an entrepreneurship focus	
	Assessment 1: Portfolio activities	
Weeks 6–9	Brainstorming ideas in teams on possible activities modeled on project-based learning underpinned by the Australian Curriculum that adopted integrative STEM for their future classrooms	In a very small number of cases, preservice teachers enrolled in the Bachelor of Education (Secondary) found it difficult to think outside their disciplinary boundaries
	Assessment 2: Designing and developing a paper for a teacher professional journal	

visitors presented structured problems that the preservice teachers needed to solve. The preservice teachers were also introduced to the application of STEM in an entrepreneurial context where they were required to define and solve problems from an ill-structured context entitled *Disrupting Education*. This was a cross-faculty workshop aimed at defining problems and developing solutions to improve *education*. In the second half of the course, as part of their assessment, the PSTs had to write a paper that could be published in a teacher professional journal. In this submission, they outlined their ideas on how integrative STEM could be taught to a cohort of students using the Australian Curriculum (see https://www.australiancurriculum.edu.au/). The course sequence and issues arising from my teaching experience are shown in Table 2.1.

The major challenge to my expectations in teaching integrated STEM education related to the lack of willingness or the lack of capacity for preservice teachers to think beyond the disciplines that defined their formal education. This was an issue that I called the *silo mentality* (Kelly & Knowles, 2016), and it was most prominent with secondary preservice teachers as they had specific educational experiences across physics, chemistry, biology, earth sciences, or mathematics. To do away with a silo mentality, I maintained a concrete focus in my teaching, meaning that design and production of artifacts were emphasized together with concepts of technology, science, and mathematics. Initially contexts were very much simulated and problems well-defined, meaning that preservice teachers were not exposed to complex contexts. Overtime, the complexity of the tasks increased. The opportunity to involve preservice teachers with entrepreneurs and in the entrepreneurship workshop was one way to introduce complexity, and further experiences in real-world problem solving.

2.3.2 Teaching Through Integrated STEM Investigations: James's Story

Teaching through integrated STEM investigations is an approach adopted by James during a 13-week course taught in a Bachelor of Education program for preservice teachers specializing in secondary school science and/or mathematics teaching. Before this course, preservice teachers undertake several courses in a specialist discipline as a *major* area of study in physics, chemistry, biology, earth sciences, or mathematics. They also undertake a shorter sequence of courses to achieve a *minor* in a second discipline area. The purpose of this STEM course is to engage preservice teachers in applying critical and creative thinking, and problem-solving skills. They are encouraged to explore transdisciplinary approaches to define, investigate, produce knowledge, and/or solve an authentic problem drawing upon STEM as a collection of analytical tools.

I, James, taught the course via workshops that included a mix of short tutorials, practical STEM investigation activities, and mentoring of groups and individuals. The first 4 weeks focused on idea and problem definition, and the preparation of a research proposal that formed the first assessment item. Students then implemented their investigations with ongoing mentor support from me, and my co-teacher, a professor of mathematics education. To foreground the learning experience of doing a STEM investigation, students maintained an individual reflective journal during the course, and this forms part of the assessment, along with a formal report written by each group. The voices of *students* (i.e., preservice teachers) presented in this section draw upon individual reflective journals that are part of a formal study of preservice teachers learning experiences. That study is funded by my university, with ethics approval and participant consent. Preservice teachers are referred to in this chapter using the pseudonyms Sabrina, Mike, and Emma. These three preservice teachers were selected for the present study because their reflective journals explored personal experiences in-depth, and their journals illustrate diversity across their experiences. Diversity was important because it shows how different preservice teachers respond to the epistemic challenges of a transdisciplinary environment. The course sequence and issues arising from my teaching experience are shown in Table 2.2.

The use of an individual reflective journal was important to foregrounding the learning experiences for my preservice teachers, so that they would appreciate just *how* they were learning by engaging in the practices of STEM. Out of their journals, a number of issues are evident that inform our understanding of this emergent field of integrated STEM education and possibilities for ways of knowing. To consider some of these issues, I provide a series of journal extracts from three preservice teachers in different project groups. See Table 2.3.

From the perspective of the journal extracts in Table 2.2, the preservice teacher experiences and interpretations of their integrated STEM investigations are quite different. For me, what is most interesting about these preservice teacher experiences, evident in Table 2.3, are their approaches to the STEM investigation task,

Table 2.2 STEM investigation project

Timeframe	Course sequence	Teaching and learning issues
Weeks 1–4	Weekly workshops with explicit teaching on transdisciplinary teamwork, problem definition and validation, research design, design thinking and writing a proposal; Mentoring groups	Defining an authentic real-world problem and seeing STEM connections
		Formulating questions to investigate a problem from different disciplinary perspectives
	Assessment 1: Research Proposal	
Weeks 5–11	Group project work and brief tutorials focused on topics such as data analysis and report writing	Teamwork
		Understanding own learning experiences through STEM
		Identity as a STEM Teacher
Weeks 12–13	I mentored individuals on writing their manuscripts	
		Contributing to transdisciplinary work
	Assessment 2: Reflective Journal	Recognizing connections between different STEM perspectives
	Assessment 3: Report	

Table 2.3 Extracts from preservice teacher journals

Pseudonym and timeframe	Journal notes
Sabrina Week 2	"I still have trouble trying to comprehend the approach promoted within this unit [course]; the idea of wicked problems especially, although not in any way confusing or revelatory, is repellent to me as a maths major, as wicked problems are so deliberately avoided in most maths subjects. Controlled experiments with all factors accounted for (or else dismissed from the discussion for ease) are my strong preference, but I wonder if they [controlled experiments] should be avoided for that very reason"
Sabrina Week 3	"This week we added another group member (Agnes), which balances out our maths major/science major split; lucky for us, as this experiment seems much heavier on biology than maths…"
Sabrina Week 6	"Not that I don't like science, but I hate science"
Mike Week 2	"This misconception evoked feelings of discomfort and made me feel disinterested as I hadn't previously been allocated this much flexibility"
	"Although we didn't have as much transdisciplinary diversity as some of the other groups, we did however bring forth differing personalities, experiences and views which I deemed integral to our success"
Emma Week 2	"This week we had to form our groups and start planning our project. While groups are supposed to be as interdisciplinary as possible my group of consists of all Maths majors, however Jean has an English minor, Sally has a Geography minor and I have a Science minor which we feel will bring some different perspectives to our project regardless of the topic. Something that we all wanted to do regardless of our project plan was to design something that mattered. For me, what's really important to consider at the moment is rubbish in our oceans and finding small ways we can reduce that"

their interpretation of the notion of transdisciplinary inquiry, the influence of their own disciplinary identities, and the recognition of what they bring to the task from personal perspectives. For example, Sabrina questions the purpose of engaging in complex, authentic problems that I described during a tutorial by introducing the notion of *wicked problems* (Rittel & Webber, 1973). Wicked problems are complex, ill-structured problems that may change over time, even after preliminary solutions have been developed. For Sabrina, the idea of wicked problems was not "confusing or revelatory." As her teacher, I would agree with this, but I would also suggest that there is a difference between thinking about something as an idea or concept, and engaging in it as a real phenomenon via enacted and lived experiences. For Sabrina, the idea of engaging in a wicked problem was not relevant to her because she was Mathematics major and "wicked problems are so deliberately avoided in most maths subjects." In the context of an integrated STEM investigation, Sabrina seems to interpret the project as requiring different discipline experts to contribute from their discipline only. This reflects a multidisciplinary rather than a transdisciplinary interpretation of integrated STEM. In addition, her interpretation of mathematics as a discipline that avoids the complexity of wicked problems may reflect experiences of procedural and decontextualized pedagogies in her own learning of mathematics as a discipline.

Sabrina's avoidance of the complexity of authentic problems that I had encouraged groups to engage with was further evident in her contentment to do a "controlled experiment," despite the fact that it "seems much heavier on biology than maths." In the context of this course where students could determine any problem they desire, Sabrina seemed satisfied at week 3 with being part of what she perceived as a simple "controlled experiment." However, by week 6 Sabrina had reached a point where she was learning about the complexity of controlled experiments. This complexity was enhanced by the learning context of a high school-style laboratory investigation with rudimentary and very dated equipment. Doing an inquiry in this context meant that students would become aware of the difference between disciplinary idealizations of *control* in a "controlled experiment," and what is actually possible in the lived reality of a school setting. Such differences also occur in real-world scientific research where science is commonly described as *messy* (McLaughlin & MacFadden, 2014). Sabrina's realization of this complexity was evident to me in her frustration during the workshops, where I recall during our interactions that she expressed her dislike for science because scientific methods were tedious and time consuming. She also wrote about this in her journal, evident with her week 6 response, "Not that I don't like science, but I hate science." Despite these emotional experiences, Sabrina and her group engaged with this project at an exceptional level. The technologies they developed as part of their emergent scientific methods involved an array of design subprojects to achieve scientific outcomes acceptable to the group. Sabrina's experience illustrates the complexity of learning through an integrated STEM investigation and how learning is not simply about applying formal analytical epistemologies drawn from idealized disciplines. To understand the localized site of knowledge formation where Sabrina was located, we would need to understand much more about the social interaction and the emotional and cognitive experiences that took place. This may be best analyzed using synthetic philosophical orientations.

A different approach to this project was adopted by Mike, who identified as a mathematics major and science minor preservice teacher. In his week 2 entry Mike refers to his "misconception" about how to proceed in this course, because as his teacher, I had not given any topics or problems to address. Instead, I taught an entrepreneurial technique for defining and validating their own issues or problems of interest (cf. Ries, 2014). Mike noted how this "evoked feelings of discomfort and made me feel disinterested as I hadn't previously been allocated this much flexibility." This is an important reflection because it indicates a source for disengagement from STEM because of the complexity introduced by the open-endedness of this project. Having designed the course for students to *experience* complexity by evoking these learner emotions, I would have been disappointed if Mike had responded differently. As the teacher, I addressed these feelings with Mike and other students by actively discussing how they felt, and explaining that such feelings are normal. I also discussed how they would need to be aware of this as part of their learning to become future STEM teachers. Having the reflective journal as part of their assessment was an excellent tool in this project, because I was able to direct students to reflect actively on their feelings as part of the learning experience. Managing their feelings and persevering with the project was part of the epistemic fluency preservice teachers experienced by learning through their authentic and complex integrated STEM investigation. In this way I was engaging them in a synthetic epistemological experience through an integrated STEM investigation involving an array of analytical epistemological traditions.

The second part of Mike's reflection in Table 2.3 addresses his understanding of transdisciplinarity and diversity in his group. Most of Mike's group was mathematics majors or minors and on his assessment of their formal disciplinary education he stated that, "We didn't have as much transdisciplinary diversity as some of the groups." He then continues with, "We did however bring forth differing personalities, experiences and views which I deemed integral to our success." Mike's observation of his group's diversity is interesting and important, because it evidences a recognition that diversity is not just about disciplinary knowledge or disciplinary epistemological beliefs. Instead, personal interests, gender, ethnicity, and life experiences, as broader informants of identity, may play a greater role in bringing transdisciplinary diversity to integrated STEM investigations as a way of knowing in educational contexts.

This idea of personal diversity in understanding transdisciplinary, integrated STEM was further evident in the reflection by Emma who also considered her groups' formal disciplinary composition. However, Emma reflected that "Something that we all wanted to do regardless of our project plan was to design something that mattered." This was a sentiment also reflected by each of her peers who were all Mathematics majors. Emma's group decided to develop their STEM project around a shared passion for recycling and care for the environment, without regard for their disciplinary backgrounds. This took their investigation down a pathway that involved design thinking, computational thinking, mathematical reasoning, and integration with scientific and everyday knowledge. The product of this project was a smart phone application for educating local students about recycling on the university campus. The engagement of Emma's group was primarily around an authentic and complex problem that the group passionately shared as a common interest to satisfy their desire to "design something that mattered." Like Mike and Sabrina, Emma experienced emotions about the open-ended and complex character of the wicked problems I was asking them to consider and engage with. The emotions were similar, as Emma did describe *apprehension* in another part of her journal; however, the way she and her peers dealt with those emotions was very different from Sabrina. The complexity of these emotional experiences and their interplay with performed practices and identity are areas requiring further investigation in various STEM education contexts.

In this section, I have highlighted the *experiences* of my preservice teachings while they *performed* and formally reported on an integrated STEM investigation project. In effect, this section is my reflection on my teaching experience with these preservice teachers and it illustrates how I have positioned them to actively engage in epistemic fluency across both analytical and synthetic ways of knowing. In the brief samples of data above, preservice teacher references to disciplines and their acknowledgment of transdisciplinary analytical ways of knowing are evident. Simultaneous with analytical ways of knowing, their act of reflecting on experiences foregrounds synthetic ways of knowing such as emotional experiences, and the interactions of these experiences with analytical choices: For example, one group followed their *passions* to find a topic (i.e., "something that mattered"), and another attempted to *avoid* negative feelings about complexity by choosing to do a *controlled experiment*. Both experiences contributed to the way they produced analytical STEM knowledge, which required a degree of epistemic fluency across analytical traditions and synthetic orientations. Importantly, the reflective journal also situated me as the teacher in relation to my preservice teachers, because their experiences became a topic for in-class dialog. In this way I was able to develop in my preservice teachers an awareness about themselves as STEM educators and how their *experiences* of integrated STEM were important for understanding the ways they came to learn, and form knowledge, *through doing* their projects. Based on my experiences as the teacher, I have performed and experienced epistemic fluency throughout teaching my integrated STEM course. Importantly I have attempted to lead my preservice teachers through their own epistemic journeys. This was possible by the structure of my teaching where I required preservice teachers to formally account for their project performance in a STEM report, while also reflecting upon *how* that performance was experienced as a series of synthetic learning experiences. In summary, epistemic fluency may be enriched in the teaching of integrated STEM with purposeful use of reflection to raise self-awareness of synthetic ways of knowing simultaneously with the performance of formal analytical ways of knowing. The approach I have described above for tackling these diverse epistemologies in integrated STEM education is offered as a start point for understanding, teaching, and exploring further the nature of epistemic fluency in initial teacher education courses.

2.4 Discussion: Re-imagining Integrated STEM Education

At the beginning of this chapter, we problematized integrated STEM in initial teacher education by exploring the need for epistemic fluency as an analytical skill. We have described how the STEM disciplines may be integrated through a confluence of formal analytical epistemologies. Our analyses of teaching experiences and some preservice teacher learning experiences across the courses of Vinesh and James suggest that initial teacher education may reinforce the boundaries between the STEM disciplines. These boundaries may be viewed as part of the professional identities of preservice teachers and give rise to what Vinesh calls *silo mentality*. As one of our preservice teachers reflected on both her identity and her discipline (i.e., Sabrina) it became evident that preservice teacher identities may be shaped not just by disciplinary knowledge but by the way that knowledge is taught, that is, pedagogy (cf. Nagdi, Leammukda, & Roehrig, 2018). While procedural understandings of disciplinary knowledge are one form of understanding, the contextualization of disciplinary knowledge within transdisciplinary, authentic, real-world, and problem-based environments provides a broader form of understanding. The broader understanding afforded by an integrated STEM approach is applied in an initial teacher education context to build teacher capacity for making connections between STEM and real-world contexts.

To achieve broader understandings of disciplinary knowledge, we point toward synthetic orientations to epistemology that enable ways of knowing through integrated STEM to be learned as part of localized contexts. Such approaches may be described as indifferent, agnostic, or predisciplinary because they enable epistemology to be understood across the analytical boundaries that define the silo mentality. For example, school students doing an empirical scientific inquiry into the effect of heat on water currents would typically described the *facts* about the equipment they use and the observations they make to reach a conclusion. Such a description would account for their analytical way of knowing that they would call *science*. But we could also study what they *actually* did during the particular occasion of their scientific practices, in terms of conversations, bodily actions, and evidence of emotional experiences (see Davis & Bellocchi, 2018). These particular bodily actions and experiences would not be typically attributed to the formation of knowledge in science, because they are glossed over and taken for granted by people in their moment-to-moment lived experiences. The study of such particularities is an example of a synthetic way of knowing that could be applied to any STEM investigation regardless of the analytical ways of knowing that may be attributed by people to the formation of knowledge. As such, a synthetic way of knowing may be applied across disciplinary boundaries to more fully understand taken-for-granted and localized synthetic epistemologies through which analytical epistemologies are typically performed.

For this reason, we may say that synthetic epistemological orientations are embedded in context that involve cognition, social phenomena, personal idiosyncrasies, and the creativity of localized contingencies such as situated cultural practices and emotional experiences. As we have seen from our data, such lived experiences of STEM investigators may influence their decisions to engage with one problem or another. This lived experience may influence their choices to engage with one method or another, to engage more deeply, or to disengage altogether. As such, the performance of epistemic fluency as a lived phenomenon during particular occasions of integrated STEM investigation needs to be understood better than it is currently.

2.5 Conclusion and Implications

This study has implications for future research into integrated STEM pedagogy for schools and higher education, including science and engineering faculties. Re-imagining integrated STEM as something greater than its constituent concepts within traditional boundaries and analytical epistemologies requires something more than is provided by conventional and procedural STEM education. New pedagogical orientations maybe needed, so that preservice teachers are immersed in the experience of integrated STEM and the complexity of real-world context, ill-structured, and wicked problems. This requires an understanding of integrated STEM epistemologies from synthetic philosophical perspectives where learning *as* and *through* context may be experienced. Understanding *how* STEM phenomena, social interaction, embodied cognition, social cognition, emotional experience, and localized epistemology are intertwined in the performance of integrated STEM investigations is important for preservice teachers learning. It is important if preservice teachers are to appreciate fully what transdisciplinary integrated STEM could be, and the possibilities for who they and their future students may become by *learning through* integrated STEM.

Acknowledgments We wish to acknowledge the STEM Education Research Group and the School of Teacher Education and Leadership of the Faculty of Education, Queensland University of Technology, for funding that partly supported this work. Any opinions, findings, and conclusions, or recommendations expressed in this article are those of the authors and do not necessarily reflect the views of the Queensland University of Technology.

References

Allchin, D. (1999a). Do we see through a social microscope? Credibility as a vicarious selector. *Philosophy of Science, 66*, S287–S298.

Allchin, D. (1999b). Values in science: An educational perspective. *Science & Education, 8*, 1–12.

Bellocchi, A. (2017). Interaction ritual approaches to emotion and cognition in science learning experiences. In A. Bellocchi, C. Quigley, & K. Otrel-Cass (Eds.), *Exploring emotions, aesthetics and wellbeing in science education research* (pp. 85–105). Dordrecht, The Netherlands: Springer. https://doi.org/10.1007/978-3-319-43353-0_5

Bellocchi, A., King, D., & Ritchie, S. M. (2016). Context based assessment: Creating opportunities for resonance between classroom fields and societal fields. *International Journal of Science Education, 38*, 1304–1342. https://doi.org/10.1080/09500693.2016.1189107

Buldt, B., Lowe, B., & Muller, T. (2008). Towards a new epistemology of mathematics. *Erkenntnis, 68*(3), 309–329.

Chalmers, A. (2013). *What is this thing called science?* Brisbane: University of Queensland Press.

Collins, R. (2004). *Interaction ritual chains*. Princeton, NJ: Princeton University Press.

Darling-Hammond, L., Barron, B., Pearson, P. D., Schoenfeld, A. H., Stage, E. K., Zimmerman, T. D., … Tilson, J. L. (2015). *Powerful learning: What we know about teaching for understanding*. Wiley.

Davis, J. P. (2017). Emotions, social beings & ethnomethods: Understanding analogical reasoning in everyday science classrooms. In A. Bellocchi, C. Quigley, & K. Otrel-Cass (Eds.), *Exploring emotions, aesthetics and wellbeing in science education research* (pp. 121–140). Dordrecht, The Netherlands: Springer. https://doi.org/10.1007/978-3-319-43353-0_7

Davis, J. P., & Bellocchi, A. (2018). Objectivity, subjectivity and emotion in school science inquiry. *Journal of Research in Science Teaching, 55*, 1419–1447. https://doi.org/10.1002/tea.21461

Davis, B., & Sumara, D. (2006). *Complexity and education: Inquiries into learning, teaching, and research*. New York, NY: Routledge.

Durkheim, E. (2008). *The elementary forms of the religious life* (C. Cosman, Trans.). Oxford: Oxford World's Classics. (Original work published 1912).

English, L. D. (2016). STEM Education K-12: Perspectives on integration. *International Journal of STEM Education, 3*. https://doi.org/10.1186/s40594-016-0036-1

Fan, L., & Zhu, Y. (2007). Representation of problem-solving procedures: A comparative look at China, Singapore, and US mathematics textbooks. *Educational Studies in Mathematics, 66*, 61–75. https://doi.org/10.1007/s10649-006-9069-6

Hong, Y.-C., & Choi, I. (2011). Three dimensions of reflective thinking in solving design problems: A conceptual model. *Educational Technology Research & Development, 59*, 687–710. https://doi.org/10.1007/s11423-011-9202-9

Huq, A., & Gilbert, D. (2017). All the world's a stage: Transforming entrepreneurship education through design thinking. *Education and Training, 59*, 155–170. https://doi.org/10.1108/ET-12-2015-0111

Kelly, T. R., & Knowles, J. G. (2016). A conceptual framework for integrated STEM education. *International Journal of STEM Education, 3*. https://doi.org/10.1186/s40594-016-0046-z

King, D., & English, L. (2016). Engineering design in the primary school: Applying stem concepts to build an optical instrument. *International Journal of Science Education, 38*, 2762–2794. https://doi.org/10.1080/09500693.2016.1262567

Liberman, K. (2007). *Dialectical practice in Tibetan philosophical culture: An ethnomethodological inquiry into formal reasoning*. Lanham, MD: Rowman and Littlefield Publishers.

Liberman, K. (2013). *More studies in ethnomethodology*. Albany, NY: SUNY.

Markauskaite, L., & Goodyear, P. (2016). *Epistemic fluency and professional education: Innovation, knowledgeable action and actionable knowledge*. Dordrecht, The Netherlands: Springer.

McLaughlin, C. A., & MacFadden, B. J. (2014). At the elbows of scientists: Shaping science teachers' conceptions and enactment of inquiry-based instruction. *Research in Science Education, 44*, 927–947. https://doi.org/10.1007/s11165-014-9408-z

Nadelson, L. S., & Seifert, A. L. (2017). Integrated STEM defined: Contexts, challenges, and the future. *The Journal of Educational Research, 110*, 221–223. https://doi.org/10.1080/00220671.2017.1289775

Nagdi, M. E., Leammukda, F., & Roehrig, G. (2018). Developing identities of STEM teachers at emerging STEM schools. *International Journal of STEM Education, 5*, 1–13. https://doi.org/10.1186/s40594-018-0136-1

Noble, T., DiMattia, C., Nemirovsky, R., & Barros, A. (2006). Making a circle: Tool use and the spaces where we live. *Cognition and Instruction, 24*, 387–437. https://doi.org/10.1207/s1532690xci2404_1

Núñez, R. E., Edwards, L. D., & Matos, J. F. (1999). Embodied cognition as grounding for situatedness and context in mathematics education. *Educational Studies in Mathematics, 39*, 45–65.

Quality Assurance Agency. (2018). *Enterprise and entrepreneurship education: Guidance for UK higher education providers*. Retrieved August 30, 2018, from http://www.qaa.ac.uk/docs/qaas/enhancement-and-development/enterprise-and-entrpreneurship-education-2018.pdf?sfvrsn=15f1f981_8

Ries, E. (2014). *The lean startup: How today's entrepreneurs use continuous innovation to create radically successful businesses*. New York, NY: Currency.

Rittel, H. W. J., & Webber, M. M. (1973). Dilemmas in a general theory of planning. *Policy Sciences, 4*, 155–169.

Sanders, M. E., & Wells, J. G. (2010). *Virginia Tech, Integrative STEM Education Graduate Program*. Retrieved from http://web.archive.org/web/20100924150636/http://www.soe.vt.edu/istemed

Shernoff, D. J., Sinha, S., Bressler, D. M., & Ginsburg, L. (2017). Assessing teacher education and professional development needs for the implementation of integrated approaches to STEM education. *International Journal of STEM Education, 4*, 1–16. https://doi.org/10.1186/s40594-017-0068-1

STEM Task Force Report. (2014). *Innovate: A blueprint for science, technology, engineering, and mathematics in California public education*. Dublin, CA: Californians Dedicated to Education Foundation.

Takeuchi, M. A. (2016). Friendships and group work in linguistically diverse mathematics classrooms: Opportunities to learn for English language learners. *Journal of the Learning Sciences, 25*, 411–437. https://doi.org/10.1080/10508406.2016.1169422

Valiela, I. (2001). *Doing science: Design, analysis, and communication of scientific research*. New York: Oxford University Press.

Van Manen, M. (1997). *Researching lived experience: Human science for an action sensitive pedagogy*. Walnut Creek, CA: Left Coast Press.

Watson, A. D. (2015). Design thinking for life. *Art Education, 68*, 12–18.

Wickman, P.-O. (2017). Back to the drawing board: Examining the philosophical foundations of educational research on aesthetics and emotions. In A. Bellocchi, C. Quigley, & K. Otrel-Cass (Eds.), *Exploring emotions, aesthetics and wellbeing in science education research* (pp. 9–37). Dordrecht, The Netherlands: Springer. https://doi.org/10.1007/978-3-319-43353-0_2

Chapter 3
Towards a Production Pedagogy Model for Critical Science and Technology Interventions

Gabriela Alonso Yanez, Kurt Thumlert, Suzanne de Castell, and Jennifer Jenson

Abstract While initiatives advancing STEM education are pervasive within the global landscape of educational reform today, STEM discourses and reforms largely fail to articulate or enact theoretical and epistemological shifts that critically conceptualize the impacts of science and technology in bio-physical and social worlds. The urgency to adopt STEM reforms in North American schools and to "train" students for competitive twenty-first century "knowledge economies" has resulted in an uncritical embrace of underlying STEM narratives, in turn foreclosing critical discussion, alternative models, and new perspectives on how we might *do* science and technology education differently. In this chapter, we review critical literature in science education in order to unpack the dominant narratives of preparation, progress, competition, and innovation that drive STEM pedagogies today. We draw upon critical sustainability studies (CSS) to articulate new axiological orientations for repositioning science and technology learning. In conjunction with CSS, we articulate the opportunities of "production pedagogy" theories and practices which

An earlier version of this chapter appears in future: the journal of policy, planning, and futures studies. This chapter has been substantially revised.

G. Alonso Yanez (✉)
Werklund School of Education, University of Calgary, Calgary, AB, Canada
e-mail: galonsoy@ucalgary.ca

K. Thumlert
Faculty of Education, York University, Toronto, ON, Canada
e-mail: kthumlert@edu.yorku.ca

S. de Castell
Faculty of Education, University of Ontario Institute of Technology, Oshawa, ON, Canada
e-mail: suzanne.decastell@uoit.ca

J. Jenson
Digital Languages, Literacies and Cultures, Department of Language and Literacy Education, Faculty of Education, University of British Columbia, Vancouver, BC, Canada
e-mail: jennifer.jenson@ubc.ca

P. Sengupta et al. (eds.), *Critical, Transdisciplinary and Embodied Approaches in STEM Education*, Advances in STEM Education,
https://doi.org/10.1007/978-3-030-29489-2_3

provide a critical framework for revisioning science and technology education—not as developmental system that prepares students for some preordained future—but rather as a dynamic vehicle that can situate learners in agentive roles *now*, in the present, using real-world tools in authentic sociotechnical contexts and communities.

Keywords Science education · STEM learning · Critical sustainability · Knowledge production · Social action

Policy and initiatives advancing STEM education are pervasive within the global landscape of educational reform. While STEM policy documents provide rationale for ratifying new curricula that can deepen students' engagement in the sciences, mathematics, engineering, and technology fields, STEM education discourses largely fail to translate innovations in policy into innovations in pedagogy, and neglect, as well, theoretical and epistemological advances in conceptualizing the impacts of science and technology in physical, social, and symbolic worlds (Rudolph, 2008; Zeidler, 2016). At the same time, the urgency to adopt and implement STEM reforms in North American schools has resulted in an uncritical embrace of underlying STEM aims and purposes, in turn foreclosing dissent, critical discussion, alternative models, and new perspectives on how we might "do" science, and (STEM) education, differently.

In schools, standardized assessments and a dependence upon routinized curricular forms mediate and mitigate efforts to critically engage and rethink STEM education, particularly under conditions where governmental policy and corporate/media pressures rapidly enroll policy makers, educators, and students onto STEM "bandwagons" (Carter, 2016; Sharma, 2016; Zeidler, 2016). At the same time, STEM reform initiatives continue to abstract science and technology practices from wider sociotechnical contexts and narrowly delimit STEM learning to the developmental acquisition of decontextualized "technical skills," compartmentalized disciplinary content knowledge (Krug & Shaw, 2016). In the rush to implement STEM reform policies in schools, STEM education aims are adapted for, and translated into, quite traditional pedagogies, conventional curricular forms, and standardized assessments (Henderson & Dancy, 2011; Sanders, 2008). Consequently, reform efforts may maintain STEM learning in discrete (non-communicating) disciplinary worlds, as well as perpetuate a "scientific rationalist approach" to curriculum and science studies (Gough, 2015) that further separates science and technology education from wider social worlds—from the dramas, living controversies, and critical social justice and ecological questions of our time.

Historically, science education reform initiatives have been mobilized as responses to national security "crises," workplace "supply shortages," and "economic downturns" (Krug & Shaw, 2016). Today, rationales for STEM education are similarly framed around concerns to address (re)emerging economic crises and perceived deficits in worker skills, accompanied by an instrumentalized and entrepreneurial language set in *preparing* students for the professional dispositions required to succeed—and to innovate—in increasingly competitive and globalized twenty-first century "knowledge economies" (Olssen & Peters, 2005). Absent from dominant STEM discourses, however, are critical analyses or a problematization of the broader social, economic, or ecological implications of these rationales, goals, and related science practices, nor questions about whose interests, or what kinds of interests, are actually being served (Habermas, 1968; Harding, 1991; Tan & Calabrese Barton, 2018).

Indeed, overshadowed by the rhetoric of educational "preparation" are the increasingly urgent ecological, ethical, and social justice exigencies that require us to critically rethink education in general, and STEM education in particular, not as instructional programs that developmentally prepare students for some preordained future, but rather as dynamic vehicles that situate learners in critical, agentive roles *now*—in the present—enabled to negotiate real-world creative challenges, controversies, and dilemmas using real-world tools in authentic sociotechnical contexts (Luke, Sefton-Green, Graham, Kellner, & Ladwig, 2017; Thumlert, 2015).

By contrast, what frequently passes as self-evident in STEM education discourses is the exigency to reform and refine the very mechanisms of "preparation," the pathways and curricular processes that would developmentally equip students for STEM fields, professions, and futures. In this chapter, we first interrogate the discourses, curricular practices, and narrative pathways that STEM reforms *anticipate* for learning actors. In short, how are students, through induction into STEM education's preparatory systems, uncritically incorporated into a curricularly prepackaged and "ready-made science" order (Latour, 1987; Visvanathan, 2006) and, with that, into unsustainable science and technology values, roles, and epistemologies.

We then sketch out a *pedagogical* model that might recompose science education and learning in an explicitly critical key, where students are positioned instead to engage science and technology questions and controversies and, further, take up agentive positions in reconfiguring their own individual and collective futures, here and now. We begin with a review of critiques of STEM from within contemporary science education theory. We then draw upon critical sustainability studies (CSS) to articulate new practical and axiological orientations for repositioning science and sustainability practices today. In conjunction with CSS, we then articulate the opportunities of "production pedagogy" theories and models (de Castell, 2010; Smythe, Toohey, & Dagenais, 2016a, 2016b; Thumlert, de Castell, & Jenson, 2015) that provide a pedagogical framework for doing (science) education differently, as well as modelling democratic science practices: means of engaging the problems of the public, and the politics of science, through critical making and media interventions. We conclude with reference to one working model that, by conjoining critical sustainability with authentic artifactual and media production today can help us rethink and *redo* science and technology education.

3.1 Contesting the Hegemony of Current STEM Reforms: Critical Science Education

Critical work in science education has contested the "hegemony" of STEM discourses (Sharma, 2016) as well as interrogated State, corporate, and educational policy directives that pressure actors in schools to rapidly adopt STEM initiatives (Zeidler, 2016). This body of work includes critiques of the neoliberal discourses and economic narratives embedded in STEM educational policies, discourses that widely underwrite reform efforts in schools (Carter, 2016; Sharma, 2016; Weinstein, Blades, & Gleason, 2016). For example, in a comprehensive review of key governmental, educational, and business documents advancing STEM education initiatives, Krug and Shaw (2016) identify three dominant narratives that inform and drive STEM policy—"progress, innovation, and global competitiveness"—narratives that link STEM education reform to economic purposes and corporate (technocapitalist) aims (p. 185). At the same time, related crisis narratives—narratives of "declining empire" (Sharma, 2016), school failure, "skills gaps," and scarcity in "human capital" and "STEM-ready workers" (Carter, 2016)—insistently reify STEM education as singular panacea to the very problems these discourses define. Zeidler (2016) describes this as the "STEM deficit model" (p. 12), where the deficits, gaps, and declines articulated in dominant STEM discourses can be ameliorated by more STEM investment, more reform, and more enhancement of existing STEM curriculum and policies.

The deficit model described by Zeidler (2016) seamlessly dovetails with a complementary technocapitalist narrative: "STEM as societal salvation" (Weinstein et al., 2016). This latter narrative envisions STEM education as a vehicle for innovation and technical solutions (e.g., solar panels, robotics) to global problems, while fitting science practices, innovation, and even "sustainability" within technical-rational and neoliberal frameworks. Here, dominant STEM discourses and practices in education, when they engage contemporary ecological and social matters, typically do so in terms of an instrumental solutionism where science is embedded in narratives that celebrate *homo faber* innovation—and where technical innovations feed forward into cycles of *homo economicus* consumption and exchange (Gough, 2015; Lyotard, 1984; Weinstein et al., 2016; Zouda, 2016).

Left unaddressed in STEM (etiological) "deficit models" and (eschatological) "salvation" narratives are problematic contradictions: above all, the very same neoliberal discourses that rationalize and promote STEM education as solution/salvation also implicitly naturalize free markets, the free movement of commodities, and the free investments of global (human) capital in largely unregulated, low-wage labor sites (i.e., in "developing nations") (Noble, 2018; Troncoso, 2018). In turn, this economic liberalization puts increasing competitive pressure on North American job markets—as well as global physical systems—through the liberalized "flows" of (human) capital and the networked "offshoring/outsourcing" of STEM jobs.

These operations devalue wages globally and further deteriorate labor conditions, increasing the precarity of already precarious twenty-first century jobs in

North America and elsewhere. In these contexts, STEM (education) narratives, fueled by neoliberal discourses of competition, demands for "national leadership" in innovation, and the closing of the so-called STEM skills gap, ultimately amplifies—and tautologically ensures the reproduction of—the very crises and deficits that STEM education would presume to "solve." Further, STEM "solutions" are articulated in terms of expert technical-rational innovations designed to, on the one hand, optimize the system that develops markets and generates economic growth and, on the other, remedy or "fix" the growing ecological/environmental consequences that exfoliate from these same global processes and relations (see Weinstein et al., 2016).

Critical educational researchers have thus called for transformations in, or "disruptions" of, dominant STEM (education) practices, with recent calls not just to contest economic narratives, but to rethink the very narratives, roles, and practices of "normal" STEM science (Bencze & Carter, 2015; Reiss, 2003). Researchers signal that STEM education, regulated by traditional assessments, curricular forms, and positivist epistemologies, selectively ignore socio-scientific issues and disengage science education from worlds outside of schools, as well as insulate STEM education from a reflexive interdisciplinary critique of the complex impacts of science and technology (Gough, 2015; Zeidler, 2014, 2016).

At the same time, critical researchers in science education share common ground in their resistance to a tacit educational embrace of "value-neutral" technology tools and methods, claims to objectivity and impartiality, along with idealized "views of the nature of science," static principles, and abstract "scientific knowledge" that continues to inform STEM curriculum, instruction and assessments (Rudolph, 2008). For example, as Barrett (2006, 2007) states, environmentally focused science education, within traditional settings where curricular objectives are imposed *upon* students, frequently negates opportunities for students to work, engage, and perceive themselves as meaningful collaborators in situated environmental, societal, and technological processes, or as agents capable of enacting change. In these contexts, STEM education often celebrates—and developmentally prepares students for—the very same technical-rational epistemological orientations to "Nature" that have, arguably, contributed to the precarity of our contemporary Anthropocene (Escobar, 2015; Houston, 2013). If STEM reforms uncritically equip learners for ready-made science dispositions, skills, and roles, they also generate "anticipatory regimes" (Adams, Murphy, & Clarke, 2009; Amsler & Facer, 2017) that delimit possible futures: what is thinkable and doable as "science."

While calls to connect STEM education to environmental issues, sustainability, and "real-world problems" are clearly audible (Krug & Shaw, 2016), more and more researchers invite us to, further, fundamentally challenge the very *construction* of science in STEM education, and to "tinker" with, and even "critically disrupt or displace" dominant STEM methodologies "toward eco-social justice" orientations (Sharma, 2016) or STS interventions that critique the inherited languages and procedures of "expert" science (Alonso-Yanez, 2018; Datta, 2018; York, 2018). STS researchers have long signalled more radical opportunities to "remake science" (Latour, 1987), articulating "post-normal," and critical science and technology models that at once resist the alignment of STEM with economic and technocratic goals, while also challenging inherited epistemologies and curricular norms/forms that reproduce "business as usual science" learning practices in schools (Amsler & Facer, 2017; Stengers, 2018). Saliently, these authors demand that the emerging practices of science research and knowledge-making be made procedurally inextricable from questions of "ethical responsibility," "transparency," and public "participatory governance."

One way to challenge the hegemony of STEM discourses, pedagogies, and policies is to draw attention to critical sustainability studies (CSS). In the next section, we explore CSS as a model for science and technology education that may very well help us articulate more critical and interdisciplinary pedagogies.

3.2 Critical Sustainability Studies: A Novel Framework for STEM Education

Critical sustainability studies (CSS) emerged in the 1990s as an interdisciplinary field of inquiry concerned with defining the limits, growth, and the inherently political nature of the organizing principle of "sustainable development" that many at the time pursued and embraced as a model for reconciling environmental challenges and science practices with imperatives of social and economic development (Rose & Cachelin, 2010; Springett, 2005).

CSS challenged the "rationality" of the capitalist paradigm that underlies the notion of sustainable development by calling for an examination of the economic, political, social, technological, and environmental forces that foster or impede rigorous, critical orientations to sustainability, and narrowly define the boundaries and meanings of "sustainability" itself. Briefly, early conceptions of sustainable development or eco-development were originally brought forward by Indigenous communities and local social groups (Springett, 2005). This original perspective embraced goals of equity and social justice in radical new ways, and was based on Indigenous rights and local community organizations' "bottom-up" models, means and models developed to fight back against emerging neoliberal policies that prominently affected local communities by applying market mechanisms to bio-physical environments.

Over time, the notion of sustainable development was hijacked by corporate discourses and modified to represent "greener" neoliberal-inspired eco-business and large-scale environmental transformations labelled as "eco-efficient" or "win-win" for all stakeholders. These conceptions helped define the emerging capitalist paradigm of "sustainability," and legitimated new forms of colonization and imperialism shaped by the pursuit of resources, expanding markets, and land management goals. At the same time, "sustainability" projects were, and continue to be, mobilized through rhetoric of "crisis" intervention and, with that, a top-down transnational solutionism governed by (non-local) experts (Alonso-Yanez, Thumlert, & De Castell, 2016; Ball, Owen, & Gray, 2000; Sterling, 2001).

CSS thus emerged as a response to the disappointments and tragedies of development assistance initiatives that had ignored local conditions, cultures, and the capacities of marginalized actors—initiatives that translated bio-physical "crises" into development opportunities legitimated by (STEM) salvation narratives (Alonso-Yanez et al., 2016). In short, CCS was a reaction against neoliberal practices advanced under the banner of "sustainable development." The starting point for a critical analysis of sustainability is the idea that unsustainability, in all its manifestations, is a result of powerful economic scientific and technological systems of domination and the influence of the institutional structures that support those systems. In this sense, sustainability, as a set of ecological discourses, has been subsumed and largely aligned with neoliberal discourses and technocapitalist interests designed to further entrench dominant global economic orders and relations.

Increasingly, however, CSS initiatives have captured scientific and popular attention, providing interdisciplinary research with a productive, critical entry point for the in-depth study of society–nature relations and the sociotechnical impacts of technology innovations in context-specific landscapes. CSS has led to a better understanding of how environmental, economic, and societal change processes are dynamically interconnected, and illuminate where contradictions in dominant "sustainability" paradigms emerge (Folke, 2006; Ostrom, 2009). Further, CSS has also explored how the historical and geographical expansion of capitalist society–nature relations has led to persistent social inequalities through technocratic governance instruments that replicate and reinforce social and environmental justice disparities (Balvanera et al., 2017; Bolin & Kurtz, 2018; Díaz et al., 2015).

Saliently, CSS critiques share many of the same concerns identified by educators critical of STEM reforms, as examined in the previous section, including the ways the economic narratives of "progress, innovation, and global competitiveness" are interwoven within "sustainable development" ideologies, and how ecological crises are rhetorically positioned to be "best addressed" by technocapitalist "solutions," or through top-down forms of neoliberalized governance and external coordination (Alonso-Yanez et al., 2016).

Alternatively, a vast body of literature is emerging, across diverse intellectual traditions and geographical locations, on the role of education in activating an informed and more critical citizenry capable of participating in, and making decisions about, current global and local problems. Much of this work focuses on issues involving science, technology, society and the environment (Pedretti & Nazir, 2011), as well as on the limitations of instrumentalist orientations of science, technology, and environmental education (Sterling, 2001). In Latin America in particular, clusters of educators, ecological-economists and environmental scientists have promoted efforts for education to foster in students the capacities to make considered, contextually informed decisions and actively "monitor technology" innovations — and what Latin American scholars have called "mercenary techno-science" (http://www.etcgroup.org/). Mercenary techno-science is understood as scientific and technological knowledge production subjugated to the ends of capital accumulation and profit (Toledo, Garrido, & Barrera-Bassols, 2015) or to the needs of the State in maximizing the efficient operations of social systems where scientific innovations are always already commodities within circuits of exchange and, as such, disconnected from critical values, questions of social justice, or alternative moves and models for practice and action (Lyotard, 1984).

It is fair to say that there is much theoretical work that has provided a vocabulary for critical discussions that might transform STEM teaching and learning today. Often however, the theoretical value of this literature, as robust as it is, remains unrealized in pedagogy and educational discourses. We suggest that CSS, as a field of inquiry, sets out ontological and epistemological presuppositions that stand as counter narratives to the conventional teaching of science and technology, as well as to STEM reforms. For one, CSS, as a field of inquiry and practice, refuses to enroll people (students) in, or prepare them for, science narratives and practices that are, we argue, fundamentally unsustainable. Further, CSS makes a strong contribution to understanding the complexities of socio-ecological and socio-technological systems by grasping the inherently political nature of negotiating sustainable futures, rather than silencing or coopting these debates.

Challenging the commoditization of scientific knowledge, CSS inquiry also emphasizes the immediate/local use value of scientific inquiry and making in relation to the needs and concerns of local actors and communities—from matters of governance to simply doing science and technology practices differently, according to the situated and self-defined interests of actors/learners. CSS demands consideration of who is most affected by economic models premised on unsustainable "sustainability" models that accelerate resource exploitation and wealth accumulation. It also interrogates the technocentric salvation narratives advanced in corporate, governmental, and STEM policy sectors—from romantically "solving" worldwide ecological crises to having the last word on debates surrounding "progress," from geoengineering and genetic modification to artificial intelligence and the very meaning of sustainability itself.

CSS thus provides a conceptual framework with which to begin analyzing the array of sociotechnical and scientific issues and to thereby disrupt "business as usual" orientations to STEM education. The framework also offers a means to evaluate the pedagogical approaches and educational practices most suited to, on the one hand, developing critical more democratic forms of engagement and action necessary for human and ecological survival and, on the other, bringing science and technology practices out of the world of experts and into everyday negotiation and meaningful use by diverse actors. These approaches and practices can work in the realm of the "not-yet possible" and challenge current formal education logics in modern capitalist societies (Amsler & Facer, 2017).

CSS problematizes the declared objective of scientific research and innovation policies in the global North, which are aimed at promoting STEM disciplines so as to enhance each nation's competitiveness in the global market. And here, CSS activates critical orientations that can transform the of "doing" science, for example, by acknowledging scientists' and engineers' societal responsibility to engage in continuous and widespread consultation with diverse publics including local and non-traditional knowledge keepers (as capable co-inquirers and sovereign partners); by offering transparency to the public about uncertainties, unknowns, and assumptions, and inviting critical debate; by using extended peer communities of stakeholders to assess the quality and value of the scientific knowledge that is produced, and who benefits from this knowledge (Craye, Funtowicz, & Van Der Sluijs, 2005; Funtowicz & Ravetz, 2003).

Extrapolating these considerations to educational contexts, it is possible to envision how CSS frameworks might create spaces for students to interrogate discourses of techno-scientific innovation, while also enacting science practices and critically inquiry shaped by different values and different orientations to knowledge and artifactual making, including immediate and self-defined "use value" (de Castell, 2016). For example, in discussing current responses to climate change advanced by governments and industries, CSS invites students to interrogate the world-altering engineering technologies portrayed as bringing unquestioned benefits. They have an opportunity to reconsider these purported benefits in terms of their potential to amplify inequalities, displace populations, threaten cultures, and harm or even eradicate physical environments (Klein, 2015).

Critical sustainability, as educational practice, thus invites modes of inquiry and learning within science, mathematics, and engineering that are socially situated, participatory, and openly "political, and not neutral" (Springett, 2005, p. 147). This perspective destabilizes the ideal of narrow technical specialization as the central goal for STEM learning and further invites closer attention to landscapes of power where capital, race, gender, and access are factors in reproducing unequal conditions that affect both bio-physical realms and social futures. It also invites educators to question the historical and cultural development of science and technology, as well as the uneven and exploitive global distribution of resources and labor in the world today. More importantly, CSS invites students to act as "skeptical agents" (Springett, 2005 p. 157), encouraging them to question the dominant narratives informing STEM education reforms today.

In the following sections, we explore production pedagogy as a means to resituate and transform science education today. As we will show, production pedagogy can activate the epistemological and axiological orientations of critical sustainability studies and enable us to situationally, through action and doing, challenge the narratives and anticipated futures embedded in dominant STEM narratives and curricular forms.

3.3 Production Pedagogies: Making and Engaging Meaningful Social Action

In contrast to the discourses of "progress" and "preparation" that underwrite the "schooling" of STEM education today, production pedagogies are premised on the view that people learn best, and learn most deeply, through design and making things that address learners' *present* needs and purposes: real-world objects and technology artifacts that have social worth, that have immanent use value, and therefore *matter* to their makers. Production pedagogies offer an interdisciplinary pedagogical orientation where learning actors are supported to engage real-world research challenges and design competences, using real-world tools, "through the making of authentic cultural artefacts—with correspondingly authentic audiences enabled to witness such acts of knowledge production" (Thumlert et al., 2015, p. 797).

Production pedagogy research suggests that student work should start off with an active engagement that connects situated exploration and research to authentic "production work"—the process of doing and making followed by an iterative course of critical reflection and theorization (de Castell, 2010) where students consider the impacts, and take responsibility, for what they have made. Production pedagogies invite students to present or publish what they create through material or networked interventions that transcend the world of standardized schooling assessments, connecting to worlds and communities outside of schools. Production pedagogy understands that making is, before anything, a process that must be "located within and subordinated to meaningful social action" where the production of socially valued "things" is integral to educational activity and "critical thinking is built into [it]" (de Castell & Jenson, 2006, pp. 240–246).

Whereas constructionism is based on a *learning theory*, closely informed by a Piagetian constructivism that centers on a theory of cognition and developmental learning, production pedagogy is not a theory of learning. Production pedagogy is just that, a pedagogy that untangles "making" from cognitivist and school-driven aims and purposes. Constructionism conceives of "student-centered" making in terms of a "reconstruction" of anticipated knowledge structures through the "making" process. School-bound social interactions and affective investments, while valorized, are nevertheless yoked and subordinated to the school-defined *learning objectives* to be ultimately acquired through making: abstract structures and "cognitive gains" which may have very little to do with learners' own purposes for producing. In these contexts, recent uses of constructionist theories of learning are frequently synchronized with, or subordinated to, the purposes of school orders, skills assessments, as well as dominant (STEM) narratives (Thumlert et al., 2015).

Indeed, constructionist theories of learning and attempts to apply it to school settings have done very little in the past 50 years to transform anything but discourse in education, even in technologically mediated education. As Kafai (2006) argued, education remains focused on "instructionist" pedagogies, rather than on constructionist ones (Jenson, Black, & de Castell, 2018), and even less so on pedagogies of production (de Castell, 2016). Since Papert's foundational work, constructionist models—and educational policies more generally—have been increasingly subsumed by means-ends educational discourses and techniques, where constructionist making in schools is (re)positioned, more conventionally, in terms of "equipping" students with technical knowledge and skills and "preparing" them for "participation in the STEM-related workforce of today or the future". Indeed, against the backdrop of the dominant narratives driving STEM education (examined above), constructionist making in schools—from robotics to digital game making—has been subjected to instrumentalized ends, where the "instructionist" and "technocentric" orientation to making critiqued by Papert long ago (1987) returns to drive what passes for "constructionist learning" in much contemporary theory and practice.

However, by privileging meaningful social action—and the making of socially valued things (as codetermined by learning actors)—production pedagogies directly challenge a jargon of authenticity that characterizes much educational discourse today, particularly when and where terms like "the social" refer to in-class sociality and "authenticity" itself refers to educational practices and "makers' activities that, far from being sociotechnically authentic, are subject to curricular mediation, are (re)contextualized for assessment purposes (where the objects of making itself are assessed against technocentric scales devised by STEM policy makers and schooling systems), or are materially translated into prepackaged "maker kits" where the outcomes of construction are uniformly predetermined in advance (Pinto, 2015). Here, the "authentic" ends and aims of production may have very little to do with the critical purposes of makers themselves.

Production pedagogies go beyond the making of school-bound "objects to think with" (Papert, 1980) to making social-valued objects *to do* with. By implication, production pedagogies link self-determined critical inquiry to forms of genuinely authentic production and can thus operate as material and/or media interventions in wider public spaces: from artifacts co-produced by learners and communities to address actual needs and purposes to student-designed documentary films, learner-produced digital research journals (Zseder, 2016), community food security projects and small-scale, local investigations where, simply put, it is the learners' concerns, and co-emerging questions and aims, that animate inquiry, knowledge-making, and artifactual design.

Further, through production using technology tools and communicational media, it is argued that students not only learn more deeply but, significantly, build "participative status" in cultural practices as they make and do (Thumlert et al., 2015, p. 797). In science and technology contexts, this translates to a more public framework for doing science, one that models and enacts much needed participation in the public governance of science and technology in democratic societies today.

Production pedagogies can transform science and technology learning, instating a focus on students' worlds—where knowing, designing, and making are embedded in the social contexts and communities where inquiry and making itself occurs (de Castell, 2010). Central to production pedagogy is an engagement, too, with external actors, models, and communities of practice: sociotechnical sites and resources that shift learning into unfinished worlds and always unfinalizable futures. Informing this view, feminist theories of science (Clarke & Olesen, 2013) suggest that the objects of inquiry, and the knowledge and things being made through learner-directed research and making, do not need to be romantically or heroically conceived as "solutions" to "crises"; rather, they simply begin as interventions that matter to their makers. Here, the "exchange value" of abstract scientific knowledge (as commodity) is replaced by local "use value" (de Castell, 2016) and reconnected with local users and communities, in ways that resonate with the more "grounded" and ethical orientations to scientific practice as articulated in critical sustainability studies.

In line with CSS, production pedagogies invite students to critically reflect on the personal purposes and theoretical premises that inform what they research and make, to take responsibility for what they make, and to understand the very purposes for engaging in a process of scientific investigation—or for intervening in states of affairs that matter to makers. Using real-world technology tools and methods instead of curricularized surrogates, students engage in *different* processes of innovation that connect local and extra-local situations and respond to the interests and values of the communities the students themselves belong to, are involved in, or imagine as possible.

This approach to learning ensures that students' research designs are contextually relevant and connected to a present fascination, need or purpose that makes sense to them, in their present situations. A key pedagogical lesson of this approach is that students are asked to engage in tasks that they can, and want to, actually achieve: this is a significant aspect of the work since it (re)engages learners, and brings students in "contact with what they themselves can accomplish" (de Castell, 2010), inviting a different kind of assessment—a process of self-referential appraisal of the products that they create, and of their wider impacts or ecological effects they may have in their world, whose ends are served, and who is involved (or excluded) in public debates and governance matters.

For production pedagogy researchers, sociomaterial interventions in the world that are less "school-bound" are operationalized: learners co-define questions and propose trajectories of inquiry for themselves through technology and materials-centric exploration (McBride, 2017), and in terms of what is, or what emerges for them as, significant to their interests, passions, or public concerns: questions about what kind of world we will inhabit, how to make sense of things, and what we might do. Simply put, learners are invited to become the scientists of their own interdisciplinary endeavors: building theories, testing them and reflecting on results and relationships, and creating new knowledge not anticipated in advance (Nolan, 2009).

Production pedagogies are thus grounded in social contexts and material localities, but are shaped, as well, by global concerns and relationships, and are thus always in principle "connected" or "connectable" to possible sites of intervention, exploration, and action outside of schools. This orientation to education, informed by CSS, offers an alternative to discipline-narrowed STEM pedagogies (Krug & Shaw, 2016) that drive students to solve contrived "school-based" problems: problems abstracted from contexts and mediated by standardized assessments, by representations and texts (rather than materials and communities), or through the staging of "science" as a spectacle of "expert wizardry" (McBride, 2017; Nolan, 2009).

STEM pedagogies today, in efforts to developmentally prepare students for knowledge economies and STEM fields, we argue, not only actively discourage students from seeing themselves as participants in science practices and creators of their own knowledge (Nolan, 2009), but also commoditize the competences to be learned in accordance with the future "exchange value" of those competences, as fitted to neoliberalized narratives of "workplace skills" (Weinstein et al., 2016). Here, as a means to a predetermined and always distant end, "preparation" itself is assumed to be the "motivation" for learning (de Castell, 2016) and, through incremental development towards some professionalized "specialization," students are endlessly equipped with skills and knowledge about states of affairs over which they themselves have neither any critical agency or embodied competence (de Castell, Jenson, & Thumlert, 2014). Here, learners are not only alienated from the use value and immanent pleasures of their own learning, making, and designing, they also are insulated from critical reflection as they are procedurally inducted into dominant STEM narratives, epistemologies, and values.

By contrast, the "outcomes" of production pedagogies are, we suggest, not just the potentially richer acquisition of competences, but, more importantly, authentic affective investment in the very processes and products of what is being made and done—and authentic precisely because productive action serves students' purposes in *their* social worlds, in their lives (de Castell, 2010). As *pedagogy*—by contrast with a "better" theory of learning—when learners take embodied roles as researchers, designers, and makers, engaging problems and stakes critically—that is, stakes and needs students have agency in identifying, interpreting, and co-defining—students more directly engage discourses, technology tools, methods, and actions in ways that enable them to *do* science differently, that is, perform practices and tell their own narratives of science and scientific "doing" that actively challenge, or disrupt instrumentalized schooling enterprises: the sequenced and segmented scientific "activities" that too often characterize STEM education's developmental aims.

By refusing to developmentally "prepare" students for STEM futures and induct them into predetermined "anticipation regimes" (Adams et al., 2009) driven by dominant STEM narratives, we see production pedagogies as an intentionally disruptive vehicle—not for anticipating the future—but for enabling learners to "remake the present" (Adams et al., 2009, p. 260).

3.4 An Example of Production Pedagogies Within STEM Learning

In this section, we show what production pedagogy in science education looks like, and how it challenges dominant STEM narratives. This example of production pedagogy is from a recent course in the Faculty of Education at the University of Calgary, Canada, where the values of CSS inform inquiry and making. The course was designed as an entry point for students to understand, engage with, and participate in public knowledge-making interventions, with three core components: The activities in the course (a) involved the use of technology, (b) connected (interdisciplinarily) with science, mathematics, and engineering (c) utilized the networked resources of online communities, which supported access to information and models, as well as provided membership in public spaces of informal inquiry and learning. In this course, pre-service teachers engaged in teaching and learning activities directed to the production of "socially valued" artifacts to be shared with various audiences (the community of learners within the course and online and/or local learning communities). Students were invited to participate in one task that helped them develop networks and competences relevant to their personal and professional interests, aspirations, and goals as educators of science, technology, and mathematics.

The project began by inviting pre-service teachers to produce something that was meaningful to them, that addressed a need or passion, and that entailed engaging with, and being apprenticed by, broader communities of practice involved in creating similar cultural/technology products. For developing their projects, the students were asked to research and engage an online community who might support their design work or reciprocally learn from the student as part of a community of practice. We (First Author) only required students to record field notes and capture critical reflections on learning processes and community interactions.

In the second phase of the project, students learned about and from the online community, and how they might contribute to the community knowledge-sharing dynamic (e.g., the student–teacher would contribute a free-to-use asset that the community might build upon and use). Finally, in relation to purposeful making, students created a prototype or artifact that expressed their own interests in relation to science and technology and contributed knowledge to the community.

One student chose to work on a prototype for a tilt-shift lens camera that connected with his passion for photography and architecture. He described his project as follows:

> The tilt-shift lens was a common photography product that was typically used in architectural photography as a way to ensure vertical and horizontal lines were true. With the emergence of digital photography editing programs, vertical and horizontal lines can be corrected after the photograph has been taken. As such, usage of the tilt-shift lens has been limited to more creative applications. Some of these more creative applications that photographers now use the lens for is to produce unique focus in portraits and miniature landscape scenes. However, the "Lensbaby" variations of today are extremely expensive, starting at a price of about $1500.
>
> Searching online, not surprisingly, I found a community of enthusiast photographers that have experimented and documented the process for creating your own tilt-shift lens for much less money. Empathizing the issue of making a tilt-shift lens for limited money, I began to design my version of the tilt-shift lens. (Student project log sample)

This student engaged with the *Stack Exchange: Photography Community* and connected with online mentors and models to create a tilt-shift lens of his own. He used his old camera and a rubber tire from a toy monster truck as a movable connector piece. He documented the progress of his ongoing work via the project log and described the many failures he overcame to build and adjust the materials he was using to build the lens.

For example, he wrote about the off-road tire being too bendy; he described attaching a bicycle, duct tape, and toilet plunger to the rubber tire to make it strong enough to hold the lens and still be malleable enough to be modified. Finally, he contributed to the Stack Exchange "Hot Questions" website section with a series of images he took, entitled "Miniature world: A unique perspective of buildings," along with a set of instructions to create a tilt-shift lens for under $30.

Initially, the student expressed that the production task seemed disconnected from its application for science teaching. However, through ongoing reflections, the teacher-candidate modeled for himself, and recognized, the valuable learning processes, purposes, pleasures, and outcomes of the production pedagogy model, as well as its implicit collaborative social linkages with others, and with dynamic spaces and applications in the world.

For example, in his reflections, the student identified himself as a member of the lens-making community and recognized the level of expertise (commensurate with the experience and mentorship provided to the community) that group members had and shared in the online group. Furthermore, he articulated that the activity allowed him to have a genuine voice in what the next steps comprised as there were no specific steps to follow in order to build the lens, but merely guidelines in what a tilt-shift lens should be able to achieve. This allowed the student to choose the items and materials he needed and to engage or repurpose them through a materials-centric (McBride, 2017) work process—and with substantially more agency, as these production processes were driven by self-and-community-defined interests, needs, and use value (de Castell, 2016).

Once completed, the student closely analyzed the *Science Program of Studies* curriculum resource—and through that he made clear connections between his project and the more official STEM curriculum objectives he enacted through the production pedagogy (such as engineering, the science of light and optical devices, and situated "design thinking"). The student describes how his own understanding of science and technology changed, using the example of his experience making the shift-tilt lens and taking photographs:

> By making a lens, I learned a lot about technology. I have never seen the inside of a lens before, even though I use the camera on my phone all the time. After I took the lens apart, I discovered that there was a lot more going on than I first thought. I was most challenged when I had to blueprint my new lens design because I was confused about how the angles would work.
>
> It was obvious that when I changed the angle of the lens, I could make some interesting effects. For example, when I bent the lens down, it would create a miniature effect on the scene I was photographing. Another technique I used a lot was bending the lens so it would only focus on certain things and makes other things blurry.
>
> What this tells me about technology is that it is actually rooted in basic principles of science. For example, my Image # One (McMahon Stadium): In this image, I held the camera above the line of sight, and tilted the lens down towards McMahon Stadium. This was done to give the stadium a model-like effect. In my Image # Three (The #9 Bus), I tilted and compressed the lens to convey a sense of motion. Because we were also pointing the lens downward, objects acquired a "miniature" quality. All of these are examples of principles that are essential to an understanding of the science of light and how optical technologies work and how we can use them to predict the effects of changes in designs, alignments, and compositions of images. Once I figured that out, I could do some interesting things with science of light and photographs and I could understand how the angles worked. (Student project log sample).

This project, "Capturing reflections: Meaning-making through a digital lens-making experience," allowed the student to, through production pedagogy, challenge the "school as usual" framework, where science teaching and learning is dominantly mediated by prescribed outcomes, uniform instructional processes, and similarly uniform assessments. Furthermore, this production pedagogy additionally supported a critical connection to place and community—both virtual and concrete—and invited the student to learn and, most importantly, develop and expand the stories of the places of the communities he visited during walkabout trips with his camera. This deeper engagement provided a different kind of knowledge and understanding of the ways in which science and technology connect us, and how we can create and communicate stories and revision the very purposes and impacts of science and technology learning.

3.5 Final Thoughts

In this chapter, we have signalled new theoretical grounds and critical models for what science education, mobilizing production pedagogies, might look like, particularly when we eschew grand narratives of expert science and corporate techno-solutionism. In the model described above, production pedagogy redraws and contextualizes science and technology education in three significant ways: (1) by giving learners flexibility and agency to participate in processes of inquiry, meaning-making, and design work through action and interaction within material and social worlds, (2) by engaging in a process of learning where students' own interests, passions, and personal purposes have a legitimate place and are fruitful and rich sites of actionable knowledge and "youth-led engagement" (Ho, Clarke, & Dougherty, 2015), and (3) by inviting (in this case) future teachers to continually pose the question: "how might we activate *this* kind of pedagogy in education today?" This latter question is fundamental in challenging current instrumentalist values and pedagogies underpinning the schooling of STEM.

As an account of production pedagogy, this kind of work generates novel relations to science and technology, providing contexts and contingencies for embodying science practices outside of standardized STEM curriculum, and for engaging with the communities that are affected by science practices, and for enacting/communicating different stories—alternative narratives—about science practices. One result is that specialist orientations to doing science driven by STEM "crisis and salvation" narratives are replaced by more immediate and multimodal investigations that emerge and evolve in relation to interests, purposes, materials, and collective forms of "community intelligence" and community needs.

And while the making of a tilt-shift lens itself may not heroically "solve" any global environmental crises, the model of action, interaction, and making refuses the commoditization of knowledge and skills and demystifies science as an expert practice or professionalized role within neoliberal economies of innovation and performativity. This is a very different orientation to science and technology learning as it focuses less upon abstract conceptual principles, propositional knowledge, or textual representations about what things "are," and instead provides insight into new relations about what things "do," and what can be done with them, thus offering more tangible, materials-centric, and embodied ways of learning and making (McBride, 2017).

Above all, we argue that it is problematic today to invoke STEM education—or any education—as a vehicle for "preparation," particularly when students are being prepared for, and inducted into, dominant narratives and practices that sustain unsustainable futures. In contrast, we reconceive science and technology production pedagogies as a means of enabling students to engage, in the present, with present "matters of concern" (Latour, 2004), invested in both the processes and social stakes of science and technology practices and futures. We contend that current STEM narratives and educational practices shaped by neoliberal logics and technocapitalist fantasies, even if based on hard science and "hard evidence," foreclose in advance opportunities for envisioning alternatives. One way to challenge the dominant narratives of STEM reform is to attend to the socio-political domains of science and technology through locally situated and meaningful production pedagogies. In educational contexts, we suggest that science-based production pedagogies, informed by the ethos of critical sustainability studies, may empower students to take agentive roles in relation to inquiry, knowledge- and artifactual-making, where actors *do* science differently in their own communities, in always unfinished worlds.

References

Adams, V., Murphy, M., & Clarke, A. E. (2009). Anticipation: Technoscience, life, affect, temporality. *Subjectivity, 28*(1), 246–265.

Alonso-Yanez, G. (2018). Exploring curriculum for science education: Lessons from a Mexican biosphere reserve. *Journal of Education for Sustainable Development, 11*(2), 89–101.

Alonso-Yanez, G., Thumlert, K., & De Castell, S. (2016). Re-mapping integrative conservation: (Dis) coordinate participation in a biosphere reserve in Mexico. *Conservation and Society, 14*(2), 134.

Amsler, S., & Facer, K. (2017). Contesting anticipatory regimes in education: Exploring alternative educational orientations to the future. *Futures, 94*, 6–14.

Ball, A., Owen, D. L., & Gray, R. (2000). External transparency or internal capture? The role of third-party statements in adding value to corporate environmental reports1 1. *Business Strategy and the Environment, 9*(1), 1–23.

Balvanera, P., Daw, T. M., Gardner, T. A., Martín-López, B., Norström, A. V., Ifejika Speranza, C., … Kittinger, J. N. (2017). Key features for more successful place-based sustainability research on social-ecological systems: A Programme on Ecosystem Change and Society (PECS) perspective. *Ecology and Society, 22*(1), 14.

Barrett, M. J. (2006). Education for the environment: Action competence, becoming, and story. *Environmental Education Research, 12*(3–4), 503–511.

Barrett, M. J. (2007). Homework and fieldwork: Investigations into the rhetoric-reality gap in environmental education research and pedagogy. *Environmental Education Research, 13*(2), 207–223.

Bencze, J. L., & Carter, L. (2015). Capitalists' profitable virtual worlds: Roles for science & technology education. In P. P. Trifonas (Ed.), *International handbook of semiotics* (Vol. 1 & 2, pp. 1197–1212). Dordrecht: Springer. https://doi.org/10.1007/978-94-017-9404-6_57

Bolin, B., & Kurtz, L. C. (2018). Race, class, ethnicity, and disaster vulnerability. In *Handbook of disaster research* (pp. 181–203). Cham: Springer.

Carter, L. (2016). Neoliberalism and STEM education. *JASTE: Journal for Activism in Science and Technology Education, 7*(1), 30–41.

Clarke, A. E., & Olesen, V. (2013). *Revisioning women, health and healing: Feminist, cultural and technoscience perspectives*. New York: Routledge.

Craye, M., Funtowicz, S., & Van Der Sluijs, J. P. (2005). A reflexive approach to dealing with uncertainties in environmental health risk science and policy. *International Journal of Risk Assessment and Management, 5*(2–4), 216–236.

Datta, R. K. (2018). Rethinking environmental science education from Indigenous knowledge perspectives: An experience with a Dene First Nation community. *Environmental Education Research, 24*(1), 50–66.

de Castell, S. (2010). Exquisite attention: From compliance to production. *Language and Literacy, 2*(2), 4–17.

de Castell, S. (2016). A pedagogy of production: An introduction. *New Media Modules* [Online]. Retrieved May 1, 2018, from https://vimeo.com/181978126

de Castell, S., & Jenson, J. (2006). No place like home: Sexuality, community, and identity among street-involved "queer and questioning" youth. *McGill Journal of Education, 41*(3), 227.

de Castell, S., Jenson, J., & Thumlert, K. (2014). From simulation to imitation: Controllers, corporeality, and mimetic play. *Simulation and Gaming, 45*(3), 322–355.

Díaz, S., Demissew, S., Carabias, J., Joly, C., Lonsdale, M., Ash, N., … Bartuska, A. (2015). The IPBES conceptual framework—Connecting nature and people. *Current Opinion in Environmental Sustainability, 14*, 1–16.

Escobar, A. (2015). Decrecimiento, post-desarrollo y transiciones: una conversación preliminar. *Interdisciplina, 3*(7), 217–244.

Folke, C. (2006). Resilience: The emergence of a perspective for social–ecological systems analyses. *Global Environmental Change, 16*(3), 253–267.

Funtowicz, S., & Ravetz, J. (2003). Post-normal science. In International Society for Ecological Economics (Ed.), *Online encyclopedia of ecological economics*. Retrieved from http://www.ecoeco.org/publica/encyc.htm

Gough, A. (2015). STEM policy and science education: Scientistic curriculum and sociopolitical silences. *Cultural Studies of Science Education, 10*(2), 445–458.

Habermas, J. (1968). *The idea of the theory of knowledge as social theory*. Cambridge, UK: Polity Press.

Harding, S. (1991). *Whose science? Whose knowledge? Thinking from women's lives*. Ithaca, NY: Cornell University Press.

Henderson, C., & Dancy, M. H. (2011, February). *Increasing the impact and diffusion of STEM education innovations*. In Invited paper for the National Academy of Engineering, Center for the Advancement of Engineering Education Forum, Impact and Diffusion of Transformative Engineering Education Innovations. Retrieved from http://www.nae.edu/File.aspx

Ho, E., Clarke, A., & Dougherty, I. (2015). Youth-led social change: Topics, engagement types, organizational types, strategies, and impacts. *Futures, 67*, 52–62.

Houston, D. (2013). Crisis is where we live: Environmental justice for the Anthropocene. *Globalizations, 10*(3), 439–450.

Jenson, Black, & de Castell (2018). Digital game-design: effects of single sex groups on student success. In M. Ciussi (Ed.), *Proceedings of the 12the European Conference on Game Based Learning*, Sophia Antipolis, France, 4–5 October 2018. pp. 258–265.

Kafai, Y. B. (2006). Playing and making games for learning: Instructionist and constructionist perspectives for game studies. *Games and Culture, 1*(1), 36–40.

Klein, N. (2015). *This changes everything: Capitalism vs. the climate*. Nueva York: Simon & Schuster.

Krug, D., & Shaw, A. (2016). Reconceptualizing ST® E (A) M (S) education for teacher education. *Canadian Journal of Science, Mathematics and Technology Education, 16*(2), 183–200.

Latour, B. (1987). *Science in action: How to follow scientists and engineers through society*. Harvard University Press.

Latour, B. (2004). Why has critique run out of steam? From matters of fact to matters of concern. *Critical Inquiry, 30*(2), 225–248.

Luke, A., Sefton-Green, J., Graham, P., Kellner, D., & Ladwig, J. (2017). Digital ethics, political economy and the curriculum: This changes everything. In *Handbook of writing, literacies, and education in digital cultures*. New York, NY: Routledge.

Lyotard, J. F. (1984). *The postmodern condition: A report on knowledge* (Vol. 10). University of Minnesota Press.

McBride, M. C. (2017). *Tangible inquiries: A study of aroma materials and sources in the built and botanical environments in Grasse, France* (Doctoral dissertation). Ryerson & York Universities. Retrieved from https://yorkspace.library.yorku.ca/xmlui/handle/10315/34467

Noble, S. U. (2018). *Algorithms of oppression: How search engines reinforce racism*. NYU Press.

Nolan, J. (2009). Scope for the imagination: Locating constructionism in science simulations. In *RICE: Reggio Inspired Care & Education conference*. Winnipeg, Manitoba, 7 May 2009.

Olssen, M., & Peters, M. A. (2005). Neoliberalism, higher education and the knowledge economy: From the free market to knowledge capitalism. *Journal of Education Policy, 20*(3), 313–345.

Ostrom, E. (2009). A general framework for analyzing sustainability of social-ecological systems. *Science, 325*(5939), 419–422.

Papert, S. (1980). *Mindstorms: Children, computers, and powerful ideas*. Brighton, UK: Harvester Press.

Papert, S. (1987). Information technology and education: Computer criticism vs. technocentric thinking. *Educational Researcher, 16*(1), 22–30.

Pedretti, E., & Nazir, J. (2011). Currents in STSE education: Mapping a complex field, 40 years on. *Science Education, 95*(4), 601–626.

Pinto, L. E. (2015). Putting the critical (back) into makerspaces. In *Monitor: Publication of the Canadian Centre for Policy Alternatives* (May/June: 2015) [online]. Retrieved April 4, 2018, from https://www.policyalternatives.ca/publications/monitor/putting-critical-back-makerspaces

Reiss, M. J. (2003). Science education for social justice. In C. Vincent (Ed.), *Social justice, education and identity* (pp. 153–165). London: RoutledgeFalmer.

Rose, J., & Cachelin, A. (2010). Critical sustainability: Promoting pedagogies of placefulness in outdoor education. *Education*.

Rudolph, J. L. (2008). Historical writing on science education: A view of the landscape. *Studies in Science Education, 44*(1), 63–82.

Sanders, M. (2008). STEM, STEM education, STEMmania: A series of circumstances has once more created an opportunity for technology educators to develop and implement new integrative approaches to STEM education championed by STEM education reform doctrine over the past two decades. *The Technology Teacher, 68*(4), 20.

Sharma, A. (2016). The STEM-ification of education: Zombie reform strikes again. *JASTE: Journal for Activism in Science and Technology Education, 7*(1), 24–51.

Smythe, S., Toohey, K., & Dagenais, D. (2016a). Video making, production pedagogies, and educational policy. *Educational Policy, 30*(5), 1–31.

Smythe, S., Toohey, K., & Dagenais, D. (2016b). Video making, production pedagogies, and educational policy. *Educational Policy, 30*(5), 740–770.

Springett, D. (2005). 'Education for sustainability' in the business studies curriculum: A call for a critical agenda. *Business Strategy and the Environment, 14*(3), 146–159.

Stengers, I. (2018). *Another science is possible: Manifesto for a slow science*. Hoboken, NJ: Wiley.

Sterling, S. (2001). *Sustainable education: Revisioning learning and change (Schumacher briefings)*. Cambridge, UK: UIT Cambridge Ltd.

Tan, E., & Calabrese Barton, A. (2018). Towards critical justice: Exploring intersectionality in community-based STEM-rich making with youth from non-dominant communities. *Equity & Excellence in Education, 51*(1), 48–61.

Thumlert, K. (2015). Affordances of equality: Ranciere, emerging media and the 'New Amateurs'. *Studies in Art Education, 56*(2), 114–126.

Thumlert, K., de Castell, S., & Jenson, J. (2015). Short Cuts and Extended Techniques: Rethinking relations between technology and educational theory. *Educational Philosophy and Theory, 47*(8), 786–803.

Toledo, V. M., Garrido, D., & Barrera-Bassols, N. (2015). The struggle for life: Socio-environmental conflicts in Mexico. *Latin American Perspectives, 42*(5), 133–147.

Troncoso, S. (2018, January 23). *Workers are the heart of the algorithm*. Retrieved August 31, 2018, from https://blog.p2pfoundation.net/workers-are-the-heart-of-the-algorithm/2018/01/29

Visvanathan, S. (2006). Alternative science. *Theory, Culture & Society, 23*(2–3), 164–169.

Weinstein, M., Blades, D., & Gleason, S. C. (2016). Questioning power: Deframing the STEM discourse. *Canadian Journal of Science, Mathematics and Technology Education, 16*(2), 201–212.

York, E. (2018). Doing STS in STEM spaces: Experiments in critical participation. *Engineering Studies, 10*(1), 66–84.

Zeidler, D. L. (2014). Socioscientific issues as a curriculum emphasis. Theory, research, and practice. In N. G. Lederman & S. K. Abell (Eds.), *Handbook of research on science education* (Vol. 2, pp. 697–726).

Zeidler, D. L. (2016). STEM education: A deficit framework for the twenty first century? A sociocultural socioscientific response. *Cultural Studies of Science Education, 11*(1), 11–26.

Zouda, M. (2016). Deconstructing STEM: A Reading through 'the postmodern condition'. *JASTE: Journal for Activism in Science and Technology Education, 7*(1), 70–83.

Zseder, N. (2016), Production pedagogy: A student-produced case study. *New Media Modules*. Online Video Document. Retrieved from https://vimeo.com/178117276

Chapter 4
Imagining the Sustainable Future Through the Construction of Fantasy Worlds

Beaumie Kim, Stefan Rasporich, and Diali Gupta

> *If you all look the same, there is less judgment by appearance or orientation.*
>
> *—Sherri, Conversation with Stefan*

Abstract Advocating the humanistic assumptions (i.e. values, norms, practices) in the discourse of sustainability education, we discuss Stefan's journey with his eighth grade students in planning and building a sustainable village in a fantasy world. As a form of a self-study, Stefan engaged in scholarly dialogues with two other authors based on his reflections on this project and conversations with the students. The students created sustainable villages and languages in *Minecraft* or other media through their interactions with their imagined land, environment, and characters. We investigated how Stefan's approach helped learners take critical positions in imagining a sustainable society. Here we discuss three student groups' emergent designs in relation to Stefan's teaching practices. We argue that humanistic and aesthetic approaches to learning may help students take their own positions regarding societal norms, values, and practices.

Keywords Sustainability education · Critical pedagogy · Aesthetic experience · Environmental literacy · Minecraft · Self-study

---

B. Kim (✉) · D. Gupta
Werklund School of Education, University of Calgary, Calgary, AB, Canada
e-mail: beaumie.kim@ucalgary.ca

S. Rasporich
Calgary Arts Academy, Calgary, AB, Canada

P. Sengupta et al. (eds.), *Critical, Transdisciplinary and Embodied Approaches in STEM Education*, Advances in STEM Education,
https://doi.org/10.1007/978-3-030-29489-2_4

In response to the Sustainable Development Goals set out by the United Nations, UNESCO created key competencies and learning objectives to promote Education for Sustainable Development (Rieckmann, Mindt, & Gardiner, 2017). One of the competencies is focused on "the ability to question norms, practices and opinions; to reflect on own one's values, perceptions and actions; and to take a position in the sustainability discourse" (Rieckmann et al., 2017, p. 10). Such critical positions seek fundamentally different approaches from what has been emphasized in sustainability education, that is the mimicking of the real-world science and engineering using models and simulations (e.g. Pallant & Lee, 2017). The approaches of using real-world scientific data and models are important for learners to understand the complexity of the dynamic systems. On the other hand, we support the argument that humanistic assumptions (i.e. values, norms, practices) should become much more central in the discourse of sustainability education (Feinstein & Kirchgasler, 2015; Stables & Scott, 1999).

In this chapter, we discuss Stefan's journey with his eighth grade students in reflecting on their own assumptions about what we look like, how we communicate, and how we live and interact with others and the environment through a creative process. Stefan teaches in an art immersion charter school in western Canada, whose curricula engages students in arts in all aspects of their learning. The project described in this chapter, co-led with a resident artist, challenged the students to plan and build a sustainable village set in a fantasy world. When given a chance to imagine a sustainable world, what kind of societies might students envision? As quoted above, Sherri and her teammates assumed that equality mattered in a sustainable society and imagined that their villagers would not have visible features of gender or race.

As a form of a self-study (Loughran, Hamilton, LaBoskey, & Russell, 2004), Stefan and two co-authors engaged in scholarly dialogues around his reflections and conversations with the students on this project. Through these dialogues, we investigated how Stefan's approach helped learners take critical positions in imagining a sustainable society, such as its governance, justice, renewable energy, water resource management, farming, language, cultural history, immigration, trade and arts. This study shows that the sustainable villages and languages created by the students in *Minecraft* or other media were a result of their interactions with their imagined land, environment, and characters. We argue that humanistic and aesthetic approaches to learning may help students take their own positions in relation to the societal norms, values, and practices in imagining sustainable futures, and "understand how (and why) our scientific knowledge and technological and artistic endeavour are historically and culturally situated" (Stables & Scott, 1999, p. 152).

4.1 Theoretical Perspectives: Critical Pedagogy, Aesthetic Experience and Environmental Literacy

Curriculum is a value-laden selection from the culture and informs the pedagogy of a classroom (Osbourne, 1991). How can a curriculum simultaneously share the established knowledge and acknowledge its value-ladenness? Acknowledging that the accepted disciplinary, societal or cultural knowledge is not necessarily objective or neutral is an important step towards taking a critical perspective (Osbourne, 1991). Giroux (1981) suggested that when linking students' lived experience with classroom experiences, they discover how they give meaning to the world and how such meaning can be used reflectively to identify or discover the sources and limits of knowledge. When students' knowledge, experiences and visions of the world become the curriculum, their learning may create actions for social change (Osbourne, 1991).

Critical pedagogy practiced in classrooms, therefore, allows for student voices to be heard (Giroux, 1981; Osbourne, 1991). In such a classroom (or society), the members recognize and rely on mutual interests, and engage in continuous readjustment of their practices (Dewey, 1916). When students' interest-driven inquiries, decisions and open dialogue in the classroom meet the interest of the curriculum, the students become critical agents that make their knowledge meaningful, critical and ultimately emancipatory (McLaren, 1989; Osbourne, 1991). At the same time, critical pedagogy emphasizes becoming attentive and responsible to the interests and experience of others, which fosters compassionate and ethical social relations (Giroux, 2011).

The learners' experience upheld with critical pedagogy has a strong relevance to Dewey's account on aesthetic quality of ordinary experience. It is the experience of "soaring beyond the immediate confines of one's experiences, entering into a critical dialogue with history, and imagining a future that would not merely reproduce the present" (Giroux, 2011, p. 155). It is the experience of recognizing and admiring "the conditions and factors that make an ordinary experience complete" and finding "ourselves faced with a problem rather than with a final solution" (Dewey, 1934, p. 12). Such engagement may lead to the consciousness and intervention in reality (Freire, 2002). As the students confront the problems in the world and with the world, they visualize a problem with various interconnections to other problems by placing them all in a holistic context, which results in the development of new understandings.

Understanding environment and sustainability should take into account such critical and aesthetic approach. Greene (1978) relatedly emphasised an epistemology that allows learners to draw from various disciplines and critically question the excessively human-dominated world. Dewey (1934) challenges our scientific approach by stating that "[n]ature is kind and hateful, bland and morose, irritating and comforting, long before she is mathematically qualified or even a congeries of 'secondary' qualities like colors and their shapes" (p. 16). We need to understand that our history, culture and knowledge have been shaped through the dialogue and interactions with the natural environment (Dewey, 1934; Stables & Scott, 1999), and it is essential to deepen this dialogue to sustain our world. The idea that human constructions and their sustainability come from the environments is actualized by Canadian indigenous architect, Douglas Cardinal. Cardinal (1998) called architecture "a living process", through which the forms evolve from the land in harmony with nature. Dewey (1934) similarly saw that art is an effort to expand our own lives by using the materials of nature.

In this chapter, following their views on considering aesthetic experience and the land as part of our living process, we explore the humanistic and aesthetic response to nature and environment as well as human knowledge and construction as a possible means to achieve critical pedagogy in the classroom. We specifically explore the approach of engaging learners in the design of artefacts. We see student-led decisions, dialogues, ideas and artefacts as designs that emerge as part of critical pedagogy in classrooms. Learners express their values and ideas through their artefacts, which are being shaped through their sociocultural interactions (Kim, Tan, & Bielaczyc, 2015). Thus when a classroom, built upon the premises of critical pedagogy, encourages learners' creation of artefacts, learners and teachers engage in purposeful interactions using objects and materials. As the learners engage in design practices, they have their "aesthetic moments" when their ideas cease to be mere ideas and become embodied in the objects (Dewey, 1934).

In advocating for learners' designs using computational technologies, Resnick, Berg, and Eisenberg (2009) emphasized learners' personalization of their forms beyond their functions, thus paying more attention to their "aesthetics". Although their contribution to seeing scientific instruments as learners' means of communication and expression was significant, their use of "aesthetics" was limited to how it looks and feels as personal creations. More recently, Farris and Sengupta (2016) used the Deweyan sense of aesthetic experience, focusing on how learners' interest-driven pursuits of using computational technologies transform these materials into expressive means. Such pursuits show the continuity and unity of learning experience, as demonstrated by Farris and Sengupta (2016). Scholars similarly have recognized the importance of allowing learners' self-initiated detours or "personal excursions" as termed by Azevedo (2006) and subversions from the proposed activities (e.g. Kim & Ho, 2018) with technologies. In our work, we suggest that the learners' designs transform the materials into objects that embody ideas. At the same time, we assume that their activities emerge as they interact with their own designs, especially in this project with the land they imagined. These are not merely detours from the planned activities, but emergent core activities, through which they engage in "a constant unveiling of reality" (Freire, 2002, p. 81) or "seeing more" (Higgins, 2008, as discussed in Farris & Sengupta, 2016) by problematizing what seems natural or inevitable in our world (Giroux, 2011).

4.2 The Study: Imagining Sustainable Futures with Students

The project of creating sustainable villages was proposed by a resident visual artist Jeff Eisen and co-developed with Stefan to incorporate an interdisciplinary approach including elements of Arts, Social Studies, Language Arts and Science. The planning process of this art immersion school (grade 4–9) begins with the collaborative relationship between a teacher and an artist, in addition to the collaborations between teachers of core disciplines. Students in this school are dominantly Caucasian and born in Canada. They come from all levels of socioeconomic status, but based on what Stefan has observed, the parents tend to be liberal or entrepreneurial, and often work in artistic and creative sectors. In some cases, students come from different schools where they could not fit in very well. Stefan has not observed much of a trend in terms of the gender balance, as it has varied from year to year.

4.2.1 Design of Study

Using a form of a self-study approach (Loughran et al., 2004), Stefan, as a teacher-scholar, engaged in dialogues with researchers. As an insider of this project, Stefan openly examined his practice by having conversations with the students and soliciting different perspectives from the university researchers (Samaras, 2010). The conversations with the researchers, who were not present in Stefan's classroom, included the kinds of questions that Stefan posed to his students and the interpretations of his taken-for-granted practices as a teacher in this art immersion school. The data collected in this study were largely Stefan's notes during students' presentations and his conversations with them. Stefan also collected available student artefacts. We (Stefan and the researchers) also took notes together while discussing Stefan's class and students' artefacts.

Stefan realized that his lessons always evolve as students engage in designing various art forms with their ideas and interest. In this project, he saw how each student was finding personal relevance, how the groups were helping others develop new skills, and how students were using their creativity and imagination to make curricular connections in an emergent design process. In order to better understand how design could be powerful for student learning, Stefan selected three groups to study the variations and richness of the designs they exhibited.

While discussing different student projects, Stefan reflected that his choice would acknowledge his specific interest such as innovative language construction or specific inquiries generated by learners. However, he also considered groups that were disengaged or subverted from their proposed activities. He felt that *"each group dynamic will provide a framework to better redesigns of similar future projects."* As the conversation with researchers progressed, Stefan inquired into how he was and could be better supporting students' emergent designs to deepen their understanding and relationships with the rich knowledge about environments and society.

4.2.2 Project Context

The grade 8 students worked in groups to design sustainable villages for 500 residents in a post-apocalyptic world using their own choice of media (e.g. *Minecraft*, websites, drawings or physical sculptures). This project lasted about 6 weeks. The students had to show the village layout (e.g. agriculture, community planning, positioning of energy devices) and demonstrate the output, efficiency, and capacity of devices to sustain the village. The renewable energy resources had to be part of the environment these 500 survivors would settle into, and produce enough power for them to sustain life and other comforts. The process started with an introduction to engage the students visually. Jeff, the resident artist, presented a website on a variety of sustainable power sources (i.e. geothermal, solar, biomass, wind, tidal and hydroelectric; see Fig. 4.1) and highlighted the visual artistry in order to capture the imagination of the students.

At this art immersion school, the focus on learning is not through specific disciplinary content in the curriculum. Instead, the approach to teaching and learning is to support students develop their skills in arts, sciences, and social sciences while addressing broader topic areas (see Table 4.1). Jeff had an idea to explore fantasy worlds in the futuristic art form of The Venus Project by Jacque Fresco, which demonstrated a village surrounded by water, the source of hydropower (thevenusproject.com). Stefan then introduced indigenous perspectives by tying in human relationships with the land in constructing culture, living environments and languages. These perspectives are based on the Canadian curriculum and are drawn from Douglas Cardinal, the Canadian Indigenous architect's plan for the Kamloops band that provides an indigenous perspective on sustainable design, highlighting wastewater and

Fig. 4.1 The introductory visual for the project by Jeff

Table 4.1 Examples of the art immersion curricular objectives for the project

	Art form: Futuristic design in *Minecraft* or other forms of media	Social studies 8	Science 8
Objective 1 (skill)	Engage in a constant creation and destruction process that affirms the redesign process	Generate creative ideas and strategies in individual and group activities	Demonstrate sensitivity and responsibility in pursuing a balance between the needs of humans and a sustainable environment
Objective 2 (skill)	Use the *Minecraft* palette of blocks as an artist's tool, or use other forms of media in creative ways to incorporate an artist's tool	Evaluate ideas, information and positions from multiple perspectives	Appreciate that scientific understanding evolves from the interaction of ideas involving people with different views and backgrounds
Objective 3 (skill)	Collaborate as a team of artists in *Minecraft* creative mode or different media	Demonstrate skills of compromise and devise strategies to reach group consensus	Work cooperatively with team members to develop and carry out a plan, and troubleshoot problems as they arise
Objective 4 (skill)	Interpret the concept of sustainability through a digital or analogue art form	Evaluate choices and the group's progress in problem-solving, then redefine the plan of action as appropriate, using networks to brainstorm, plan and share ideas with group members	Seek and apply evidence when evaluating alternative approaches to investigations, problems and issues
Objective 5 (knowledge)	Use vocabulary related to technical *Minecraft* or other media-associated communication and post-apocalyptic themes of cultural reconstruction	Review year-long historical study of the cultures of Aztecs, Spanish, Edo Japan and Renaissance Europe in terms of hierarchy, urban planning and indigenous culture	Understand the concepts related to sustainability, renewable energy sources, fresh and saltwater systems

cohabitation with animal populations (djcarchitect.com). The idea of creating languages emerged through the conversations with the researchers, which Stefan chose to explore with students. It was agreed that the new language would also be a reflection of the environments the survivors were in.

Students were given the opportunity to select a partner they felt comfortable with, and then put into larger groups of 4–5 out of their pairings, matching them up in a way that often mixed gender, research skills and proficiency in *Minecraft*. They were then tasked with a brainstorming phase, in which they drew out their initial ideas with a large piece of paper and show accountability for their early creative process (see Fig. 4.2 for an example). After the brainstorming, they entered into the

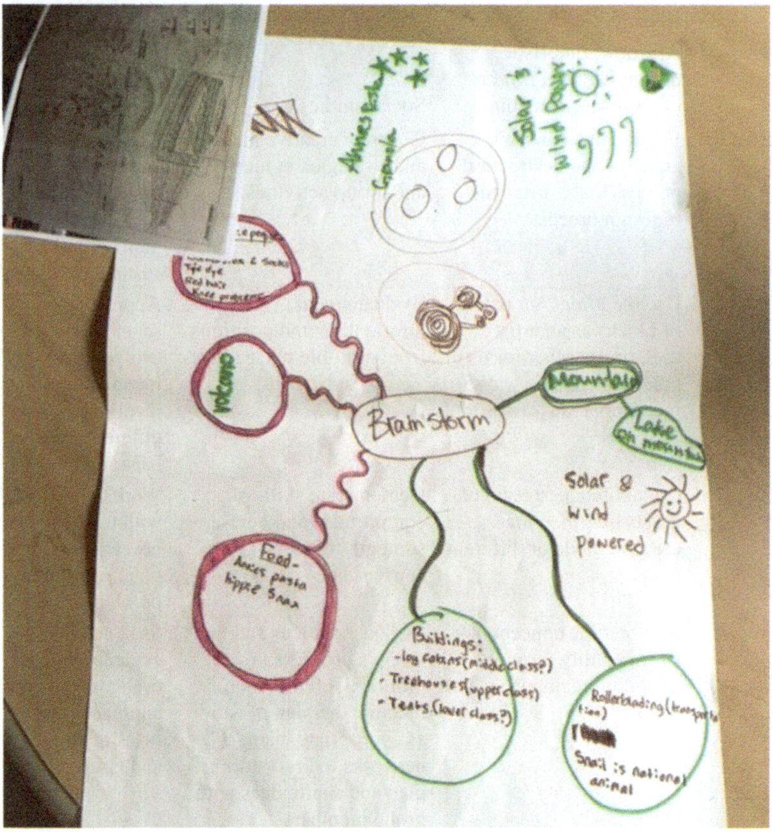

Fig. 4.2 Students' brainstorming on paper

first development stage. Most students began to work in *Minecraft*, while some started creating websites, paper maps, or new language codes. At different times, Stefan brought in language construction materials such as the sample phrase sheets for Dothraki and High Valerian from David Petersen's book *The Art of Language Invention*, as well as a phrase sheet of Blackfoot from the school field trip to the local Museum's exhibit, *Nistapiisiini*, meaning "our way of life". The exhibit showed how language is embedded in culture and led Stefan to further explore this topic. He found that in indigenous philosophy actions taken by anyone are for the benefit of the community which clarified why there is no word in Blackfoot for "thank you". He further interpreted that such actions were unlike any transactions between people in a commerce-driven society.

Stefan also reflected upon the assessment that he conducted through a mid-term check. He asked the students to share their work and the rest of the class served as an audience who had to give positive feedback to their peers. The emphasis on positive feedback was to reinforce their choices as artists and be more specific and relevant in their comments as an audience. This provided a chance for the students to become aware of each other's work, draw inspiration from it and check their own progress. In our conversations about the exchange of positive feedback, Stefan discussed how Pixar Animation Studios follows a similar process with the assumption that affirming good ideas are most important in a creative process. Stefan also liked the exercise as it fostered active listening skills during the feedback process which in turn provided positive energy to the presenters and highlighted the strengths of their work.

The idea of a cultural "grand entry" emerged later in the project, where students could create both a song and dance indicative of the culture of their world, and possibly their language. Students further developed and refined their sustainable villages, bringing in websites or deeper exploration of the language creation beyond the simple alphabet codification, as they rehearsed for their music and dance numbers. The final presentation started with a "grand entry" outside on the school compound with a huge parade where everyone simultaneously did their music and dance and then entered the school one by one to make their way to the final presentation area. The presentations themselves were extrapolations of their mid-term check with a small portion devoted to their music and dance, as well as receiving positive feedback.

4.3 Findings: Constructing Fantasy Villages

Learners encounter sociocultural norms or practices that include gender, identity or even languages in their classrooms and lifeworlds. Extending Freire's (2002) notion of learners as critical interventionists, we may notice the learners' "moments of resistance and tension" (Dewey, 1934, p. 15) towards our existing norms and practices as they engage with their own designs and develop ways of viewing the world. In the process of constructing sustainable villages, students explored different ways to communicate their ideas. For example, some students engaged in coding to maintain *Minecraft* server integrity and created a tiered student permissions system. Here, we introduce the works of three groups out of the eleven groups that worked on the project. Stefan chose these three groups because their projects were unique and might inspire us to see the critical roles of our work. These three groups constructed fantasy villages named Caveia, Plumatopia and Whistle-Whistle-Click. In the following examples, we observe learners' interesting and creative but sometimes conflicting accounts on what would be ideal, sustainable societies through their designs. We illustrate their work based on Stefan's notes on student presentations and conversations with students and screenshots of *Minecraft* fantasy villages. All students' names are pseudonyms.

4.3.1 Interacting with the Darkness of Lava Tube in Caveia

The Caveia group with five members (Brenda, Cornelius, Kerry, Michael, and Sally) conceptualized their village and its language and constructed the village in *Minecraft.* Stefan saw this group's strengths as their construction in *Minecraft* evolved over time. In their presentation, they explained that the survivors found a big cave, which was a dried lava tube where gas had been trapped after a volcanic eruption. The survivors then founded a society within it. Sally explained that some ideas were rather randomly suggested. For example, they had the ideas of acid rain versus lava tube forming a cave and voted for the lava tube.

The choices they made for the environments influenced their subsequent decisions. The development of a language based in part on echolocation was inspired by the darkness and how bats in caves evolved an extra sense in this environment. Michael, during his conversation with Stefan after the project was finished, remembered that the ideas of Caveia actually started from exploring different ways of communicating: adopting bats' echolocation led them to choose a cave as a good post-apocalyptic environment for the settlement. The healthcare system dealt with those facing issues of limited sunlight and set up recovery areas in tubes that received sunlight. The division of labour in the village was also determined by the need for farming above ground, hunting near the top opening, and the creation of artificial sunlight for the lower farms (see Fig. 4.3).

They created the history of this village on which they continued to build its culture and situated the sustainable development considering possible environmental resources. In Stables' and Scott's (1999, citing Schama, 1995) term, they are developing "a cultural history of landscape" (p. 150) in a small scale in this fantasy setting. With the chosen environmental setting, they decided that hydroelectric energy from rainwater and an underground river would be Caveia's main energy resources. Geothermal and solar energy would supplement or act as backups when needed. These explorations reflected the curricular objective (see Table 4.1), as in "demonstrate sensitivity and responsibility in pursuing a balance between the needs of humans and a sustainable environment". Using the flooding, the students were able to find a balance with the forces of nature and their needs as humans in this imagined society.

The work of the Caveia group also demonstrates how their decisions and visions became the curriculum (Osbourne, 1991) and how this student-led curriculum lends itself to students' "aesthetic moments" of their objects embodying their ideas (Dewey, 1934) in this class. What initially appeared to be randomly generated ideas of being in the dark cave were elaborated upon with the ideas of rain and flooding. Understanding the environment they created within the *Minecraft* world (i.e. rain and flood) and creating solutions to these emergent world variables became the student-led curriculum. The whole society was built on the frequent flooding, and Michael, who was most proficient in *Minecraft* programming in this group, created a system where the underground rivers would actually rise and recede to generate power. He also created a working elevator that was much needed for the villagers to

Fig. 4.3 Caveia: (**a**) elevator to the small opening to the surface; (**b**) vertical farming

connect to the ground. For Michael, figuring out the technology side of *Minecraft* became an important part of his curriculum, i.e. creating and managing the team's own server to fully control their creative work. Other groups then joined in to help building Caveia and were inspired to explore this aspect of *Minecraft* to construct their villages. In Dewey's (1934) terms, students were inspired to participate in this aesthetic moment and culminating process of transforming materials into objects that embodied their ideas.

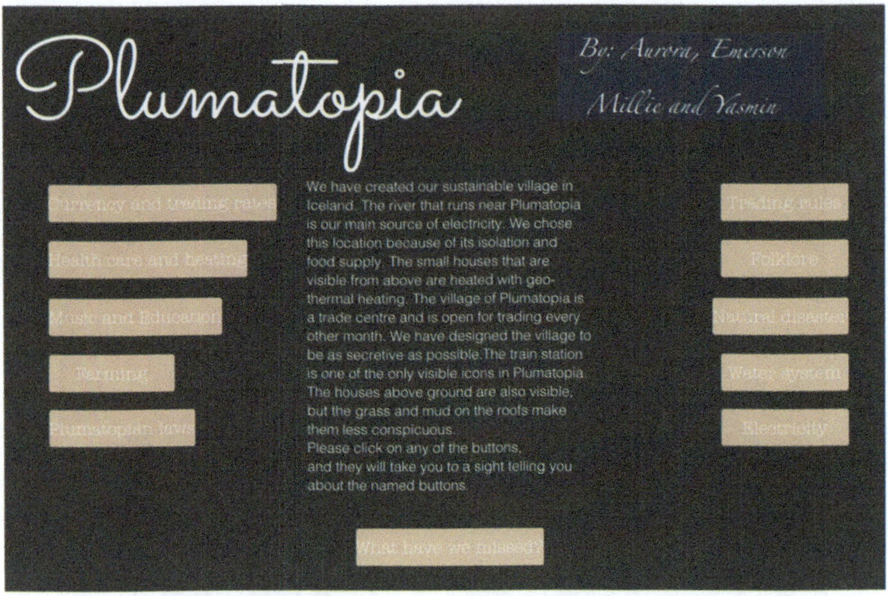

Fig. 4.4 The main page of Plumatopia group's website

4.3.2 Gift of Telepathy for the Villagers of Plumatopia

Aurora, Emerson, Millie, and Yasmin conceptualized and created Plumatopia in *Minecraft*, and explained this hidden society in detail on their website (see Fig. 4.4). Stefan chose this group because they had the most extensive backstory about the founder, Pluma, which gave many important trajectories for their designs. Their world in *Minecraft* and the Plumatopian's ways of living described in their writing were the result of the development that spanned about 300 years since Pluma founded the group. Plumatopia's history was embedded within the existing world history starting from the late 1800s and evolved in seclusion, instead of a futuristic post-apocalyptic setting as the project parameters initially stated.

They explained in their website that they situated their world in Iceland near Reykjavik, and "pluma" refers to the hot steam coming off the venting geothermal areas. Their story goes that Pluma, who was speech-impaired and a social outcast young female in her city, found a home in a cavern and created a new village. Other outcasts started to come and live in this city. This background of being an "outcast" without vocal communication and living in a hidden dark cavern provided a setting for emergent designs in order to sustain this community. The Plumatopia group came up with the concept of a telepathic language and its origin: After Pluma's death, her spirit put a spell on the villagers to have the gift of telepathy without the ability to vocalize.

During the conversation with Stefan, this group mentioned that they wanted to "step outside of the box" and "do the opposite" of what they were familiar with (e.g.

how people could interact when they do not have some senses). To design a sustainable village with these premises, they had to constantly unveil the reality of living in these environmental and social conditions they imagined (Freire, 2002), which yielded various new ideas (i.e. telepathic language, trade language, seclusion as a protection).

The underground cave was the primary consideration for the sustainability features of the village within the social hierarchy, which had evolved for the last 300 years. Their cave was under the river, where they imagined a connecting tube to catch fish for food. They also suggested having a transparent opening towards the river, so that they would have some natural light illuminating the village. Geothermal energy being the main source of sustenance, the monarch of Plumatopia lived in the deepest and most protected underground area that is also the warmest. Families were placed next in the hierarchy, and were close to the warmth of the geothermal areas. This group created water filtration tanks as well as safety precautions in case of earthquakes, which would be the most devastating to their village.

Plumatopia is described as a secluded society never revealed to the traders on the surface who arrive at a train station. They are brought blindfolded from the train station to the cave and through a labyrinth until they meet at the underground marketplace (see Fig. 4.5). As an isolated society, their telepathic language developed without the influence of other languages and thus they required a trade language when the marketplace was active. This language employs signs, a simplified version of the telepathic/visual glyphs so that traders could learn and communicate. One of the group members explained how Aurora led the creation of the sign language as an intermediary between Plumatopians and the traders:

> *Aurora was in charge of it... ((omitted)) it would be hard for new people to understand what was going on so she came up with a way for them to sign so there was an intermediary, ((omitted)) challenged herself to a sign language that would be unique and novel, and then taking a word and morphing it so they could be mute, speak telepathically, but also sign.*

The considerations around building a marketplace in Plumatopia show how they evaluated ideas, information and positions from multiple perspectives (see Table 4.1). They considered how trades in their hidden and strict society with an unfamiliar language could happen based on the environmental and social constraints they designed. Their marketplace embodied Pluma's backstory and the history of Plumatopians who went through conflicts, tensions, and struggles. In Dewey's (1934) terms, the marketplace represents the form that this group arrived at as a stable equilibrium, which is aesthetic.

This group looked into gender issues through reproductive rights and identified that the villagers would have a delayed self-identification of gender. Hence all girls had a protective internal "flower" that would guard against accidental pregnancy until they could identify their genders. They imagined that a seed would grow into a flower as the person grows. The students talked about the problems of sexual assaults and youth pregnancy. They wanted to express their perspective on safety, using Pluma's voice as their voice, by creating this protective system and countering the discourse of blaming the youth. Stefan saw this group's work as powerful when

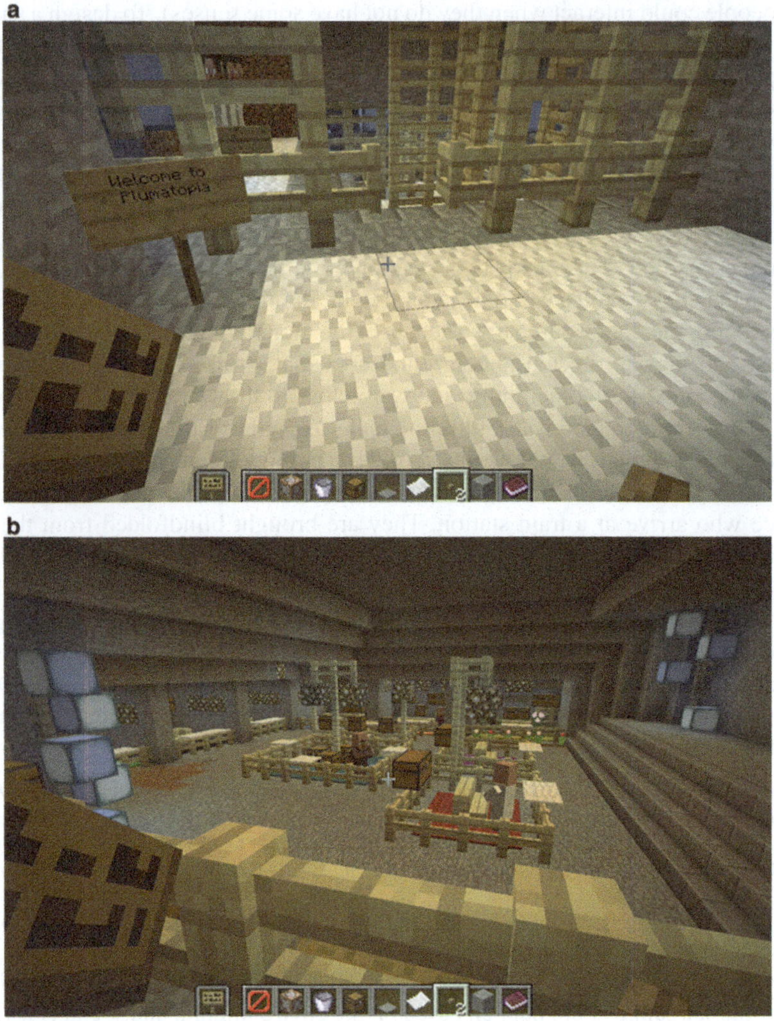

Fig. 4.5 Plumatopia: (**a**) entrance; (**b**) underground marketplace

thinking of a new kind of sustainability where a person finds their full identity and connects the ideas expressed in the classroom to their own lives. Stables and Scott (1999) reminded us to acknowledge "that human history and natural history are mutually implicated" (p. 152). Such emancipatory, rights-focused, and humanistic approach to protecting ourselves and our youths, questioning binary notions that exist in our society including gender, may lead to expanding such humanistic approaches to other entities in the environments (e.g. water being alive, therefore, having rights; Peltier, 2018).

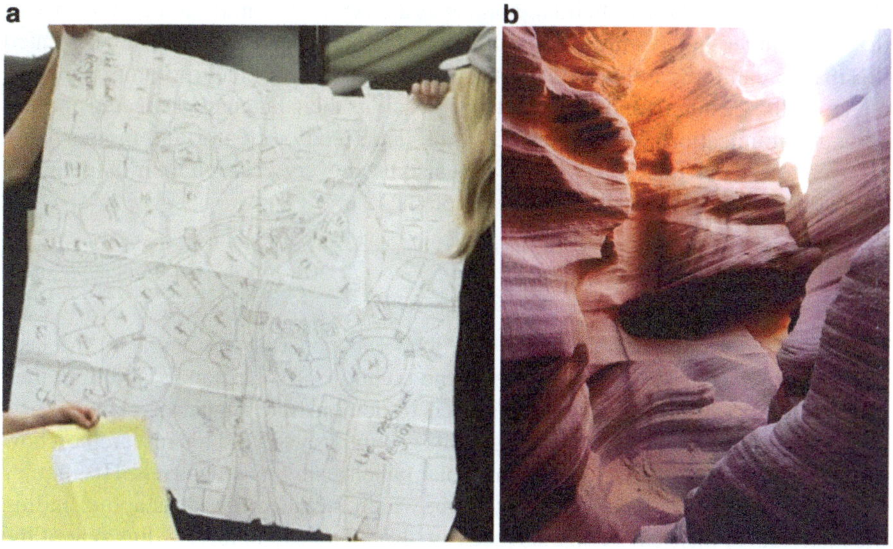

Fig. 4.6 Whistle-Whistle-Click: (**a**) village map, modelled after (**b**) Slot Canyon (photo by Kibeom Lee)

4.3.3 *Speaking Through Whistle-Whistle-Click*

Corrie, Eamon, Stan, Treyvon and Jacquie imagined a village that resembles Slot Canyons in Utah, USA (see Fig. 4.6) and oral traditions of indigenous people in this region. They explained that Whistle-Whistle-Click is what they called themselves, which could include who they are, where they live, and how they speak. To Stefan, this group's work showed how students explored personal inquiries into the sustainability of the imagined or real society and how their inquiries could emerge in conversations with the environments. Starting with an art form, Stefan focused on the students' constant process of developing, carrying out, evaluating and redefining their plans of actions as an essential skill. The inquiries that this group carried out in the course of their project indicate such constant process.

They were inspired by looking up the terrain on Google Earth, and decided that their survivors would live in the caves of canyons. This group also convinced Stefan to let them focus on mapping out the village, energy systems and other living environments on paper rather than constructing it in *Minecraft* (see Fig. 4.6). Even though people live in caves in Whistle-Whistle-Click, they have a plenty of sunlight in the canyons unlike the underground lava tube of Caveia. Therefore, sun became an important starting point of their ideas for energy source and other aspects of the society.

One of the inquiries was related to the language they started to construct, which is reflective of the environment of canyons under the big sky. They used sounds that would travel and bounce in the canyons. In fact, they called themselves

Whistle-Whistle-Click as their language was a mixture of sounds and gesture: whistle twice, finger gun gesture with both hands, and click with tongue. They indicated this in a textual form as "[whistle×2, figure guns, click]". By incorporating whistles, clicks, trills, gestures, and even body percussion, their language reflected indigenous communications, deeply rooted in the land. God or creator and mother were referred to as "hak tok click", which was also used everywhere in their language as a greeting and a praise. Corrie further explained,

> *... when written it was handwriting and very connected. Instead of a letter representing a part of word it depended on pronunciation. So letter 'e' was pronounced in a couple of ways, trill itself was its own word to mean 'praise'...*

One of the group members, Treyvon, who was generally not engaged in class, pursued a personal inquiry project into the creation of solar panels and how they functioned. It started from Treyvon's need to demonstrate academic calibre as he was having some trouble contributing to his group. Treyvon did his research on how to construct photovoltaic cells to create solar panels using raw materials. His inquiry enriched the social culture of their village, which included the ideas about finding the materials in their environments (see Fig. 4.7 for an example). Finding one of the minerals became the rite of passage for the society's youths to become adults: They need to travel for days to a natural spring, mine and collect sassolite and return to the village to extract Boric Acid, the rarest mineral.

Whistle-Whistle-Click's work showed the start of embracing indigenous perspectives that Cardinal (1998) used in his approach to new architecture and the aesthetic approach to imagining the culture as the product of "cumulative interaction with environment" (Dewey, 1934, p. 28). That is, the new creation is rooted in the land, harmonizes with the land, and changes its people and their culture. At the same time, its meaning and use evolve with the land and its people. Interestingly, both Caveia and Whistle-Whistle-Click used the dripping of water from stalactites to make sense of harmonizing with the environment and evolving their culture. For Caveia, the water was made to drip into different pool frequencies to generate music. For Whistle-Whistle-Click, the stalactite dripping eroded a pattern that was interpreted as a giant upside-down tree, which was believed to connect them to the creator (the root) who trimmed their paths. These scenarios provide a glimpse of the humanistic approach to entities as water to have independent agency (Peltier, 2018) to shape the culture.

3. Phosphorous

Phosphorus is the second most abundant mineral in the human body, only inferior to Calcium. It appears most densely in the bones and teeth (85% percent, to be exact). [Whistle x2, Finger Guns, Click] extracts the phosphorus directly from their deceased members. The Phosphorous is then purified via heat, and stored until ready for use.

Fig. 4.7 An example of a type of mineral needed and how they would find it

4.4 Discussion and Conclusions

We believe that the students work described above demonstrates Stables and Scott's (1999) position on critical environmental literacy that "human history, knowledge and action are ongoing dialogue with the natural environment, with this dialogue itself responsible for human identity" (p. 152). Through their continuous dialogue with their imagined environment, their designs of sustainable villages emerged and evolved. This design process, therefore, forced them to give meaning to the world they were creating and in turn, imagine how their choices have consequences for the villagers, the environment, and their interactions. Students came up with ideas to create renewable energy sources in their villages, and tie in relevance to present global problems. Their choices of natural settings shaped the lifestyle, the culture, and the needs of the society which all bear connections to social issues pertaining to the real world. Through this study, we identified four main areas of reflection to continue the journey of critical pedagogy as an educator and as researchers.

4.4.1 Emergent Designs and Perspectives on Sustainability

In this project, there were several emergent designs that helped Stefan to think much more deeply about indigenous perspectives on living in harmony with the environment. It also provided trajectories for students to become aware of the varying influences that shape their current and future lifeworlds. For most of the groups, *Minecraft* served as "possibility spaces" (Gee, 2008) to create and experiment with the imagined conditions of the environments. The way that blocks replace or destroy the empty space of *Minecraft* resembles the students' emergent design process in this project. Students often created a "backstory" or "origin" for their world—that it must go through a phase of destruction for a new world to emerge. Placing the notion of "sustainability" within a post-apocalypse also implies a total falling apart of an economic system of commerce/profit motive. By framing this project in a survivalist landscape, without need for commerce, there is an emancipatory approach to claiming the word "sustainable" in a purer form that includes greater emphasis on harmony with the land and its "entities". For many students, it opened new spaces within a classroom to reframe knowledge and understanding (Parker, 2013) and imagine the possibility of rebuilding with limited resources.

On the creation of languages, a noteworthy feature of most groups was that they removed language as understood in the present context, which is often considered a source of power and social control. Although their language constructions for the fantasy villagers were mostly conceptual and preliminary, their ideas show the social and cultural roles the languages play and how they are value-laden. Plumatopia's language, for example, privileges signs and telepathy over sounds, and makes speaking with the tongue abnormal. This becomes the "aesthetic moment" of creating a language "object" that embodies their ideas related directly to their lives and subverts accepted social constructs. Gender identification was an issue that emerged as students in their constructed language experimented with specific language usage tied to an individual's gender identity. Some of the groups looked at gender from the perspectives of appearances and norms accepted in society and came up with suggestions that actually looked into societal problems or issues such as pregnancy. They established new societal laws for Plumatopia, for example, where villagers have a delayed self-identification of gender. Thus we found that students were looking into problems that deal with issues of equity and identity (Freire, 2002).

We suggest that engaging students in such emergent design practices help learners to take their own positions in relation to the societal norms, values, and practices within the environment they live in. While creating sustainable villages situated in their own imagined landscape, its resources, history and culture, land, water, and the sun became living entities and their construction (or architecture) became "a living process" (Cardinal, 1998). In the current self-study of the project, students brought in some critical perspectives in different ways from their own interest (e.g. youth justice, gender, race, cultural traditions coming from the land) in addition to creating sustainable energy sources. Their work expands the meaning of sustainability beyond the contemporary implications of economic ties to sustainable developments to how we (i.e. humans, other animals, and environments) live together in harmony. In the future project, Stefan plans to have initial conversations with students to identify the issues that exist in the world that prevent us from living in harmony from this expansive view of sustainability, and come up with possible goals and benchmarks, similar to how they chose the landscape and energy sources in the course of their projects.

4.4.2 Creating a Narrative of a Protagonist

Most of the groups had some narratives or folklore associated with certain cultural practices or artifacts. Specifically, Stefan connected the story of Pluma, who took actions against the existing ways of living that gave her hardships, with Joseph Campbell's (1949/2008) articulation of a hero's adventure. Elaborating on how Plumatopia developed a backstory, Stefan would like to further explore how students could write narratives that cast one of their villagers' journey as a hero. The backstory is a way of generating empathy for the protagonist—an essential part of the screenwriting in the first act—so that the audience is rooting for them to achieve their goal. By giving the hero an "undeserved misfortune" at the beginning of the story, the audience is more engaged in witnessing them achieve an ultimate "poetic justice" (Vogler, 1991/2007). By creating a world and situating a protagonist's narrative within it, learners need to unpack the environments, obstacles and encounters throughout the journey. At the same time, the experience of creating the narratives of the protagonists, who transform from ordinary or unfortunate persons in unjust situations into heroic characters, would help the students empathize and identify with the characters and see themselves as protagonists (Rasporich, 2007) of future social actions. We suggest that creating a narrative of the village with a protagonist and other characters in addition to creating the details of the sustainable development would help learners to take a more humanistic approach to their project and take similar actions as a protagonist in their lifeworlds. By advocating critical pedagogy of becoming conscious, attentive, and responsible to the experience of others, students may seek "social relations which foster a mix of compassion, ethics, and hope" (Giroux, 2011, p. 83).

4.4.3 Pursuing the Ongoing Dialogue

The sharing of their evolving designs also engaged students in an ongoing dialogue and co-designing of the learning experience within the classroom environment. Dialogue in this context refers to their reflection on how to transform the imagined reality of fantasy villages and what course of action to seek (Freire, 2002), such as the limited exposure to sunlight for the villagers of Caveia. Students were engaged with a commitment to transform or affect the issues they felt needed attention. They presented their ideas and designs multiple times in the course, thus opening up the opportunity to explicitly recognize innovative aspects of others' work and reflect on their assumptions. These were also critical moments in the pedagogy where the students evaluated their designs, heard other students' interpretations of their work, and drew inspiration from each other's "aesthetic moments" of creating expressive objects (Dewey, 1934).

Along a similar line of becoming attentive to others' experiences, students may explore their own intentions of their work more consciously in the future project by including artist statements during the mid-term feedback session. They will have identified the expected audience's feelings or actions, which they can compare against the audience' actual responses. At the same time, Stefan realized that there could have been many opportunities to have conversations around current events and social systems in the world. For example, all the groups in the current study had some conflicting ideas of villagers being equal or escaping oppressions together with being born into social classes or evolving into a hierarchical society. This might have been influenced from their recent Social Studies curriculum connection, but such students' decisions could become their own curriculum (Osbourne, 1991) to explore why hierarchies emerge in societies, how we can find resemblance in our natural environments, and if there is any necessity to form a class system in our society. The experience of such dialogue becomes aesthetic not by shunning the "moments of resistance and tension" but by "cultivating them, not for their own sake but because of their potentialities, bringing to living consciousness an experience that is unified and total" (Dewey, 1934, p. 15).

4.4.4 Teaching as Emergent Design

As a form of a self-study, our intention was to reflect on Stefan's practice from the lens of critical pedagogy. Giroux (2011) stressed how Freire's work of critical pedagogy is about seeing beyond what appears to be natural or inevitable state, challenging what are assumed to be common senses, move into the critical conversations about creating a different future. Through our conversations, we realized how Stefan engages in a continuous redesign of his lessons and how his teaching itself is an emergent process. His teaching as emergent design is only possible through the ongoing dialogues with other adults, such as Jeff, the resident artist, the researchers, who bring perspectives from their own disciplines, and the students, who bring their interest and concerns. His assumptions are challenged by students and other adults while he could also challenge their assumptions. In this project, the spoken and unspoken inquiries of Stefan, Jeff, the students, and the researchers were intermingled and informed the curriculum of the class. Students were not only creating their inquiries into sustainable energy sources for their villages (e.g. hydroelectric in Caveia, solar panel fabrication in Whistle-Whistle-Click) but also juxtaposing various worldviews on geophysical conditions and how they could harness different forms of energy. In fact, the idea of constructing language came about through the conversations with several university researchers, and the creation of song and dance with the grand entry emerged through the conversations with students.

In the future implementation of the project and his teaching in general, Stefan would like to make these inquiries as conscious efforts in the classroom. He would also like to explicitly acknowledge the new aspects of the project and make expectations fully transparent to the students, so that the students could become mindful of how things emerge in their classroom. Students will similarly pursue their inquiries into the sustainable energy sources or other aspects of their villages or current affairs. This study demonstrates how engaging students in the designs within an art form may integrate and expand the curriculum above and around disciplines, and shows the possibilities of arts immersion as a transdisciplinary and humanistic approach to sustainability education. Our aesthetic responses to nature, environment, and landscape alone may not be enough to understand and promote environmental sustainability (Stables & Scott, 1999), and we argue that approaching learning from an art or design standpoint may help appreciate the plurality of our views.

Acknowledgements We thank Jeff Eisen for his work on this project and Stefan's students who eagerly participated in this project and had further conversations with Stefan in this regard. We would also like to thank the reviewers and editors who gave helpful feedback on the earlier versions of this chapter.

References

Azevedo, F. S. (2006). Personal excursions: Investigating the dynamics of student engagement. *International Journal of Computers for Mathematical Learning, 11*(1), 57–98. https://doi.org/10.1007/s10758-006-0007-6

Campbell, J. (1949/2008). *The hero with a thousand faces* (3rd ed.). Navato, CA: New World Library.

Cardinal, D. (1998). Architecture as a living process. *Canadian Journal of Native Education, 22*(1), 3–9. https://doi.org/10.2752/175145213X13760412749953

Dewey, J. (1916). *Democracy and education: An introduction to the philosophy of education*. New York: Macmillan.

Dewey, J. (1934). *Art as experience*. London, UK: Penguin.

Farris, A. V., & Sengupta, P. (2016). Democratizing children's computation: Learning computational science as aesthetic experience. *Educational Theory, 66*(1–2), 279–296. https://doi.org/10.1111/edth.12168

Feinstein, N. W., & Kirchgasler, K. L. (2015). Sustainability in science education? How the Next Generation Science Standards approach sustainability, and why it matters. *Science Education, 99*(1), 121–144. https://doi.org/10.1002/sce.21137

Freire, P. (2002). *Pedagogy of the oppressed*. New York: The Continuum.

Gee, J. P. (2008). Learning and games. In K. Salen (Ed.), *Ecology of Games* (pp. 21–40). Cambridge, MA: The MIT Press.

Giroux, H. A. (1981). *Ideology, culture and the process of schooling*. Philadelphia: Temple University Press.

Giroux, H. A. (2011). *On critical pedagogy*. New York: Continuum.

Greene, M. (1978). *Landscapes of learning*. New York: Teachers College Press.

Higgins, C. (2008). Instrumentalism and the clichés of aesthetic education: A Deweyan corrective. *Education and Culture, 24*(1), 6–19. Retrieved from https://www.jstor.org/stable/10.5703/educationculture.24.1.7

Kim, B., & Ho, W. (2018). Emergent social practices of Singapore students: The role of laughter and humour in educational gameplay. *International Journal of Child-Computer Interaction, 16*, 85–99. https://doi.org/10.1016/J.IJCCI.2018.01.001

Kim, B., Tan, L., & Bielaczyc, K. (2015). Learner-generated designs in participatory culture: What they are and how they are shaping learning. *Interactive Learning Environments, 23*(5), 545–555.

Loughran, J., Hamilton, M. L., LaBoskey, V. K., & Russell, T. (Eds.). (2004). *International handbook of self-study of teaching and teacher education practices*. Dordrecht: Kluwer Academic.

McLaren, P. (1989). *Life in schools*. Toronto: Irwin.

Osbourne, K. (1991). *Teaching for democratic citizenship*. Toronto: Our Schools/Our Selves Education Foundation.

Pallant, A., & Lee, H.-S. (2017). Teaching sustainability through system dynamics: Exploring stocks and flows embedded in dynamic computer models of an agricultural land management system. *Journal of Geoscience Education, 65*(2), 146–157. https://doi.org/10.5408/16-169.1

Parker, J. K. (2013). Critical literacy and ethical responsibilities of student media production. *Journal of Adolescent and Adult Literacy, 56*(8), 668–676.

Peltier, A. (2018, March). *An opening address for the UN General Assembly at the launch event of International Decade for Action "Water for Sustainable Development" 2018-2028*. New York: United Nations Headquarters.

Rasporich, S. (2007). *The learner as protagonist.* Conceptual paper. Retrieved from http://wikieducator.org/The_Learner_as_Protagonist

Resnick, M., Berg, R., & Eisenberg, M. (2009). Beyond black boxes: Bringing transparency and aesthetics back to scientific investigation. *The Journal of the Learning Sciences, 9*(1), 7–30. Retrieved from http://www.tandfonline.com/doi/abs/10.1207/s15327809jls0901_3

Rieckmann, M., Mindt, L., & Gardiner, S. (2017). *Education for sustainable development goals: Learning objectives*. Paris, France: UNESCO.

Samaras, A. P. (2010). Explorations in using arts-based self-study methods. *International Journal of Qualitative Studies in Education, 23*(6), 719–736. https://doi.org/10.1080/09518390903426212

Stables, A., & Scott, W. (1999). Environmental education and the discourses of humanist modernity: Redefining critical environmental literacy. *Educational Philosophy and Theory, 31*(2), 145–155. https://doi.org/10.1111/j.1469-5812.1999.tb00381.x

Vogler, C. (1991/2007). *The writer's journey: Mythic structure for writers* (3rd ed.). Studio City, CA: Michael Wiese Productions.

Chapter 5
Preservice Teachers' Perceptions of STEAM Education and Attitudes Toward STEAM Disciplines and Careers in China

Wenjing Li and Feng-Kuang Chiang

Abstract The term STEM, an acronym for science, technology, engineering, and mathematics, was coined the US National Science Foundation in the 1990s. It is considered to be an efficient way for promoting economic growth and international competitiveness. Educators have recently integrated arts with STEM, calling this effort STEAM. The Chinese government has established a set of policies to enhance the development of STEAM education in K-12, thus increasing the necessity for qualified teachers. However, barriers continue to inhibit the preparation of qualified teachers proficient in STEAM foundational knowledge and instructional approaches. In this chapter, we report and discuss preservice teachers' current perceptions of and attitudes toward STEAM education. We also highlight the correlation between their perceptions and attitudes. The results indicate that most participating preservice teachers did not attend lectures or training related to STEAM. Most preservice teachers' perceptions of STEAM education remained superficial and many showed minimal interest in STEAM education. On the other hand, the participants expressed varying attitudes toward each of the five disciplines within STEAM. We argue for better teacher education opportunities that meaningfully implement China's policy for STEAM education.

Keywords STEAM education · Preservice teachers · Attitudes toward STEAM disciplines and careers

W. Li
Faculty of Education, School of Educational Technology, Beijing Normal University, Beijing, China

F.-K. Chiang (✉)
Department of Educational Technology, College of Education, Shanghai Normal University, Shanghai, Shanghai, China
e-mail: fkchiang@shnu.edu.cn

P. Sengupta et al. (eds.), *Critical, Transdisciplinary and Embodied Approaches in STEM Education*, Advances in STEM Education,
https://doi.org/10.1007/978-3-030-29489-2_5

5.1 Introduction

With the growth of global competition, science, technology, engineering, and mathematics (STEM) education has been considered as an efficient method for promoting economic growth and international competitiveness (Guyotte, Sochacka, Costantino, Walther, & Kellam, 2014). The acronym STEM was first coined by the US National Science Foundation in 1986. Over the past few decades, the focus on STEM education has been gradually established worldwide. Many countries, including China, have formed educational policies to promote STEM education.

In the recent years, education researchers have begun to pay significant attention to supporting students' creativity and innovation through their educational experiences at all levels. Trilling (2009) noted that the arts offer a crucial method for cultivating creativity. The integration of the arts with STEM disciplines is referred as the STEM-to-STEAM movement, which is considered to foster creativity and innovation. The concepts of STEM and STEAM are not universally agreed (Breiner, Harkness, Johnson, & Koehler, 2012). In the context of Chinese education, however, many researchers agree that STEAM does not refer to any of the independent disciplines of mathematics, science, computers, music, or arts, but one interdisciplinary and comprehensive education involving knowledge and methodology from at least two disciplines (e.g., Jiang, 2017; Xie, 2017; Ye & Yang, 2017). Teachers from various disciplines work together to solve a problem to break down barriers between disciplines, connect with and influence each other, create interactions, and foster an increasingly comprehensive understanding regarding any topic (Pernecky, 2016).

Chinese educational leaders have established a set of policies and organizations to enhance the development of K-12 STEAM education. In the *13th Five-Year Plan* (2015), ratified by the Ministry of Education of the People's Republic of China, it was proposed that the exploration of a new model for education, such as STEAM, can stimulate students' creativity and innovation. In 2016, the Research and Development Center of Educational Equipment of the Ministry of Education made STEAM education crucial agenda. A STEAM School Alliance was established in 2016 to promote STEAM education, coordinated by the Educational Committee of Chaoyang District, Beijing. This alliance comprises 35 K-12 schools that have successfully conducted several activities, such as STEAM expert lectures. Despite these efforts, barriers continue to inhibit the preparation of adequate qualified teachers proficient in STEAM foundational knowledge and instructional approaches.

In China, "STEAM" is not an officially recognized disciplinary category, thus indicating that teachers in China cannot obtain qualifications as STEAM teachers. Therefore, schools typically have teachers of different subjects, such as science, mathematics, art, and computer science teachers, collaborate and design the STEAM class together. Teachers themselves must find ways to gain interdisciplinary teaching skills and integrate different subjects into the STEAM class. We therefore use the term "preservice STEAM teacher" to indicate undergraduate students majoring in teaching specific subjects with the intention and potential to teach STEAM classes in K-12 schools after their graduation.

Many teachers in China are accustomed to the traditional test-oriented teaching programs and therefore may resist the new STEAM education model. Those who teach in rural or Western schools in China have few opportunities to participate in professional development trainings and further education (Li, 2015). Moreover, several factors, including the lack of time and resources and work pressure, have discouraged rural teacher development (Zhu & Zhou, 2009). Therefore, we believe teachers in different regions of China do not have equitable access to teacher education and professional development on innovative pedagogical approaches, especially in the context of an integrated method of teaching STEAM education.

We believe that preservice teachers' perceptions and attitudes toward STEAM education and careers are important consideration in order for us to adequately train future STEAM teachers. This involves both understanding their opinions about STEAM, and helping them develop deep conceptual understandings of STEAM education, including connections between and across disciplines, so that we can help them feel increasingly comfortable in teaching STEAM disciplines in an integrated manner.

5.2 Literature Review

5.2.1 STEM and STEAM Education

The concept of STEM was first introduced in *Undergraduate Science, Math, and Engineering Education* (1986) by the US National Science Foundation. Johnson (2013) defines STEM as "an instructional approach, which integrates the teaching of science and mathematics disciplines through the infusion of the practices of scientific inquiry, technological and engineering design, mathematical analysis, and 21st century interdisciplinary themes and skills" (p. 367). This concept focuses on the learning processes and skills in the content whereas Tsupros, Kohler, and Hallinen (2009) focused on the interdisciplinary approach to learning in real-world context.

Recent years have seen a surge of interest in integrating arts with STEM education, including China. Compared to STEM education, STEAM education pays increased attention toward the cultivation of students' artistic attainment. One argument is that while science education tends to focus on students' inductive thinking, arts education emphasizes students' divergent thinking, which helps them to explore more potential solutions (Li & Lv, 2018). When inductive thinking is combined with divergent thinking, individual creativity and high-level problem-solving skills are developed (Lamore et al., 2013).

Moreover, the arts can provide new perspectives and can even help students to better visualize their subject matter at hand, which in turn can help students to improve their understanding of science and thus continue exploring it (Li & Lv, 2018). Spelke (2008) noted that learning art influenced the understanding of mathematical concepts among children and adolescents aged 5–18. Resnick, Berg, and Eisenberg (2000) noted that if students were encouraged to creatively design and construct their own scientific instruments, they were more likely to develop a deeper engagement with science.

5.2.2 Visualizations and STEAM Education

Educational researchers have shown that visualizations can support learning in several ways. For example, students can maintain their attention and motivation toward STEAM through visual representatives (Cook, 2006); connections between ideas and representations can sometimes be missed through the use of text alone (Mayer et al., 1996; Peeck, 1993), and graphs can be used effectively to present data, illustrate abstract concepts, and organize complex sets of information (Roth, Bowen, & Mcginn, 1999). Cognitively oriented studies have indicated that graphs can serve as composites of individuals' cognitive abilities and skills (Roth et al., 1999) and can reveal unseen or intangible phenomena that cannot be detected or experienced directly (Buckley, 2000).

Moreover, quick visualizations requiring the illustration of graphs in a short period may help to provide effective visual frameworks, through which future STEM educators can improve STEM knowledge internalization (Radloff & Guzey, 2016). Along these lines, Bybee (2013) presented nine potential perspectives of STEM education by using graphs that ranged from viewing STEM as a single subject to viewing it as transdisciplinary.

In summary, the vague definitions of STEM and STEAM education do not allow preservice teachers to fully understand STEAM education; however, we believe that the use of visualizations can help preservice teachers develop deeper understandings of both STEM and STEAM education. In addition, preservice teachers' attitudes toward STEAM might also affect their motivation to teach STEAM. To understand these and related concerns more deeply, this research will investigate preservice teachers' perceptions of STEAM education through both textual and visual responses and through their attitudes toward the five disciplines and careers in STEAM.

5.2.3 Research Questions

We investigated the following research questions:

1. What is the overall perception of STEAM education among preservice teachers? How do they perceive the connection among the five disciplines, as revealed by their visualization of STEAM education?

2. What are preservice teachers' attitudes toward STEAM disciplines and careers in STEAM areas? Do STEAM dispositions and career interests differ based on attributes, such as a teacher's gender or major study area?
3. What other opinions do preservice teachers have about STEAM education—such as what types of abilities should be necessary for STEAM teachers and whether they have suggestions for policymakers to help develop STEAM education?

5.2.4 *Methodology*

We adopted quantitative methods to derive general answers from a large group of people (Radloff & Guzey, 2016). By using a cross-sectional survey, the study data were collected over 2 weeks from the spring college semester of 2017 (March 2017). The participants were 485 students studying at a normal university in Beijing and indicated a high potential for becoming STEAM teachers. We adopted convenience and stratified sampling to gather data. We distributed printed questionnaires in each class. Of the distributed questionnaires, 379 were returned (validity rate, 78.1%).

5.2.5 *Instruments*

The survey comprised of five parts. Part one involved collection of demographic data regarding several personal characteristics, including gender, age, major, and grade. Part two was about preservice teachers' basic perceptions regarding STEAM education. The participants were requested to provide both textual and visual responses. They were asked to illustrate a diagram explaining how they visualize STEAM education by using the letters S, T, E, A, and M. Part three included the career interest questionnaire (CIQ), adapted from a longer instrument originally developed for a native Hawaiian studies project that promoted STEM interest in Hawaii (Bowdich, 2009). We modified some of the items to increase the focus on STEAM education, rather than STEM education. The participants answered the questionnaire using a Likert-type instrument, composed of nine items on three scales. Items 1–3 documented the participants' perceptions of being in an environment supportive of STEAM careers, such as "My family is interested in the STEAM courses I take." Items 4–6 analyzed participants' intent to pursue a career related to STEAM, such as "I will make it into a good college and major in an area needed for a career in STEAM." Items 7–9 documented the perceived significance of STEAM careers, such as "A career in STEAM would enable me to work with others in meaningful ways" (Peterman, Kermish-Allen, Knezek, Christensen, & Tyler-Wood, 2016). The Cronbach's alpha of Part three was 0.888. Part four was about preservice teachers' other opinions regarding STEAM education. The participants answered three open-ended questions: (1) “What ignited your interest in STEAM education?” (2) “What types of abilities do you believe are necessary for STEAM teachers?” (3) “What are your suggestions for policymakers to develop STEAM education?” Part five involved the STEAM semantics survey, adapted from Christensen, Knezek & Tyler-Wood, (2014) “teachers’ attitudes toward technology” questionnaire, derived from earlier semantic differential research by Zaichkowsky (1985). The participants answered the questionnaire by using a Likert-type instrument composed of six sections representing science-, mathematics-, engineering-, technology-, art-, and STEAM-related careers. The semantic differential scales were rated on a seven-point scale, ranging from “strongly disagree” to “strongly agree” (Christensen, Knezek, & Tyler-Wood, 2014). The Cronbach’s alpha for Part five was 0.929.

5.2.6 Results

Our analysis indicates that preservice teachers’ perceptions of STEAM education and their attitudes toward STEAM disciplines and careers varied. We first report the demographic data and then provide our study findings that organized according to the three aforementioned research questions. The analyses were completed on SPSS (version 23.0).

5.2.7 Demographic Data

The data were gathered from 379 preservice teachers; of them, 108 were men and 269 were women. Among the participants, 71% were aged <20 years. Data were collected from 20 majors, most of whom were related to a STEAM fields, such as physics, chemistry, mathematics, and computer science. In this survey, 48.2% of the participants indicated their interest in becoming a teacher after graduation, whereas 40.4% were unsure. Only 11.3% of the participants did not plan on becoming teachers in the future. Most teachers indicated that they had not decided which institutes they would teach at in the future, but among the teachers who had, most were planning to teach at senior high schools. Only a few participants chose teaching in elementary and junior high schools. However, only 38.6% of the participants had a prior teaching experience. Some of them had worked as tutors for high school students, whereas some of the others had volunteered as teachers in underdeveloped areas of China. The data are presented in Table 5.1.

The participants who answered the questionnaires were from different majors, in science and arts disciplines. We also attempted to determine whether these participants planned to teach the same subjects after graduating from their majors. Therefore, we asked them which subjects they would teach if they planned to become a teacher in the future. The results indicated that most of the participants who majored in mathematics, geography, physics, chemistry, and computer science

Table 5.1 Distribution of preservice teachers based on demographic data

Gender	Frequency	Valid percentage	Age	Frequency	Valid percentage	Teaching experience	Frequency	Valid percentage
Male	108	28.6	<20	266	71.1	Yes	145	38.6
Female	269	71.4	21–25	106	28.3	No	231	61.4
Total	377	100	26–30	1	0.3	Total	376	100
Missing	2		>31	1	0.3	Missing	3	
			Total	374	100			
			Missing	5				

Grade	Frequency	Valid percentage	Future teacher	Frequency	Valid percentage	Future teaching school	Frequency	Valid percentage
Freshman	187	47.5	Yes	179	48.2	Elementary school	3	0.80
Sophomore	115	30.3	No	150	11.3	Junior High school	16	4.2
Junior	61	16.1	Not sure	42	40.4	Senior High school	161	42.5
Senior	1	0.3	Total	371	100	Not sure	199	52.5
First-year master student	21	5.5	Missing	8		Total	379	
Second-year master student	1	0.3						
Total	379	100						

wished to teach the same subjects. The participants who majored in science and technology were unsure about their future career.

Research Question 1 What is the overall perception of STEAM education among preservice teachers? How do they perceive the connection among the five disciplines, as revealed by their visualization of STEAM education?

Of all participants, 49 (12.9%) reported that they had attended lectures or training sessions regarding STEAM. A five-point Likert scale (ranging from *completely uninterested* to *very interested*) was used to measure if participants indicated interest in STEAM education. The mean score was 3.18; the related frequencies are illustrated in Fig. 5.1. The data is presented in Table 5.2. Several participants indicated no distinctive attitude; however, some indicated minimal interest in STEAM education.

To examine participants' opinions regarding how the five STEAM disciplines are related to one another, we used a nine-point Likert scale (1 = *completely independent* to 9 = *completely related*). The highest number of participants (35.1%) selected the answer "completely related." For further detail, we asked participants to describe the reasons they selected their responses about the five disciplines. The seven types of responses were obtained (Table 5.3), translated from Chinese.

Because STEAM education provides a new pedagogical approach, it differs from the traditional forms of education and has several unique characteristics. YU & HU, (2015) noted the nine primary characteristics of STEM education: interdisciplinary, interesting, experiential, applicative, cooperative, designable, artistic, practical, and technological characteristics. We considered that STEAM education also has the same nine characteristics and designed an open item to determine whether participants were aware of the characteristics of STEAM. The participants' answers are presented in Table 5.4. Many participants (47.5%) claimed they were unaware about

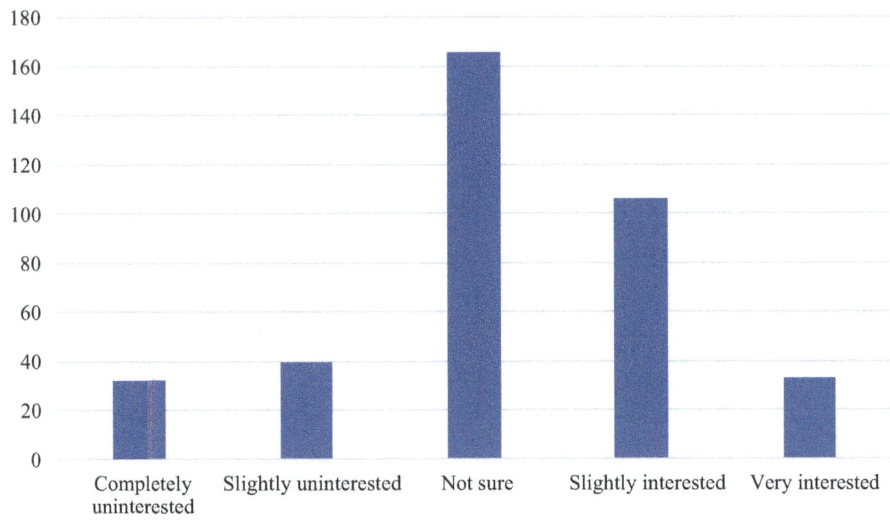

Fig. 5.1 Student' interests in STEAM education

Table 5.2 Student' interests in STEAM education

Student' interests in STEAM education	Frequency	Valid percentage
Completely uninterested	32	8.5
Slightly uninterested	40	10.6
Not sure	166	44.0
Slightly interested	106	28.1
Very interested	33	8.8
Total	377	100
Missing	2	

Table 5.3 The reasons that the participants selected responses regarding the relationship between the five STEAM disciplines

Codes	Students' quotes illustrating theme	Frequency	Percentage
All the subjects are related to each other	"There must be some connection between every two subjects" "Everything is connected to each other"	71	18.7
Five STEAM disciplines are closely related to each other	"Five STEAM disciplines are closely related to each other" "There is something in common among these five disciplines"	96	25.3
These five disciplines are connected but every discipline has its own features	"These disciplines interrelate to each other without losing their separate identities"	75	19.8
Four STEM disciplines are closely related to each other	"Science, Technology, Engineering and Mathematics are closely related to each other. But Art is so different from them" "Math is the fundamental discipline, on which science, technology and engineering are based. But art has little in common with them"	31	8.2
These five disciplines are independent of each other	"Every subject has its own identity" "Little things in common among these five subjects"	3	0.8
No idea	"I don't know" "I have no idea"	5	1.3
Others or no response	Others	98	25.9

the characteristics of STEAM education, whereas only 29.8% of the participants knew about interdisciplinary characteristic of STEAM education.

We asked the participants to illustrate how they visualized STEAM education by using the letters S, T, E, A, and M. The common types of visual representations and their individual explanations are listed in Table 5.5. The first four types match those from Bybee's (2013) codes.

Table 5.4 Participants' opinions regarding the characteristics of STEAM education

Characteristics	Students' quotes illustrating theme	Frequency	Percentage
Interdisciplinary	*"Interdisciplinary"; "comprehensive"; "contains different subjects";*	113	29.8
Designable	*"Designable";*	27	7.1
Interesting	*"Interesting"; "fun"; "attract students' interests";*	23	6.1
Practical	*"Practical"; "hands-on activities";*	19	5.0
Artistic	*"Artistic";*	7	1.8
Applicative	*"Applicative"; "application"; "Apply what you've learnt to real life";*	6	1.6
Technological	*"Technological"; "use of technology";*	6	1.6
Cooperative	*"Cooperative"; "cooperate with others"; "collaborative";*	3	0.8
Experiential	*"Experiential"; "kids get experience in learning";*	2	0.5
No idea or vacant	*"I don't know" "I have no idea"*	180	47.5

Most participants who drew a "nested" visualization selected science or arts as the dominant discipline. Only two participants selected technology as the overarching discipline. Interconnected visualizations were the most abundant visualizations. Overlapping visualizations were divided into various types. Most participants who drew this type of diagram only indicated that the disciplines were connected but did not elaborate about which factors connected them. However, some participants believed that two overarching subjects were connected by two other subjects.

Research Question 2 What attitudes do preservice teachers have toward STEAM disciplines and careers in STEAM areas? Do STEAM dispositions and career interests differ based on attributes, such as a teacher's gender or major study area?

The internal consistency of the CIQ in this study is listed in Table 5.6. CIQ can potentially serve as a measure for individual attitudes (Peterman et al., 2016). An analysis was conducted to examine participants' STEAM career attitudes. The mean and standard deviation are shown in Table 5.6. An independent samples test was performed to analyze the difference in participants' STEAM career attitudes based on the demographic variables. Table 5.7 shows the results of this analysis.

According to the data in Table 5.7, no significant differences were noted in terms of gender and teaching experience toward CIQ. However, the results indicated a difference in the CIQ between participants who attended STEM lectures or training and those who did not. We also adapted the STEAM semantic survey to analyze participants' attitudes toward five STEAM disciplines and careers.

As listed in Table 5.8, the participants' dispositions toward arts were the most positive, whereas those toward engineering were the least positive. An independent samples test was performed to analyze participants' attitudes toward STEAM disciplines and careers across gender (Table 5.9). The dispositions of male and female participants toward science, technology, engineering, and mathematics did not indi-

Table 5.5 Types of STEAM visualizations, explanations, frequencies, and percentages

Type	Visualization	Explanation	Frequency	Percentage
(a) Nested	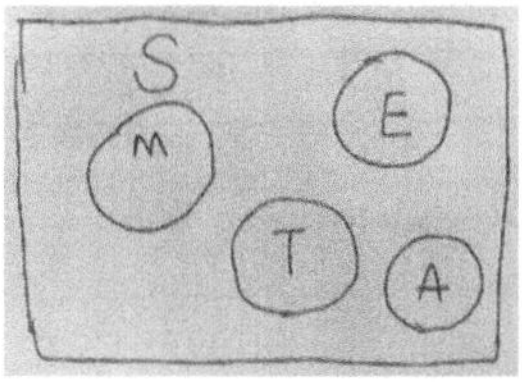	*Nested* visualizations meant that there was an overarching discipline of five STEAM disciplines	44	22.2
(b) Interconnected	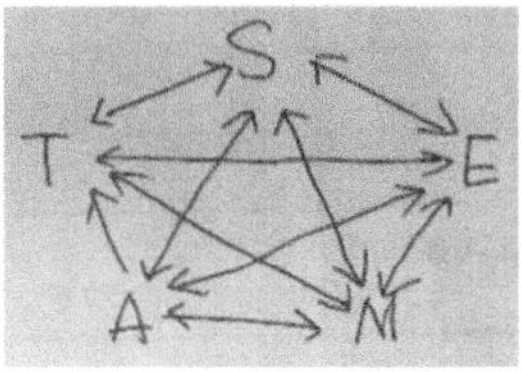	*Interconnected* visualizations showed that there was a connection between each two of the five disciplines	61	30.8
(c) Overlapping		*Overlapping* visualizations showed a Venn diagram that the disciplines were overlapping	48	24.2
(d) Sequential	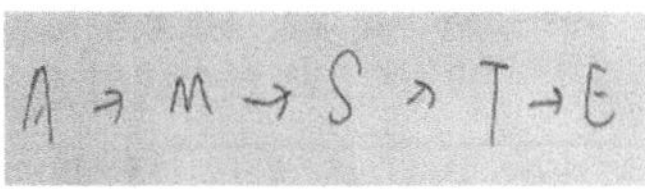	*Sequential* visualizations viewed STEAM as disciplines in a sequence	12	6.1
(e) Art independent	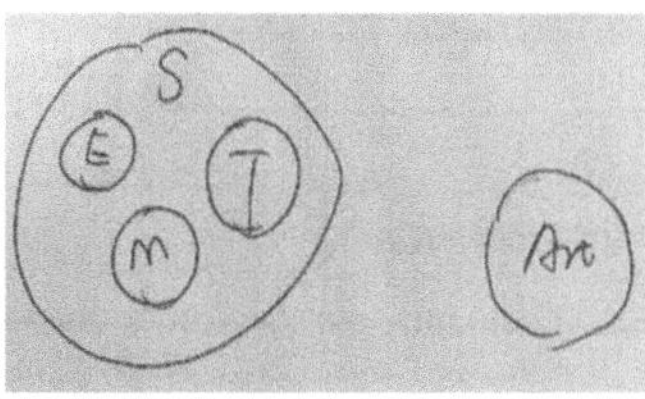	*Art independent* visualizations showed that there were connections among STEM four disciplines, but Art was independent	21	10.6
(f) Others			12	6.1
Total			198	100

Table 5.6 Cronbach's alpha for internal consistency reliabilities and the mean and standard deviations for CIQ subscales

Scale (# of items)	CIQ ($N = 379$)	M	SD
Interest (3)	0.887	3.44	1.021
Intent (3)	0.710	3.27	0.994
Importance (3)	0.825	3.68	0.960
Total (9)	0.888	3.46	0.991

Table 5.7 Significant differences in CIQ across gender, region, and whether participants attended the STEAM lectures or training sessions

CIQ			N	Mean	Std. dev	Sig
	Gender	Male	108	31.56	6.629	0.454
		Female	269	31.01	6.427	
	Teaching experience	Have teaching experience	231	30.72	6.661	0.104
		No teaching experience	144	31.85	6.204	
	Students attended the STEAM lectures or training or not	Attended the STEAM lectures or training	330	30.73	6.490	0.001
		Not attended the STEAM lectures or training	48	34.13	5.603	

Table 5.8 Means and standard deviations for STEAM semantic survey scales

	M	SD
Science	5.58	1.068
Technology	5.09	1.256
Engineering	4.62	1.270
Art	5.80	1.245
Mathematics	5.10	1.381

Table 5.9 Significant differences on STEAM semantic survey across gender

STEAM semantic survey			N	Mean	Std. dev	Sig
	Science	Male	108	28.33	5.55	0.292
		Female	269	27.69	5.25	
	Technology	Male	108	25.76	7.10	0.524
		Female	296	25.31	5.92	
	Engineering	Male	108	23.68	7.06	0.241
		Female	296	22.83	6.05	
	Art	Male	108	26.29	7.68	0.000
		Female	296	30.10	5.19	
	Mathematics	Male	108	26.21	6.62	0.226
		Female	296	25.26	7.00	

cate significant differences. However, female participants' attitudes toward arts were significantly more positive than those of male participants.

Research Question 3 What other opinions do preservice teachers have about STEAM education?

Table 5.10 summarizes the participants' opinions regarding STEAM teachers' capabilities. Numerous participants believed that teachers should have comprehensive knowledge regarding the subject. Some participants believed that they should also become experts of the subjects they are teaching. However, we classified responses 1, 2, 4, 5, 6, 7, 8, and 9 as capabilities that must be necessary for every

Table 5.10 Participants' opinions about STEAM teachers' abilities

Codes	Students' quotes illustrating theme	Frequency	Percentage
1. Have comprehensive knowledge	"Teachers must have comprehensive knowledge" "A wide range of knowledge......" "Knowledge"	92	24.3
2. Have professional knowledge	"Teachers should gain solid professional knowledge" "They must know their own teaching subject well"	42	11.1
3. Integrate interdisciplinary knowledge	"Integrate interdisciplinary knowledge" "Know how to teach different subjects together"	29	7.7
4. Innovation ability	"Innovation ability"	28	7.4
5. Be an interesting teacher	They should be interesting teachers "Teaching with fun"	19	5.0
6. Logical thinking ability	"Teachers should have logic thinking ability" "Logical mind......"	17	4.5
7. Communication ability	"Communication ability" "Communication skills" "Know how to communicate with students and other teachers"	14	3.7
8. Teaching skills	"Grab students' attention in class" "Teach the knowledge accurately"	9	2.4
9. Practical ability	"Apply theoretical knowledge to teaching practice"	8	2.1
10. Make connections between knowledge and real life	"Make connections between knowledge and real life"	2	0.5
11. Problem-solving skills	"Teachers need to gain problem-solving skills while teaching STEAM"	1	0.3
12. No idea	"No idea" "I don't know"	35	9.2
13. Vacant		68	17.9

teacher, and not only STEAM teachers. Responses 3, 10, and 11 are the capabilities that STEAM teachers should develop. Based on the data in Table 5.9, we can deduce that most participants were discussing the general abilities that are necessary for every teacher.

We asked the participants to provide suggestions for developing STEAM education in China. Many participants suggested that STEAM teachers are the key component for developing STEAM education. However, some believed that education in undeveloped rural areas should receive increased attention. Some participants further argued that STEAM education in China can adapt ideas from other countries that have well-developed and experienced STEAM education.

5.3 Discussion

5.3.1 Preservice Students Indicated Minimal Interest in STEAM and Lacked Interdisciplinary Awareness

Only some participants had attended training sessions and lectures about STEAM. Preservice teachers' familiarity with STEAM education contrasted that of in-service teachers. In a prior study, we examined in-service teachers' perceptions about STEM education and noted that most teachers had attended training sessions and lectures about STEM or STEAM. The *13th Five-Year Plan* launched by the Ministry of Education noted the significance of developing STEAM education for the first time. STEAM education is on the rise in both K-12 and higher education. Teachers have realized that education in the future will change from the traditional exam-oriented model to one that focuses on participants' comprehensive capabilities. Participants need to comprehend the knowledge they learn and connect it with aspects of real life, instead of only memorizing it for exams. However, the participant who majored in a specific area, such as physics or chemistry, concentrated on the content knowledge, rather than the educational policy. They also lacked practical experience. Because many students know little about STEAM education, they either had no opinion of STEAM education or only showed minimal interest in it.

To examine participants' opinions about the level of correlation among the five STEAM disciplines, the participants were asked to state their answers using a Likert scale, wherein "1" indicated that the five disciplines were completely independent and "9" indicated that they were completely related. Numerous participants selected "9," thus indicating that they believed that the five disciplines were related to each other.

To analyze their perceptions regarding STEAM, we asked the participants to describe why they chose their responses. We noted that many participants believed that everything in the world was related. Although that is true from a philosophical perspective, we believed that these participants did not understand the basic connotation of STEAM. Few participants (8.2%) believed that science, technology, engineering, and mathematics were related to each other, but the arts were independent from these disciplines. We also noted a similar result in the visualization of the five disciplines. Only 10.6% of the participants drew *arts-independent* visualization to explain the correlation among the five disciplines. The findings suggest that participants considered the arts to be unrelated to the other four disciplines because the other four disciplines were all science based. In China, participants in senior high school must compulsorily learn Chinese, mathematics, and English. However, they can choose to learn literature-based subjects, including politics, history, and geography; science-based subjects, including chemistry, physics, and biology; or arts-based subjects, including music, painting, and sport. Participants who select subjects from only one category are not forced to attempt the college entrance examination for the other subjects. Therefore, some participants continue to perceive the arts and science as unrelated to each other. The arts and science have often been considered different from each other (Wilson, 2012). However, artistic, cultural, and aesthetic experiences can help to create and maintain meaningful engagement with learning (Chemi, 2014). "Artistic experiences can greatly facilitate dedication, self-development, and learning and that what has a major effect on learning and creative expression is when children and young people experience great commitment and passion for what they do" (Mclellan, Galton, Steward, & Page, 2012).

Consistent with the earlier research, our analysis implied that the most common rationale for participant visualizations was that "STEAM disciplines were related to each other" (Table 5.4). Radloff and Guzey (2016) noted that mathematics has always been perceived as the predominant discipline because it is considered useful in all disciplines. However, in our research, science was selected as the discipline connecting all the other disciplines. Our findings also suggest that participants may have thought that all the other disciplines were guided by scientific principles.

We designed an open-item survey to analyze whether participants are aware of the STEAM characteristics. Numerous participants indicated that they were unaware of these, whereas some, who knew about the characteristics, only knew about the interdisciplinary characteristics.

In conclusion, most participants were unfamiliar with STEAM education and only had superficial knowledge about it. Therefore, for them, being interested in STEAM education would be difficult.

5.3.2 Female Participants' Increasing Interest in STEAM

The gender-based findings regarding STEAM careers in CIQ that female participants were less likely to pursue STEAM careers were inconsistent with those reported previously: Diao (2012) noted that in China, male college students indicate stronger interests in the realistic types of careers related to engineering and technology than do female college students. However, the author also found that the differences between male and female students' attitudes toward artistic types of careers were insignificant. Thus, the author's results were affected by the traditional implicit sex–occupational stereotype: men have more improved critical-thinking skills than do women, subsequently making them more rational and independent. Our research indicated no significant differences between the female and male students. Similar conclusions can be made regarding their semantic perceptions of science, technology, engineering, and mathematics. However, the female students indicated a significantly positive attitude toward the arts than did male students—also inconsistent with the findings of Diao (2012). Additional research determining the reasons for the increasing interest of female students in STEAM is warranted. Our participants who attended training sessions or lectures on STEAM indicated significantly positive attitudes related to finding a career in STEAM than did those who had not. For those who have minimal STEAM knowledge, being interested in working in a field they are unfamiliar with may be difficult. By contrast, they have no idea what it would be like to work in the STEAM area.

5.3.3 Preservice Teachers Were Unfamiliar with Qualities Necessary for STEAM Teachers

Many participants focused on the general capabilities that not only STEAM but also all other teachers must have. For instance, the results indicated that they considered it vital that all teachers have comprehensive or extensive professional knowledge regarding their selected areas of expertise. We admit that these skills were crucial for STEAM teachers. However, STEAM teachers and teachers who only teach one subject differ. For instance, STEAM teachers require integrated interdisciplinary knowledge. The most common STEAM teaching approaches involve project- and problem-based learning. A project or problem typically involves several subjects, which requires teachers to be proficient in various subjects. STEAM teachers are also required to teach according to the project-based learning principles. However, none of the participants discussed this. In conclusion, most participants did not possess a deep understanding about STEAM education.

5.4 Conclusion

The current data were gathered from 379 university students who indicated the potential to become preservice teachers. The findings indicated that most participants did not attend lectures or training sessions about STEAM education and indicated minimal interest in STEAM education. Moreover, most participants only had superficial perceptions about STEAM education and did not understand fundamental ideas in STEAM education, such as how the different disciplines can be connected, or the method of teaching STEAM courses. According to the STEAM policy proposed by the Ministry of Education, STEAM courses must be offered in elementary and high schools. However, similar to Chemi's (2017) recommendations, our findings also suggest that integration of STEM with the Arts should be conducted meaningfully—that is, it should not serve only as entertainment or be positioned to play only an ancillary role for other disciplines. When engaged in deep and meaningful artistic and creative experiences, we believe that preservice teachers can begin to develop a rich understanding of creative inquiry in STEM education.

A particular concern that our analysis raises goes beyond elementary and high schools, given that in our data, the understandings regarding the significance of STEAM education was lacking among the current preservice teachers who were enrolled in university courses. This suggests that universities must also offer educational experiences in STEAM education to preservice teachers.

5.5 Limitations

This study only represents the 379 participants who are majoring in the educational program, and therefore, an overgeneralization of the results should be avoided. We also were not able to interview individual students to follow-up on their survey responses. In the next research phase, we will interview preservice teachers who can serve as atypical representatives of questionnaire respondents.

References

Bowdich, S. (2009). *Analysis of research exploring culturally responsive curricula in Hawaii*. In Educational Research Association Annual Conference, Hawaii.

Breiner, J. M., Harkness, S. S., Johnson, C. C., & Koehler, C. M. (2012). What is STEM? A discussion about conceptions of STEM in education and partnerships. *School Science and Mathematics, 112*(1), 3–11. https://doi.org/10.1111/j.1949-8594.2011.00109.x

Buckley, B. C. (2000). Interactive multimedia and model-based learning in biology. *International Journal of Science Education, 22*(9), 895–935.

Bybee, R.W. (2013) The case of STEM education: challenges and opportunities. NSTA Press, Arlington.

Chemi, T. (2014). *The art of arts integration*. Aalborg: Aalborg University Press.

Chemi, T. (2017). Science and the arts: Curriculum integration, learning, and emotions in schools. In A. Bellocchi, C. Quigley, & K. Otrel-Cass (Eds.), *Exploring emotions, aesthetics and well-being in science education research* (pp. 203–218). Dordrecht: Springer Nature.

Christensen, R., Knezek, G., & Tyler-Wood, T. (2014). Student perceptions of Science, Technology, Engineering and Mathematics (STEM) content and careers. *Computers in Human Behavior, 34*, 173–186. https://doi.org/10.1016/j.chb.2014.01.046

Cook, M. P. (2006). Visual representations in science education: The influence of prior knowledge and cognitive load theory on instructional design principles. *Science Education, 90*(6), 1073–1091.

Diao, Y. (2012). *Correlation research of college students' vocational interests and its relationship with social support and coping style* (Unpublished master's thesis). Gannan Normal University, Jiangxi, China.

Guyotte, K. W., Sochacka, N. W., Costantino, T. E., Walther, J., & Kellam, N. N. (2014). STEAM as social practice: Cultivating creativity in transdisciplinary spaces. *Art Education; Reston, 67*(6), 12–19.

Jiang, F. K. (2017). STEM education for connecting formal and informal learning—Review of the 4th international STEM in education conference. *E-Education Research, 38*(02), 53–61.

Johnson, C. C. (2013). Conceptualizing integrated STEM education. *School Science and Mathematics, 113*(8), 367–368. https://doi.org/10.1111/ssm.12043

Lamore, R., Root-Bernstein, R., Root-Bernstein, M., Schweitzer, J. H., Lawton, J. L., Roraback, E., … Fernandez, L. (2013). Arts and crafts critical to economic innovation. *Economic Development Quarterly, 27*(3), 221–229. https://doi.org/10.1177/0891242413486186

Li, J. (2015). Research on the training needs of rural teachers. *China Adult Education, 7*, 92–93.

Li, G., & Lv, L. (2018). From STEM to STEAM: The Role of Arts. *China Educational Technology*, (280), 31-39+47.

Mayer, R. E., Bove, W., Bryman, A., Mars, R., Tapangco, L., & Levin, J. R. (1996). When less is more: Meaningful learning from visual and verbal summaries of science textbook lessons. *Journal of Educational Psychology, 88*(1), 64–73.

Mclellan, R., Galton, M., Steward, S., & Page, C. (2012). *The impact of creative initiatives on well-being: A literature review* (Creativity, Culture and Education Series). Newcastle upon Tyne.

Peeck, J. (1993). Increasing picture effects in learning from illustrated text. *Learning & Instruction, 3*(3), 227–238.

Pernecky, T. (2016). The epistemic foundations of event studies. In T. Pernecky (Ed.), *Approaches and methods in event studies* (pp. 3–22). London, UK: Routledge.

Peterman, K., Kermish-Allen, R., Knezek, G., Christensen, R., & Tyler-Wood, T. (2016). Measuring student career interest within the context of technology-enhanced STEM projects: A cross-project comparison study based on the career interest questionnaire. *Journal of Science Education and Technology, 25*(6), 833–845. https://doi.org/10.1007/s10956-016-9617-5

Radloff, J., & Guzey, S. (2016). Investigating preservice STEM teacher conceptions of STEM education. *Journal of Science Education and Technology, 25*(5), 759–774. https://doi.org/10.1007/s10956-016-9633-5

Resnick, M., Berg, R., & Eisenberg, M. (2000). Beyond black boxes: Bringing transparency and aesthetics back to scientific investigation. *Journal of the Learning Sciences, 9*(1), 7–30. https://doi.org/10.1207/s15327809jls0901_3

Roth, W., Bowen, G., & Mcginn, M. (1999). Differences in graph-related practices between high school biology textbooks and scientific ecology journals. *Journal of Research in Science Teaching, 36*(9), 977–1019.

Spelke, E. (2008). Effects of music instruction on developing cognitive systems at the foundations of mathematics and science. In *Learning, arts and the brain*. New York, Washington DC: Dana Press.

Trilling, B. (2009). *21st century skills: Learning for life in our times*. San Francisco: Jossey-Bass.

Tsupros, N., Kohler, R., & Hallinen, J. (2009). *STEM education: A project to identify the missing components*. Pennsylvania: Carnegie Mellon University.

Wilson, S. (2012). *Art + science now*. London: Thames & Hudson Ltd..

Xie, Z. (2017). Discrimination between creator education and STEM education. *China Modern Educational Equipment, 260*, 16–19.

Ye, Z., & Yang, Y. (2017). Innovation and integration: Rethinking of STEM education, creator education and creator course. *Jiangsu Education*, (19), 7–10+19.

YU, S., & HU, X. (2015). STEM Education and Its Model for Interdisciplinary Integration. *Open Education Research, 21*(4), 13–22.

Zaichkowsky, J. L. (1985). Measuring the involvement construct. *Journal of Consumer Research, 12*(3), 341–352.

Zhu, J., & Zhou, Z. (2009). *Continuing education and training of teachers in rural areas based on mobile learning and mobile services*. Presented at the 1st International Conference on Information Science and Engineering (ICISE), Nanjing, China.

Chapter 6
Engaging Emerging Bilingual Students in Language and Scientific Practices Through Collaborative Disciplinarily Integrated Games from a Co-operative Action Lens

Douglas Clark and Ashlyn Pierson

Abstract Recent research on science education with emerging bilinguals has emphasized the co-development of language learning and science learning. This aligns with overarching science education research and reform advocating for an increased emphasis on engaging students in disciplinary practices. We developed disciplinarily integrated games (DIGs) as an approach for engaging students in disciplinary practices of science with an emphasis on models and modeling. The goal of this chapter is to (a) propose a collaborative DIG as a rich context in which emerging bilinguals can leverage a broad range of resources and representations and (b) consider the potential promise of the proposed collaborative DIG for emerging bilinguals from the perspective of Charles Goodwin's co-operative action framework. Toward this goal, we present a collaborative DIG prototype with agent-based models as the mode of control wherein players create, trade, and elaborate on one another's code as part of gameplay. We then discuss how such a platform might engage emerging bilingual students in the practices of modeling in a manner that also increases opportunities for language learning from the perspective of the framework.

Keywords Science education · Science as practice · Modeling · Disciplinarily integrated games · Collaboration · Digital games · Emerging bilingual students · English language learners · Co-operative action

D. Clark (✉)
University of Calgary, Calgary, AB, Canada
e-mail: douglas.clark@ucalgary.ca

A. Pierson
Vanderbilt University, Nashville, TN, USA

P. Sengupta et al. (eds.), *Critical, Transdisciplinary and Embodied Approaches in STEM Education*, Advances in STEM Education,
https://doi.org/10.1007/978-3-030-29489-2_6

6.1 Introduction

All students, including emerging bilingual students, benefit from engaging in authentic disciplinary practices (Lehrer & Schauble, 2006a; Moschkovich, 2015; National Research Council (NRC), 2012). Therefore, researchers and educators need to support emerging bilingual students in appropriating science practices like modeling (Lee, 2005; NGSS Lead States, 2013). Clark, Sengupta and colleagues developed disciplinarily integrated games (DIGs) such that players' actions involve the iterative development and manipulation of scientific models and other computational and mathematicized formal representations as the core game mechanics (Clark, Sengupta, Brady, Martinez-Garza, & Killingsworth, 2015). In the current chapter, we explore the potential of collaborative DIGs to engage emerging bilingual students in the discourses and practices of science leveraging diverse disciplinary representational resources as well as students' multiple linguistic resources. To set the stage for this exploration and discussion, we first establish the need for supporting emerging bilingual students in the practices of science. We then discuss a proposed collaborative DIG as a potentially rich context in which emerging bilinguals can leverage a broad range of resources and representations from the perspective of Charles Goodwin's co-operative action framework (Goodwin, 2017).

6.2 The Need: Emerging Bilingual Students and Science Education

Contemporary research promotes engaging emerging bilingual students in language-intensive science practices to simultaneously promote language learning and science learning (O. Lee, Quinn, & Valdés, 2013; Suarez & Otero, 2014). Researchers further argue that (a) academic language proficiency is not a prerequisite for engaging in science learning, and (b) language learning can be supported through science learning because teachers can mediate linguistic bridges from informal to academic registers (Gibbons, 2003; Quinn, Lee, & Valdés, 2012). Along these lines, multiple studies demonstrate students' ability to engage in science practices using "less-than-perfect" English (Lee, Miller, & Januszyk, 2014; NGSS Lead States, 2013). Based on this research, Lee et al. argue for moving away from discrete, content-based language instruction and sheltered models. They suggest that students can develop language proficiency and understanding of science content simultaneously, and they argue that teachers should support language development in science by scaffolding students' participation in discourse (Lee et al., 2013, 2014). Lee et al. explain that this approach benefits all students, not just emerging bilinguals, because monolingual English speakers also benefit from scaffolding that bridges everyday discourse and academic discourse (Lee et al., 2014). Earlier work by our group focused on supporting emerging bilingual students engaging in inquiry by providing access to the full range of their linguistic resources including named languages and other semiotic resources (e.g., Clark, Touchman, Martinez-Garza, Ramirez-Marin, & Drews, 2012; Medina-Jerez, Clark, Medina, & Ramirez-Marin, 2007; Tate, Clark, Gallagher, & McLaughlin, 2008). Current research on supporting emerging bilinguals in science education warrants exploring approaches for engaging emerging bilingual students in the practices of scientific modeling in a manner that also increases opportunities for language learning. Our goal in this chapter is to shift our focus deeper into supporting students in connecting across disciplinary representations and resources.

6.3 Science as Practice and DIGs

Recent research and reform efforts in science education have shifted away from a focus on the accumulation of knowledge and skills toward conceptualizing science as a form of disciplinary practice (Cheng & Lin, 2015; Duschl, Schweingruber, & Shouse, 2007; Erduran & Jiménez-Aleixandre, 2007; Louca & Zacharia, 2015; National Research Council (NRC), 2012; Östman & Wickman, 2014; Pierson, Clark, & Sherard, 2017). From this perspective, science classrooms should provide students with opportunities to engage in epistemic disciplinary activity as a way of building scientific understanding (Duschl, 2008; Ford & Forman, 2006). This framing of science as practice moves beyond science education's traditional focus on developing conceptual understanding or decontextualized "scientific processes" (Lehrer & Schauble, 2006b).

Games as a medium have been demonstrated to be effective in supporting various conceptual, epistemological, and practice-focused aspects of science learning (e.g., Clark, Nelson, Sengupta, & D'Angelo, 2009; Martinez-Garza, Clark, & Nelson, 2013; Sengupta, Krinks, & Clark, 2015). Developing games that engage students in modeling from a science-as-practice perspective would seem to hold great promise. Toward these goals, Clark, Sengupta and colleagues have proposed DIGs as one such approach (Clark et al., 2015, 2016; Sengupta & Clark, 2016).

Disciplinary integration can be thought of in terms of "model types" and "modeling strategies" (Collins, 2017), which Collins et al. have termed "epistemic forms" and "epistemic games" in earlier work. Collins et al. argue that the professional work of scientists can be understood in terms of model types (epistemic forms) that are the target structures guiding scientific inquiry and modeling strategies (epistemic games) that are the sets of rules and strategies for creating, manipulating, and refining those model types (e.g., time-series analyses, system-dynamics models, and other canonical model types).

Clark et al. (2015) argued that the nature of learning experiences in DIGs could be construed as engaging with epistemic forms and games in the sense that Collins et al. have argued for. DIGs are designed to engage students in specific modeling and representational practices of developing, interpreting, manipulating, and translating across specific model types. Initial DIGs focused primarily on Cartesian change over time graphs as the formal model types through which players enacted strategies and control over the game as well as the formal model types through which the game communicated goals and challenges to the player (Clark et al., 2015, 2016). Early work by Sengupta, Clark, Krinks, Killingsworth, and Brady (2014), Sengupta and Clark et al. (2016), and Krinks, Sengupta, and Clark (2019) extended this frame by proposing and exploring the possibility of using agent-based models as the representations through which the player enacted strategies within the game. As we elaborate below, agent-based models offer a different set of affordances for engaging in modeling than Cartesian representations. For example, they support students in visualizing the phenomena they represent and ask students to consider the system from the perspective of the agent within the system rather than from an aggregate perspective only.

Essentially, all DIGs have the following characteristics: (a) formal model types for controlling the game, (b) formal model types for communicating challenges and opportunities, (c) a phenomenological representation presenting the phenomenon being modeled, (d) intermediate aggregating representations to help students bridge between formal model types, and (e) game mechanics and goals focused on engaging the player in interpreting, creating, modifying, and translating across these formal and phenomenological representations.

In the current chapter, we will consider a prototype of a collaborative DIG focusing on ecology with a core model type of agent-based modeling for the purposes of subsequently discussing the potential of such a platform for engaging emerging bilingual students in disciplinary practices of science, including discourse practices. We originally introduced and discussed this prototype along with another prototype focusing on physics in Clark, Medlock-Walton, Boquín, and Klopfer (2018) and Clark and Sengupta (2019), but we provide an overview here to support our discussion of the potential affordances for English learners.

The software, called SURGE Gameblox, was programmed by the Gameblox designer at MIT funded by the SURGE NSF grant at Vanderbilt University. It represents a reconfiguration and integration of functionality from the Gameblox and Starlogo Nova environments developed by MIT through prior funding. SURGE Gameblox projects are organized into block pages, similar to how large coding projects are separated into multiple files (Fig. 6.1). The collaboration code is designed so that each player can edit only her own code but can view the other player's code as it is being changed. The design could be revised to allow players to directly edit other players' pages, but our experience with other collaborative environments suggests that allowing such direct editing can undermine incentives for players to talk with one another because it is often easier for a player to edit a page directly than it is to explain to a collaborator how to make the changes. SURGE Gameblox thus allows pairs of players to simultaneously edit different pages within the same overarching model, facilitating discourse between players.

In the ecosystem game prototype, players attempt to get as many animals living on the two players' farms as possible (Fig. 6.2). The animals roam between the two farms, pause to eat grass, and get varying levels of energy from each type of grass. Players program where, when, and how much grass of each type to plant at their farms. Players have graphs of the seasonal changes in sunlight and rainfall (Fig. 6.2)

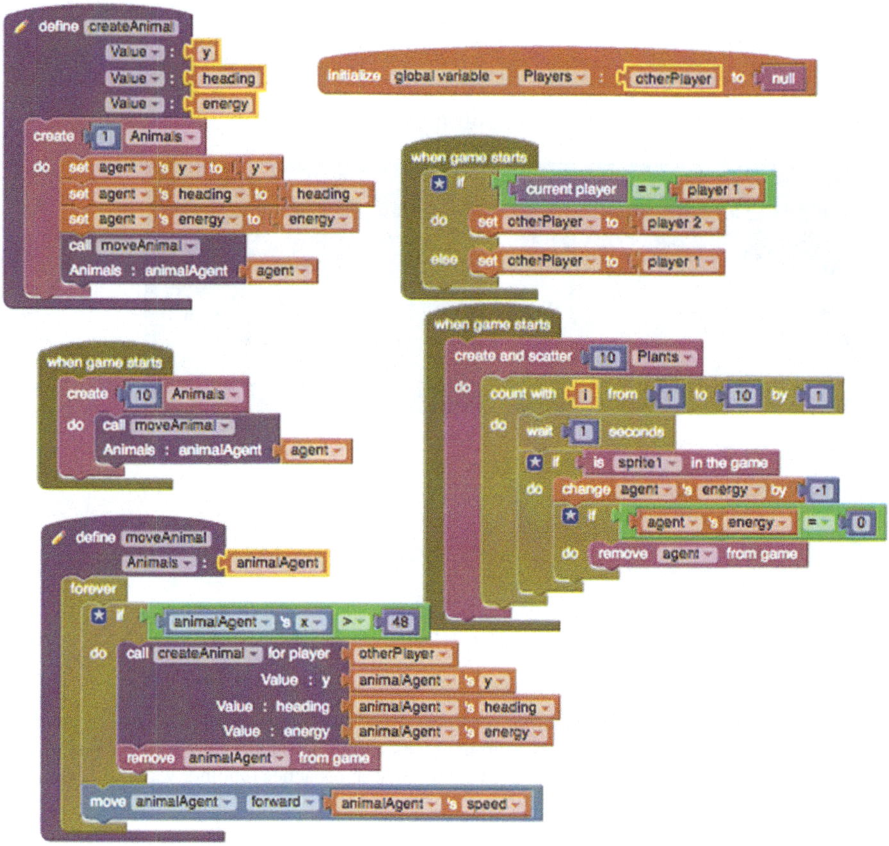

Fig. 6.1 Example of code for the underlying model. This code is visible to both players but neither player can edit it. This is the code that defines the relationships and dynamics at the heart of the challenge and thus communicates the challenge to the players

along with other data about the animal and grass types. Each type of grass does better in different climate conditions. Another important feature is a block that allows the player to get data from the graphs about particular points in time. This functionality incentivizes more complicated coding strategies, allowing the user to decide what to plant and where based on the values of the rainfall and sunlight instead of simply choosing static times and locations. This shifts the nature of computational thinking within the game to emphasize core CT ideas such as loops, functions, and conditionals (cf. Grover & Pea, 2013).

Agent-based models are thus the formal model type through which the player enacts control and strategies within the game. The game also uses formal model types to communicate challenges and goals to the player, including Cartesian graphs of scientific interactions or relationships over time, as well as agent-based model block pages with model functions that are visible but not editable by the players that provide the parameters, breeds, and functions driving the challenge. Players create

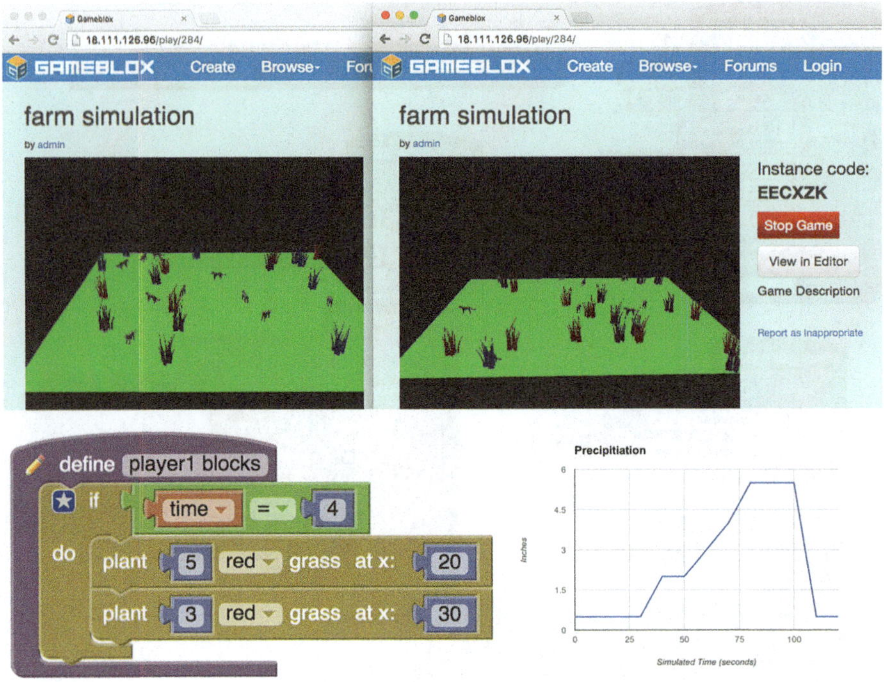

Fig. 6.2 Prototype of the game including a screenshot of the game running on the screens of the two players along with an example of a simple block a player might create to plant grass and a precipitation graph that players consult in planning their approaches (another way the game communicates challenges to the players)

their own breeds and code to interact with the code on these block pages in order to explore and potentially overcome the central challenges that define that particular game. Gameplay focuses on interpretation, translation, and manipulation of disciplinary representations.

6.4 Collaborative DIGs for ELs Through Lens of Co-operative Action

Having now set the stage, we want to explore and consider the affordances of the proposed collaborative DIG for supporting emerging bilingual students in conceptual science learning, engagement in the disciplinary practices of science, and engagement in discourses, registers, and modalities of science through the lens of the co-operative action framework (Goodwin, 2017). Toward that end, we first provide a brief overview of the framework and then consider the proposed cooperative DIG from the perspective of the framework.

6.4.1 Overview of Goodwin's Co-operative Action Framework

At its core, Goodwin's co-operative action framework proposes that people produce new actions by decomposing and reusing with transformation the resources made available by the actions of others. More specifically, the co-operative action framework posits that:

> New action is built by decomposing and reusing with transformation the resources made available by the earlier action of others. We thus inhabit each other's actions. Such co-operative action differs from cooperation in that it is not restricted to mutual aid; more crucially it provides, in the midst of action itself, a systematic mechanism for progressive accumulation with modification on all scales, from chains of local utterance, through tools, to the unfolding differentiation through time of human social groups. (Goodwin, 2017, p. 1)

Goodwin makes clear with detailed analyses and examples how these processes of co-operative action support interaction with co-present participants and with predecessors through sedimented artifacts. Goodwin argues that this set of practices for the construction of action are at the core of much of human social and cognitive life including language, social organization, tools, pedagogy, sharing experience, and the progressive differentiation of human societies and cultural worlds. This allows for the ongoing accumulation and transformation of resources by individuals, communities, and fields. This is true for language use as well as for the practices and tools that a community or society leverages. Thus, co-operative action spans spatial and temporal scales. At the smallest grain size, Goodwin describes reuse with transformation within conversational turns in mundane settings, but Goodwin also describes changes and innovations at broader grain sizes that can be viewed as learning by communities in terms of cultural and technological practices, tools, and symbol systems.

Substrates need not consist of talk or language structure. Goodwin provides examples of the ways that environmentally coupled gestures and other semiotic resources, such as prosody, allow participants to fully and effectively participate in interaction. Moreover, historical artifacts can act as substrates, facilitating co-operative action with predecessors and juxtaposing semiotic fields in particular ways to support disciplinary or community practices. Toward this end, Goodwin provides the example of a Munsell chart from archeology, which appears to be a simple color chart printed on cardboard with holes punched next to each color to allow categorization of soil by color and type. As Goodwin explains, "By juxtaposing unlike spaces, but ones relevant to the accomplishment of a specific cognitive task, the chart creates a new, distinctively human, kind of space. It is precisely here, as bits of dirt are shaped into the work-relevant categories of a specific social group, that 'nature' is transformed into culture" (p. 199). Goodwin also provides the examples of the transformation of sedimented artifact and tools by communities over time, demonstrating that a variety of substrates are available to participants for reuse with transformation in building subsequent action.

It is important to note that co-operative action provides not only a lens for analyzing action by a group, but also proposes an effective mechanism through which individuals and communities learn and change. Many of Goodwin's examples focus on the small-scale co-operative actions that occur among small groups of people within a few conversational turns, but he also highlights how co-operative action supports learning by creating pathways through which individuals become capable actors within a community and pathways through which novel ways of thinking and doing are developed. Co-operative action therein can offer pragmatic advantages when thinking about learning compared to traditional perspectives on learning through interaction such as Legitimate Peripheral Participation (LPP), for example, in the sense that LPP primarily accounts for mechanisms through which people are "reproduced" in the mold of the community of practice rather than accounting for mechanisms of change within a community. In contrast, co-operative action provides mechanisms for reproduction while also outlining mechanisms for learning that allow the community, community practices, or the individual to become something new. Co-operative action therefore outlines mechanisms though which the individual can become not only a competent member of the community but to also change that community and its practices and contribute new resources in terms of thinking, ideas, practices, and tools.

6.4.2 *Affordances of Proposed DIG from the Perspective of Co-operative Action*

From the perspective of Goodwin's co-operative action framework (Goodwin, 2017), a collaborative DIG grounded in agent-based modeling could create a multimodal learning environment with affordances for all students, especially emerging bilingual students. The proposed DIG environment allows students to build upon and transform one another's actions as well as the historical disciplinary resources provided by the DIG. The following sections explore aspects of these affordances.

Multimodality Beyond English-Language Resources First, the multimodal nature of the DIG environment allows emerging bilingual students to draw upon a variety or resources. Along these lines, we can think about the affordances provided in DIGs in terms of the multimodal semiotic fields juxtaposed within the DIG. As Goodwin states:

> By participating in [an] interactive field people with very different resources and abilities… are nonetheless able to use language, including grammatical structures that are beyond the capacities as individuals to create, to build relevant action. Moreover, within such an interactive field language typically does not stand alone, but instead is used in conjunction with other semiotic resources, including most crucially a range of quite different kinds of displays visible in the orientation in action of the participants' bodies, as well as structure in their environment. (Goodwin, 2017, p. 89)

The interactive field of the DIG thus provides a range of substrates for co-operative action beyond English-language resources, increasing opportunities for emerging bilingual students to fully participate in the game and related learning opportunities.

Meaning making in science classrooms traditionally involves multiple modalities, including action, speech, writing, symbols, and diagrams (Danielsson, 2016). Engaging with multiple modes is essential to engaging in disciplinary practices; thus, these modalities are not simply scaffolds for language development (Grapin, 2019; Moschkovich, 2015). Grapin (2019) explains that language has limitations, for example, compared to language, modalities such as diagrammatic models are better suited to representing spatial relations among components in a system. For this reason, multimodal representations like models and data visualizations are powerful conceptual tools across domains of science (e.g., Lehrer & Schauble, 2006b).

In the proposed DIG, for example, although the code blocks include English-language text, these short phrases are accompanied by color coding related to function. Moreover, students can quickly link blocks with simulated and visualized outcomes as they run their code in the game. In addition, the game provides visual representations of data, which have been shown in previous studies to be a powerful resource for emerging bilingual students (e.g., Moschkovich, 2015). Thus, the game creates opportunities for laminating dialog across multiple registers of scientific discourse. This dialog is not limited solely to verbal interaction in English—instead, dialog is supported by the multiple representational systems constituted within the game with which players can laminate verbal dialog, deictic gesture, images, disciplinary representations, and code. Whereas traditional classrooms may focus on lectures, text books, and high stakes IRE dialog led by the teacher that lean predominantly on English language resources, dialog around the DIG provides a more varied and richer substrate within a robust interactive field. Furthermore, in addition the affordances of multimodal environments for learning, examining learning from a multimodal perspective can offer a more complete and nuanced picture of what students know and can do in science, thus exposing "the rich possibilities in these students' thinking" (Fernandes, Kahn, & Civil, 2017, p. 280).

Sedimented Resources of Historical Actors Through Canonical Representations Second, the game allows students to draw upon present actions as well as the sedimented resources of historical actors through canonical disciplinary representations and tools. DIGs include modalities such as code and dynamic data visualization that are common in professional science practice, yet rare in K-12 classrooms (Finzer, Erickson, Swenson, & Litwin, 2007). These additional modalities provide students with new ways to make sense of content and express ideas. These features of the DIG provide additional modalities and access to formal registers of communication in science. From the perspective of co-operative action, these formal representations are cultural tools that have sedimented into science practice to support seeing the world in a certain way. The representation of farms, animals, and climate data within the DIG are an inscription of relevant features of the world seen through the lens of the science standards that culturally define what students should know and understand about ecological relationships. Thus, the proposed DIG supports seeing and operating on the world in a way that builds on existing cultural tools and scientific practices, providing more rich and authentic opportunities for engaging in disciplinary discourse and registers than are available in traditional classroom environments that focus more exclusively on English-language resources and pedagogical practices.

Differential Knowledge States and Collaboration Motivate Co-operative Action Third, the DIG motivates co-operative action by creating differential states of knowledge among participants. Goodwin argues that differential states of knowledge are constitutive of the structure of human action (2017, p. 101). The proposed DIG necessarily positions participants as *knowing* in terms of their own code and *unknowing* in terms of the other player's code. In this sense, to be successful in the game, students must engage in co-operative action by engaging with the other player's code.

Furthermore, while co-operative action does not require cooperation, the collaborative setting of the DIG may encourage students to build upon and transform the resources created by their peers across these differential states of knowledge. Asking students to collaboratively grapple with a problem challenges them to establish common frames of reference, resolve discrepancies in understanding, negotiate the distribution of labor, and arrive at joint understanding (Barron, 2000). Rochelle (1992) explained that in order for conceptual change to occur, students must experience convergence, which includes iterative cycles of "displaying, confirming, and repairing situated actions" (p. 237). In the context of digital games, constructive collaboration often leads to positive learning outcomes and higher levels of abstract thinking (Echeverrîa et al., 2012). One possible explanation from research in non-game settings involves the opportunity to bridge multiple perspectives in collaborative settings (e.g., Schwartz, 1995). Additionally, as explored by Farris and Sengupta (2014), providing collaboration opportunities between peers during game play can allow students to leverage discourse in order to negotiate meaning of concepts and actions within games, construct ideas, and resolve conflicts through perspectival shifts (Greeno & MacWhinney, 2006; Greeno & van de Sande, 2007, 2012; Thagard & Verbeurgt, 1998).

At the most basic level, collaboration in this setting and in other shared settings thus provides potential advantages for all students and emerging bilingual students in particular. Students playing digital learning games in the classroom rarely play alone, even in digital games that are ostensibly "single-player" games (Van Eaton, Clark, & Smith, 2015). Rather, students engage "offline" continuously and informally with their nearby peers, and students may also simultaneously engage "online" with their peers throughout the classroom and beyond in online forums, chat, or message boards if they are available (Van Eaton et al., 2015). In other words, there are rarely truly "single-player" digital game experiences for learning in a typical classroom. Furthermore, our research on engaging students in argumentation more generally demonstrates the power of engaging students collaboratively in digital environments for the quality of the argumentation and for students' opportunities to engage in argumentation (e.g., Clark, Weinberger, Jucks, Spitulnik, & Wallace,

2003; Sampson & Clark, 2006; Weinberger, Clark, Häkkinen, Tamura, & Fischer, 2007).

Concrete Engineering Goal to Focus Co-operative Action and Intersubjectivity Fourth, the DIG provides concrete engineering goals (i.e., maximize biomass in the proposed DIG) to focus co-operative action and galvanize intersubjectivity. Sengupta, Krishnan, Wright, & Ghassoul (2014, April) demonstrate the affordances of engineering goals for supporting intersubjective constructionist meaning making. Sengupta et al. build on the ideas of Suthers (2006) in terms of intersubjective meaning making occurring as "multiple participants contribute to a single composition that is the result of bringing together of interrelated interpretations" in manner that is more complex than simple establishing "common ground" in terms of shared understanding (p. 274). As Sengupta et al. explain, intersubjective meaning making is more complex than simply establishing common ground in the sense that: (a) it represents a participatory process within which beliefs are enacted and thus shared without necessarily being mutually accepted, (b) ideas can be created through interaction as well be formed by individuals before being introduced to the group, and (c) the cognitive activities can be distributed across the individuals as well as the artifacts through which they interact.

Aligning with this perspective, the collaborative game prototype highlights relevant measures for students to attend to. This shared goal could facilitate productive co-operative action because it allows students to consider features of the game in categories that are consequential to their goal, even when students' beliefs about how to achieve their goal differ. Thus, students are able to more easily evaluate each other's actions in relation to the goal, providing a shared lens for epistemic activity and supporting collaborative action. This is productive from the perspective of the co-operative action framework because tools and practices take on meaning within a community by facilitating progress toward valued and shared goals. By staking out concrete engineering goals, as opposed to more open-ended sandbox exploration goals, the proposed DIG orients co-operative action by providing a clear and shared criteria against which to interpret and evaluate actions in the game while allowing for richer intersubjective meaning making to occur by providing a shared goal that does not demand or depend upon fully aligned perspectives across the participants.

Dynamic Representations Facilitating Intersubjectivity Fifth, and finally, the agent-based models provide real-time feedback and implications in response to students' actions, which may support consensus building and achievement of intersubjectivity by participants. In agent-based models, students identify agents (individual actors within the model representing individual animals and plants), create computational rules of behavior and interaction for each type of animal or plant agent (i.e., "breed" of agent) and "run" the system. This enables the computational representation to simulate emergent whole-system behaviors through the interactions of the individual agents with one another based on the rules so that the modeler-plus-computer can reason about the agent-level and aggregate-levels of the system to build deeper understandings of causal mechanisms (Dickes & Sengupta, 2013; Dickes, Sengupta, Farris, & Basu, 2016; Sengupta & Wilensky, 2009; Wilensky & Resnick, 1999). By presenting the agent-based modeling environment through a visual, rather than text-based interface, agent-based modeling can make computational thinking more accessible to learners syntactically (Guo et al., 2016; Sengupta et al., 2015; Sengupta, Kinnebrew, Basu, Biswas, & Clark, 2013; Wilkerson-Jerde, Wagh, & Wilensky, 2015) and avoid reliance on a single semiotic language system.

With these dynamic representations, students can quickly encounter and demonstrate differences in outcomes by changing their code. Compared to other representations, like verbal or written explanations or diagrammatic models, agent-based models allow learners to test relationships between components of the model and communicate about their ideas. Thus, not only does the DIG provide a variety of resources within the substrate, it also provides and strategically juxtaposes dynamic disciplinary representations within the substrate that could support understanding, communication, and explanation. Furthermore, the real-time feedback from the DIG may also the game as a participant in interaction, adding additional and dynamic resources to the substrate (Pierson, Brady, & Clark, in press). These dynamic characteristics might facilitate intersubjectivity among players because the implications of any given proposition are made clear to all participants as the model is run. The models could therefore increase shared meaning as they dynamically "talk back" with feedback to the participants. In addition, interaction around the DIG could increase intersubjectivity by bringing into alignment students' efforts and progress in a manner that might help constrain and focus participants' interpretations. The causal and emergent outcomes of the game thus act as multimodal substrates for reuse and transformation as students continue to play and learn in interaction with the game and their peers.

6.5 Conclusions and Implications

In summary, from the perspective of the co-operative action framework, the proposed DIG has affordances in terms of learning and communication for all students. These resources could be particularly valuable for emerging bilingual students because they support participation frameworks in which no single actor requires complete mastery of all semiotic systems included within the interactive field to participate fully and effectively. Instead, participants can draw on and manipulate substrates from a range of modalities to engage in game play and science learning.

More specifically, the proposed DIG seems particularly well adapted to supporting productive interactive fields for small groups of emerging bilingual students because: (a) the multimodal nature of the environment allows emerging bilingual students to draw upon a variety or resources, (b) the structure allows students to draw upon present actions as well as the sedimented resources of historical actors through canonical disciplinary representations and tools, (c) the structure motivates co-operative action by creating differential states of knowledge among participants, (d) the structure provides concrete engineering goals to focus co-operative action and galvanize intersubjectivity, and (e) the agent-based models provide real-time feedback and implications in response to students' actions, which may support consensus building and achievement of intersubjectivity by participants.

To help clarify these proposals, it is important consider the affordances and limitations of the resources and organization at any given moment within the interactive field. At a concrete level, an obviously central semiotic field within the substrate is the programming code. This syntax is visible to the players in the form of the rules that agents follow, as well as the ongoing transformations and additions to that code added by the players. Within this semiotic field, new action is built by decomposing and reusing with transformation the resources made available by the earlier action of others. Around this core endeavor, players are also likely to engage in ongoing multimodal interaction around the game about the code, models, their intentions, and the ways in which the model is reacting to the code that they have created. The enhanced opportunities to laminate multiple semiotic systems to produce and transform meaning would seem particularly valuable for emerging bilinguals. The specificity of meaning as expressed in the code and demonstrated when the code is run also seems potentially useful in supporting generativity and shared meaning convergence amongst participants.

At a more global level in terms of the affordances of the co-operative action framework for analyzing learning and designing environments for education, future research should continue to explore the affordances of the co-operative action framework in science education and other fields. We argue that co-operative action as a framework provides potentially powerful linkages between sociocultural and cognitive theories on learning by simultaneously carefully analyzing (a) the interactions of multiple participants and artifacts both in the immediate present as well as with predecessors and (b) the detailed combination and recombination of ideas and symbols as participants decompose and reuse with transformation the resources made available by the actions of others. This latter focus of the analysis aligns well with knowledge in pieces and related resources perspectives on learning (e.g., Clark & Linn, 2013; diSessa, 1993). While these initial knowledge in pieces perspectives focused on conceptual learning in science at the level of p-prims (e.g., diSessa, 1993; Ozdemir & Clark, 2009), more recent work has broadened to focus on epistemology and disciplinary practices (e.g., Hammer & Elby, 2002), and even more recently to move beyond cognitive lenses to focus on ideology in socioscientific arenas (e.g., Philip, 2011). While not the focus of our specific work, the co-operative action framework therefore could provide a useful lens for exploring cognitive and sociocultural intersections in science education through knowledge-in-pieces related perspectives in terms of individual and group learning at the levels of ideas, epistemologies, disciplinary practices, or possibly even ideologies.

Although Goodwin proposes that the co-operative action framework applies to thinking about a wide range of human interaction from short conversational interactions to the accumulative co-operative organization of human action and tools, most of his detailed research focuses on the former. Our goals for the research and design approaches outlined here are therefore twofold: (a) to explore the affordances of DIGs for emerging bilingual students engaging in the disciplinary practices of science while simultaneously engaging in the discourses, registers, modalities of science and (b) to explore the affordances of the co-operative action framework for research in these areas. We propose that the theoretical affordances of the cooperative DIG genre and the co-operative action framework appear compelling for the reasons outlined in this chapter. Future research should explore these affordances more closely.

References

Barron, B. (2000). Achieving coordination in collaborative problem-solving groups. *The Journal of the Learning Sciences, 9*, 403–436.

Cheng, M.-F., & Lin, J.-L. (2015). Investigating the relationship between students' views of scientific models and their development of models. *International Journal of Science Education, 37*(15), 2453–2475.

Clark, D. B., & Linn, M. C. (2013). The knowledge integration perspective: Connections across research and education. In S. Vosniadou (Ed.), *International handbook of research on conceptual change* (2nd ed., pp. 520–538). New York: Routledge. https://doi.org/10.4324/9780203154472.ch27

Clark, D. B., Medlock-Walton, P., Boquín, R., & Klopfer, K. (2018). Multiplayer disciplinarily-integrated agent-based games: SURGE gameblox. In D. Cvetković (Ed.), *Simulations and gaming*. Rijeka, Croatia: InTECH Open. https://doi.org/10.5772/intechopen.73051

Clark, D., Nelson, B., Sengupta, P., & D'Angelo, C. (2009, October). Rethinking science learning through digital games and simulations: Genres, examples, and evidence. In *Learning science: Computer games, simulations, and education workshop sponsored by the National Academy of Sciences, Washington, DC.*

Clark, D. B., & Sengupta, P. (2019). Reconceptualizing games for integrating computational thinking and science as practice: Collaborative agent-based disciplinarily-integrated games. *Interactive Learning Environments*. https://doi.org/10.1080/10494820.2019.1636071

Clark, D. B., Sengupta, P., Brady, C., Martinez-Garza, M., & Killingsworth, S. (2015). Disciplinary integration in digital games for science learning. *International STEM Education Journal, 2*(2), 1–21. https://doi.org/10.1186/s40594-014-0014-4. Retrieved from http://www.stemeducationjournal.com/content/pdf/s40594-014-0014-4.pdf

Clark, D. B., Touchman, S., Martinez-Garza, M., Ramirez-Marin, F., & Drews, C. S. (2012). Bilingual language supports in online science inquiry environments. *Computers and Education, 58*(4), 1207–1224. https://doi.org/10.1016/j.compedu.2011.11.019

Clark, D. B., Virk, S. S., Sengupta, P., Brady, C., Martinez-Garza, M., Krinks, K., ···, D'Angelo, C. M. (2016). SURGE's evolution deeper into formal representations: The siren's call of popular gameplay mechanics. *International Journal of Designs for Learning, 7*(1), 107–146. Retrieved from https://scholarworks.iu.edu/journals/index.php/ijdl/article/view/19359

Clark, D. B., Weinberger, A., Jucks, I., Spitulnik, M., & Wallace, R. (2003). Designing effective science inquiry in text-based computer-supported collaborative learning environments. *The International Journal of Educational Policy, Research and Practice, IV*(1), 55–82.

Collins, A. (2017). *What's worth teaching?: Rethinking curriculum in the age of technology*. New York: Teachers College Press.

Danielsson, K. (2016). Modes and meaning in the classroom—The role of different semiotic resources to convey meaning in science classrooms. *Linguistics and Education, 35*, 88–99.

Dickes, A. C., & Sengupta, P. (2013). Learning natural selection in 4th grade with multi-agent-based computational models. *Research in Science Education, 43*(3), 921–953.

Dickes, A. C., Sengupta, P., Farris, A. V., & Basu, S. (2016). Development of mechanistic reasoning and multilevel explanations of ecology in third grade using agent-based models. *Science Education, 100*(4), 734–776.

DiSessa, A. A. (1993). Toward an epistemology of physics. Cognition and instruction, 10(2-3), 105–225.

Duschl, R. (2008). Science education in three-part harmony: Balancing conceptual, epistemic, and social learning goals. *Review of Research in Education, 32*(1), 268–291.

Duschl, R. A., Schweingruber, H. A., & Shouse, A. W. (Eds.). (2007). *Taking science to school: Learning and teaching science in grades K-8.* National Research Council Board on Science Education, Center for Education, Division of Behavioral and Social Sciences and Education. Washington, DC: The National Academies Press.

Echeverrîa, A., Améstica, M., Gil, F., Nussbaum, M., Barrios, E., & Leclerc, S. (2012). Exploring different technological platforms for supporting co-located collaborative games in the classroom. *Computers in Human Behavior, 28*, 1170–1177.

Erduran, S., & Jiménez-Aleixandre, M. P. (2007). *Argumentation in science education: Perspectives from classroom-based research.* Dordrecht, Netherlands: Springer.

Farris, A. V., & Sengupta, P. (2014). Perspectival computational thinking for learning physics: A case study of collaborative agent-based modeling. In *Proceedings of the 12th International Conference of the Learning Sciences (ICLS 2014)* (pp. 1102–1107).

Fernandes, A., Kahn, L. H., & Civil, M. (2017). A closer look at bilingual students' use of multimodality in the context of an area comparison problem from a large-scale assessment. Educational Studies in Mathematics, 95(3), 263–282.

Finzer, W., Erickson, T., Swenson, K., & Litwin, M. (2007). On getting more and better data into the classroom. *Technology Innovations in Statistics Education, 1*(1), Article 3. Retrieved from repositories.cdlib.org/uclastat/cts/tise/voll/issl/art3

Ford, M. J., & Forman, E. A. (2006). Redefining disciplinary learning in classroom contexts. *Review of Research in Education, 30*, 1–32.

Gibbons, P. (2003). Mediating language learning: Teacher interactions with ESL students in a content-based classroom. *TESOL Quarterly, 37*(2), 247–273.

Goodwin, C. (2017). *Co-operative action.* Cambridge: Cambridge University Press.

Grapin, S. (2019). Multimodality in the new content standards era: Implications for English learners. *TESOL Quarterly, 53*(1), 30–55.

Greeno, J. G., & MacWhinney, B. (2006). Learning as perspective taking: Conceptual alignment in the classroom. In *Proceedings of the 7th International Conference on Learning Sciences* (pp. 930–931). International Society of the Learning Sciences.

Greeno, J. G., & van de Sande, C. (2007). Perspectival understanding of conceptions and conceptual growth in interaction. *Educational Psychologist, 42*(1), 9–23.

Greeno, J. G., & van de Sande, C. C. (2012). Achieving alignment of perspectival framings in problem-solving discourse. *Journal of the Learning Sciences, 21*(1), 1–44.

Grover, S., & Pea, R. (2013). Computational thinking in K–12: A review of the state of the field. *Educational Researcher, 42*(1), 38–43.

Guo, Y., Wagh, A., Brady, C., Levy, S. T., Horn, M. S., & Wilensky, U. (2016). Frogs to think with: Improving students' computational thinking and understanding of evolution in a code-first learning environment. In *Proceedings of the 15th International Conference of ACM SIGCHI Interaction Design and Children (IDC 2016)* (pp. 246–254).

Hammer, D., & Elby, A. (2002). On the form of a personal epistemology. In B. K. Hofer & P. R. Pintrich (Eds.), *Personal epistemology: The psychology of beliefs about knowledge and knowing* (pp. 169–190). Mahwah, NJ: Lawrence Erlbaum.

Krinks, K., Sengupta, P., & Clark, D. B. (2019). Benchmark lessons, modeling, and programming: Integrating games with modeling in the curriculum. *International Journal of Gaming and Computer-Mediated Simulations, 11*(1), 39–50. https://doi.org/10.4018/IJGCMS.2019010103

Lee, O. (2005). Science education and English language learners: Synthesis and research agenda. *Review of Educational Research, 75*(4), 491–530.

Lee, O., Miller, E., & Januszyk, R. (2014). Next Generation Science Standards: All standards, all students. *Journal of Science Teacher Education, 25*(2), 223–233.

Lee, O., Quinn, H., & Valdés, G. (2013). Science and language for English language learners in relation to next generation science standards and with implications for Common Core state Standards for English language arts and mathematics. *Educational Researcher, 42*(4), 223–233.

Lehrer, R., & Schauble, L. (2006a). Cultivating model-based reasoning in science education. In R. K. Sawyer (Ed.), *The Cambridge handbook of the learning sciences* (pp. 371–388). Cambridge, England: Cambridge University Press.

Lehrer, R., & Schauble, L. (2006b). Scientific thinking and science literacy: Supporting development in learning in contexts. In W. Damon, R. M. Lerner, K. A. Renninger, & I. E. Sigel (Eds.), *Handbook of child psychology* (Vol. 4, 6th ed.). Hoboken, NJ: Wiley.

Louca, L. T., & Zacharia, Z. C. (2015). Examining learning through modeling in K-6 science education. *Journal of Science Education and Technology, 24*, 192–215.

Martinez-Garza, M., Clark, D. B., & Nelson, B. (2013). Digital games and the US National Research Council's science proficiency goals. *Studies in Science Education, 49*(2), 170–208. https://doi.org/10.1080/03057267.2013.839372

Medina-Jerez, W., Clark, D. B., Medina, A., & Ramirez-Marin, F. (2007). Science for ELL: Re-thinking our approach. *The Science Teacher, 74*(3), 52–56. Retrieved frpm http://www.nsta.org/publications/news/story.aspx?id=53492

Moschkovich, J. N. (2015). Academic literacy in mathematics for English learners. *The Journal of Mathematical Behavior, 40*, 43–62.

National Research Council (NRC). (2012). *A framework for k-12 science education: Practices, crosscutting concepts, and core ideas*. Washington, DC: The National Academy Press.

NGSS Lead States. (2013). *Next generation science standards: for states, by states (Appendix F—Science and Engineering Practices in the NGSS). The Next Generation Science Standards*. Washington, DC.

Östman, L., & Wickman, P.-O. (2014). A pragmatic approach on epistemology, teaching, and learning. *Science Education, 98*(3), 375–382.

Ozdemir, G., & Clark, D. B. (2009). Knowledge structure coherence in Turkish students' understanding of force. *Journal of Research in Science Teaching, 46*(5), 570–596. https://doi.org/10.1002/tea.20290

Philip, T. M. (2011). An "ideology in pieces" approach to studying change in teachers' sensemaking about race, racism, and racial justice. *Cognition and Instruction, 29*(3), 297–329. https://doi.org/10.1080/07370008.2011.583369

Pierson, A., Brady, C. E., & Clark, D. B. (in press). Balancing the environment: Computational models as interactive participants in a STEM classroom. *Journal of Science Education and Technology*.

Pierson, A. E., Clark, D. B., & Sherard, M. K. (2017). Learning progressions in context: Tensions and insights from a semester-long middle school modeling curriculum. *Science Education, 101*(6), 1061–1088.

Quinn, H., Lee, O., & Valdés, G. (2012). *Language demands and opportunities in relation to Next Generation Science Standards for English language learners: What teachers need to know.* Paper presented at the Understanding Language Conference, Stanford, CA.

Rochelle, J. (1992). Learning by collaboration: Convergent conceptual change. *The Journal of the Learning Sciences, 2*(3), 235–279.

Sampson, V., & Clark, D. B. (2006). The development and validation of the nature of science as argument questionnaire (NSAAQ). In *Proceedings of the National Association of Research in Science Teaching Conference 2006*. San Francisco, CA.

Schwartz, D. (1995). The emergence of abstract representations in dyad problem solving. *The Journal of the Learning Sciences, 4*(3), 321–354.

Sengupta, P., & Clark, D. B. (2016). Playing modeling games in the science classroom: The case for disciplinary integration. *Educational Technology, 56*(3), 16–22.

Sengupta, P., Clark, D. B., Krinks, K., Killingsworth, S., & Brady, C. (2014). Integrating modeling with games for learning Newtonian mechanics. In N. Holbert & D. Weintrop (Org), N. Holbert (Chair), & Y. Kafai (Discussant), Combining video games and constructionist design to support deep learning in play. In J. Poleman, E. Kyza, I. Tabak, & K. O'Neill (Eds.), *Proceedings of the 11th International Conference of the Learning Sciences (ICLS 2014)* (pp. 1388–1395). University of Colorado at Boulder: ISLS.

Sengupta, P., Dickes, A., Farris, A. V., Karan, A., Martin, D., & Wright, M. (2015). Programming in K-12 science classrooms. *Communications of the ACM, 58*(11), 33–35.

Sengupta, P., Kinnebrew, J. S., Basu, S., Biswas, G., & Clark, D. (2013). Integrating computational thinking with K-12 science education using agent-based computation: A theoretical framework. *Education and Information Technologies, 18*(2), 351–380.

Sengupta, P., Krinks, K., & Clark, D. B. (2015). Learning to deflect: Conceptual change in physics during digital game play. *Journal of the Learning Sciences, 24*(4), 638–674. https://doi.org/10.1080/10508406.2015.1082912

Sengupta, P., Krishnan, G., Wright, M., & Ghassoul, C. (2014, April). Mathematical machines and integrated stem: An intersubjective constructionist approach. In *International Conference on Computer Supported Education* (pp. 272–288). Cham: Springer.

Sengupta, P., & Wilensky, U. (2009). Learning electricity with NIELS: Thinking with electrons and thinking in levels. *International Journal of Computers for Mathematical Learning, 14*(1), 21–50.

Suarez, E., & Otero, V. (2014). Leveraging the cultural practices of science for making classroom discourse accessible to emerging bilingual students. In *Proceedings of International Conference of the Learning Sciences, ICLS* (Vol. 2, pp. 800–807).

Suthers, D. D. (2006). Technology affordances for intersubjective meaning making: A research agenda for CSCL. *International Journal of Computer-Supported Collaborative Learning, 1*(3), 315–337.

Tate, E. D., Clark, D. B., Gallagher, J. J., & McLaughlin, D. (2008). Designing science instruction for diverse learners. In *Designing coherent science education: Implications for curriculum, instruction, and policy* (pp. 65–93).

Thagard, P., & Verbeurgt, K. (1998). Coherence as constraint satisfaction. *Cognitive Science, 22*(1), 1–24.

Van Eaton, G., Clark, D. B., & Smith, B. E. (2015). Patterns of physics reasoning in face-to-face and online forum collaboration around a digital game. *International Journal of Education in Mathematics, Science and Technology, 3*(1), 1–13. Retrieved from http://ijemst.com/issues/3.1.1.Van_Eaton_Clark_Smith.pdf

Weinberger, A., Clark, D., Häkkinen, P., Tamura, Y., & Fischer, F. (2007). Argumentative knowledge construction in online learning environments in and across different cultures: A collaboration script perspective. *Research in Comparative and International Education, 2*(1), 68–79. https://doi.org/10.2304/rcie.2007.2.1.68

Wilensky, U., & Resnick, M. (1999). Thinking in levels: A dynamic systems approach to making sense of the world. *Journal of Science Education and Technology, 8*(1), 3–19.

Wilkerson-Jerde, M., Wagh, A., & Wilensky, U. (2015). Balancing curricular and pedagogical needs in computational construction kits: Lessons from the DeltaTick project. *Science Education, 99*(3), 465–499.

Chapter 7
Getting Beyond Functional Rationality in the Kid Coding Movement: An Agenda for the Learning Sciences

D. Kevin O'Neill

Abstract Today we are in the midst of a kid coding movement which has been heavily promoted by tech companies and is having a large influence over curriculum revisions and teaching practice in North America. This is neither the first nor the best kid coding movement that there has been. The current cultural influence of high tech companies has made educators and politicians intellectually lazy about examining why it might be worthwhile for all children to learn some computer programming, even though most will not pursue it in later life. Educators and policy-makers today appear to be driven by naïve expectations of an unending high tech employment boom, or of advanced problem-solving skills that will transfer easily to all areas of life. A research agenda is laid out that would allow learning scientists to help place the current kid coding movement on a better intellectual footing, and hopefully live up to the standard of its predecessor in the 1980s.

Keywords Coding · Computer programming · Transfer · Research · Code.org

Today the popular press, computing industry executives and public officials in both Canada and the United States appear united in belief about the urgency of teaching every child to code. This is not the first time that our countries have witnessed such a coalescence of enthusiasm about kids learning to program computers; but the current movement is not simply a rerun of the kid coding craze of the 1980s. In some important respects, I argue that today's movement is a poor imitation. Learning scientists must carry a share of the blame for this; but they could also do something about it. Below I sketch an agenda that we could pursue for the benefit of educators and society as a whole.

D. K. O'Neill (✉)
Faculty of Education, Simon Fraser University, Burnaby, BC, Canada
e-mail: koneill@sfu.ca

P. Sengupta et al. (eds.), *Critical, Transdisciplinary and Embodied Approaches in STEM Education*, Advances in STEM Education,
https://doi.org/10.1007/978-3-030-29489-2_7

7.1 Background

In the United States, the industry-funded lobby group Code.org has executed a cunning, multipronged strategy to influence the revision of state curricula in the service of their business needs. This strategy has been very successful—leading to at least 24 states making changes in related policies and laws (Singer, 2017). One Code.org-influenced bill passed in Idaho read in part, "It is essential that efforts to increase computer science instruction, kindergarten through career, be driven by the needs of industry and be developed in partnership with industry" (Singer, 2017). What children might get out of programming *besides* an eventual job was clearly of little interest to the Idaho legislators (though that outcome is somewhat dubious, as I will discuss later).

Channeling students toward employability in software-related jobs from the earliest possible age also appears to have been the driving force behind curriculum revisions in Canada. In 2016, British Columbia's Premier announced to a government-sponsored industry summit that "Every kindergarten to grade 12 student will have…the opportunity to learn the basics of coding" (Silcoff, 2016). This announcement was made in the context of a broader set of initiatives meant to strengthen the Province's tech sector and pivot away from dependency on resource industries such as mining, forestry, oil, and gas. In 2017, the government of Saskatchewan followed suit, announcing its intention to revise its curriculum to support the growth of its own comparatively small tech sector (Macpherson, 2017).

Since we are addressing *mandatory* curriculum, what we are expecting *all* students to gain from developing programming skills is a question that deserves serious examination. Yet with some notable exceptions in the learning sciences literature (Grover & Pea, 2013; Sengupta, Kinnebrew, Basu, Biswas, & Clark, 2013; Weintrop et al., 2016), few people appear to have been asking these questions recently. Instead, most discussions taking place around the current kid coding agenda concern questions about *how* to get the job done. For example, should students be introduced to coding with a "block language" that makes knowledge of syntax less necessary to achieving a working program, or should they be taught in an environment that insists on correct, hand-coded syntax from the beginning (Bennehum, 2016; Harel, 2016)? If the former, how should we manage the transition between block languages and the kind of coding that working software engineers do every day? These are worthwhile questions, but by themselves they are insufficient to guide the development of a robust and lasting, modern-day kid coding movement. Given how the bubble burst on the kid coding movement of the 1980s, robustness should be a concern to anyone involved in today's movement.

7.2 Framing the Issues

In his book *Man and Society in an Age of Reconstruction,* the sociologist Karl Mannheim laid out a useful distinction between *substantive rationality* and *functional rationality* (Mannheim, 1940). Functional rationality pursues the most efficient and effective way to achieve a predetermined goal. As Mannheim noted in his day, this kind of rationality is a hallmark of thinking in the business world and in large bureaucracies. In contrast, what Mannheim called *substantive rationality* involves examining the *relative worth* of goals. It asks, "is this what we should be doing right now?"

In my view, neither functional nor substantive rationality is inherently superior to the other. In practice, they should dialogue with and strengthen one another, in order to determine the *best* way to achieve the *most defensible* goals. However, when one stands back and takes a hard look at it, today's kid coding movement is clearly dominated by functional rationality. The question it primarily pursues is, "how can we motivate the largest possible number of kids and teachers to learn introductory programming?" In this respect, I find it inferior to the kid coding movement of the 1980s, which I lived through as a child, later studied as a learning scientist, and continue to teach my students about every year. In what follows, I want to explore why the previous kid coding movement was better than the current one has been so far, and what learning scientists can do to help today's movement live up to, and hopefully surpass, the standard of its predecessor.

7.2.1 Today's Kid Coding Movement

The worst thing that I think can be said of the kid coding movement in today's schools is that rather than promoting active investigation into the various ways in which kids might benefit from coding, it takes its ideas directly from computing industry lobby groups — especially Code.org (Singer, 2017). It has become obvious that government officials, the general public, and educators alike are ill equipped to treat their sophisticated lobbying efforts with the skepticism that they deserve. To hear the way some politicians talk about coding as an essential job skill, one would think there was no such thing as an unemployed programmer. Current government statistics show that in the USA, unemployment among computer programmers runs at a very low 1.9% nationally, though the Bureau of Labor Statistics predicts the number of positions to decline by 7 percent over the next decade (Bureau of Labour Statistics, 2017b). Further, while the national picture looks strong, there are substantial regional differences (Bureau of Labour Statistics, 2017a). National differences can be much larger. Detailed data are not readily available for Canada, but the UK currently has a high unemployment rate among computer science graduates. Heath (2016) reported that:

> The vast majority of UK business leaders and IT execs, 78 percent, told PricewaterhouseCoopers' recent Global Digital IQ Survey that a shortage of digital skills was holding their firm back. But that claim sits uneasily next to the relatively high proportion of computer science graduates struggling to find work, with 11.7 percent unemployed six months after leaving university. Compared to graduates in related disciplines — in science, technology, engineering and mathematics — their employment prospects are particularly poor. (Heath, 2016)

The industry's claims about its needs are important to notice here. In the USA where employment prospects for IT professionals are comparatively good, tech giants are not only encouraging every American child to consider a software career, but at the same time working as hard as they can to put American IT professionals out of work through abuses of the H1-B visa program (Thibodeau, 2014, 2016). This two-faced relationship with the public should give us pause.

While these issues should be of vital concern to parents, politicians, and educators, they should also interest learning scientists on a professional basis. Harkening back to Mannheim's concept of substantive rationality, the current kid coding movement as it is promoted to parents and politicians seems to lack a believable evidence-based argument about how youth subjected to the new curricula will benefit from learning programming. The case currently being made to parents and educators involves two main claims. The first is that learning coding will open doors to good jobs in the future (an assertion that there is reason to be skeptical about). The second (currently more soft-pedaled) claim is that coding helps develop powerful thinking and problem-solving skills that are broadly applicable in life. I will discuss each of these claims in some detail below.

7.2.2 The Original Kid Coding Movement

When I was a child in the 1980s, many parents and educators shared a blind faith that somehow, computers would be very important in the future. When I informed my grandfather of my intention to study computer science, I remember him saying, "I lost my job to a computer, so there might as well be someone in the family who gets a job from computers." Other adults' attitudes were similar, so despite there being very little in the way of computer science education in my local schools at this time, I was encouraged to learn as much as I could on my own. My parents could afford to support my interest, buying me a computer and enabling me to eventually become educated as a computer scientist, despite not having high enough grades in mathematics to be granted access to the two TRS-80s my elementary school owned. (High grades in mathematics had been decided upon as the key criterion for entry into the school's after-hours computer club.).

The fact that the uninformed faith of my parents was rewarded by my eventual success does not prove that their beliefs were well founded. In the 1980s, nobody had *any* idea how large and ubiquitous the computing industry would one day become. In 1984, discussing the results of a national survey of educators regarding the use of computers in schools, Henry Jay Becker expressed reasonable skepticism about the value of kids learning programming:

> ...compared with the importance of teaching other culturally valued knowledge, such as that from scientific, historical, and literary domains, how necessary is it that schools spend valuable instructional time teaching students about computers, and specifically, about how to program them in general-purpose computer programming languages like BASIC? (Becker, 1984, p. 23).

Due in part to such skepticism, Seymour Papert worked hard to develop broad rationale for kids to learn programming, publishing his ideas in a popular book that many educators read (Papert, 1980). Those days, promoters of programming for children spoke of the "powerful ideas" and general problem-solving abilities that children would acquire through programming, and this way of thinking was not confined to the literature of education. The report of an Association for Computing Machinery task force on elementary and secondary computer science curriculum stated in part that:

> A course in computer science should be oriented primarily toward teaching problem solving skills rather than being vocationally oriented (Aiken, Hughes, & Moshell, 1980, p. 172).

As this quotation shows, even computer scientists were not foolish enough in the 1980s to believe that the computing industry would provide large-scale employment. As it turns out, they weren't half wrong. Today Amazon, Apple, Facebook, Google, Netflix, and Twitter are huge companies in terms of their market capitalization and influence on culture—but they employ remarkably few people relative to their earnings (Rosoff, 2016). These companies are, in fact, attractive investments *because* they are poor employers.

Nonetheless, with the computing industry having become as large and profitable as it is, educators have gotten intellectually lazy about justifying the time and effort invested in providing coding lessons for a broad swath of children. The working assumption of many politicians and educators today seems to be that, in the words of Marc Andreessen, software will "eat the world" (Andreessen, 2011) and the computing industry will continue to expand forever.

7.3 A Role for Learning Scientists

All of this matters for learning scientists, because given the scope of Code.org's ambitions, it is obvious that most kids who are taught coding in school today will *not* one day become professional software developers. Some will code in later schooling and on the job, as a part of evolving computational practices in STEM (and other) fields (Weintrop et al., 2016). For years, learning scientists have worked toward integrating computational thinking into STEM curricula, teaching it in a discipline-situated way through computational modeling of phenomena (Sengupta, Dickes, & Farris, 2018; Sengupta et al., 2013). However, this vision is not the one that motivates the majority of people involved in today's kid coding movement—particularly not most politicians, policy-makers, and industry advocates.

While embedding computational thinking within STEM fields is a defensible approach, both student and veteran teachers eager to meet new coding mandates in the most straightforward way have seized upon materials supplied by Code.org and their peers as the easiest way possible to tick the boxes in the new curriculum standards. This dominant domain-general approach is worthy of learning scientists' attention, because while many educators assume it to be effective, it has not been rigorously evaluated with regard to what teachers and parents actually expect of it. What many teachers appear to be enacting today is essentially a curriculum of wishful thinking.

The argument that programming is a useful environment in which to learn problem-solving approaches and concepts applicable to other areas of life (even when programming is taught in the relatively isolated manner typical of Code.org's materials) was made explicitly in one of the earliest academic papers of the current coding movement, by Jeannette Wing from Carnegie Melon University (Wing, 2006). In a follow-up paper in 2010, she doubled down on her argument by stating:

> When I use the term computational thinking, my interpretation of the words "problem" and "solution" is broad; in particular, I mean not just mathematically well-defined problems whose solutions are completely analyzable, e.g., a proof, an algorithm, or a program, but also real-world problems… The educational benefits of being able to think computationally… enhance and reinforce intellectual skills, and thus can be transferred to any domain. (Wing, 2010, p. 1)

Wing's 2010 paper featured a series of anecdotes drawn from colleagues at Carnegie Mellon University, sharing how computational thinking appeared in their everyday lives: organizing LEGO blocks for later use, standing in a buffet line, and sorting sheet music. What I observe in these examples is a fallacy that has plagued learning scientists for generations, including Seymour Papert himself from time to time. The fallacy is that if a person with deep expertise can observe an analogy between two problem contexts, and can describe knowledge as applying across them, then teaching novices this knowledge in one context should enable transfer to the other.

This assumption has long roots in our culture. In centuries past, Latin and chess were taught in schools partly because it was believed (based on similar reasoning) that they developed powerful general abilities to think precisely and strategically. Unfortunately, research has shown this hopeful assumption of transfer to be false (Bransford, Brown, & Cocking, 2000). In one classic study, Gick and Holyoak (1980) examined the ability of psychology undergraduates to transfer learning between two formally identical problems under various conditions. Without careful preparation, most were unable to recognize the opportunity to transfer knowledge between the two problems (Gick & Holyoak, 1980). Even mature learners tend to be so distracted by the surface features of a problem that they do not recognize opportunities for transfer (Bransford & Schwartz, 1999). This being the case, why should K-12 students who are learning and taking part in computational thinking for the first time be expected to do better?

At the time of their publication, Wing's assertions about the broad applicability of computational thinking to domains other than computing had no research evidence behind them, and this is still lacking. Attempting to assemble such evidence is a worthwhile challenge for learning scientists, and would produce findings with large implications for policy and practice worldwide, whatever the outcomes are. Well-done research with positive outcomes could channel current enthusiasm in productive directions, while negative outcomes could potentially topple the Humpty Dumpty of naïve expectations of transfer that today's kid coding movement mostly uses as its figurehead. In the latter case, research could generate broader public interest in teaching computational thinking in the more discipline-situated manner promoted by some learning scientists.

For learning scientists hoping to empirically cast out naïve assumptions about transfer, it will be necessary to revive and update a program of research that was effectively abandoned in the 1980s. This research produced disappointingly mixed results when it went looking for evidence that kids taught coding were able to transfer problem-solving skills outside that context (Mitterer & Rose-Krasnor, 1986; Pea, Kurland, & Hawkins, 1985). In fairness though, back then researchers had a difficult time producing transfer even under carefully controlled laboratory conditions, let alone in classrooms (Bransford & Schwartz, 1999). At the end of the 1990s, psychologists still described transfer as "a problem of Gordian proportions" (Alexander & Murphy, 1999).

Today the LOGO transfer studies are simply being ignored by industry lobbyists, politicians, and educators, most of whom are too young to remember the first kid coding craze; but those studies ought to concern learning scientists because simply re-branding "powerful ideas" and general problem-solving skills as "computational thinking" does nothing to address the lack of convincing evidence that they can be routinely taught in a way that fosters transfer across domains (Grover & Pea, 2013). Without doubt, the LOGO transfer studies of the 1980s had their flaws (Emihovich, 1990; Papert, 1987). It makes sense to revisit these flaws now, and consider how learning scientists today might do better, with their much-improved theoretical and methodological toolkits.

7.3.1 Setting an Agenda

As is now well known, transfer depends importantly on the depth of the initial learning (Bransford & Schwartz, 1999). In some of the LOGO transfer studies, programming may not have been taught in as deep a way as necessary to foster transfer. Papert once pointed out how absurd it was for any serious educator to believe that a few minutes a week of drawing circles and squares on a computer would somehow change how a child thinks—though in practice, this appeared to be what thousands of teachers believed. At the same time, some of the researchers who studied transfer from LOGO problem-solving (e.g., Mitterer & Rose-Krasnor, 1986) had learned from the masters at MIT and had worked as hard as they could to make LOGO a success. They found their research results every bit as disappointing as Papert and his acolytes did (Mitterer, 2009).

Second, critics of the LOGO transfer studies have pointed out that the batteries of paper-and-pencil transfer tests administered in the studies may simply have been too different from the problem-solving that students did while coding to make transfer a realistic possibility. In effect, these tests may have constituted feats of unreasonably far transfer for the students involved. This too is a reasonable critique, particularly in light of what the ensuing decades taught us about the situated nature of knowledge and skills (Greeno, 1998; Lave & Wenger, 1991).

Third, beyond the format of the problem-solving transfer tests, learning sciences researchers have fundamentally critiqued the whole paradigm in which the earlier transfer studies were conducted. As Bransford and Schwartz (1999) pointed out, many transfer studies of the 1980s and 1990s were conducted using a "sequestered problem solving" (SQS) approach, in which the results of weeks or months of collaborative problem-solving in an information-rich environment were assessed in an environment very like a high-stakes final examination, stressing solo problem-solving in an information-poor environment. Any skills or knowledge a student may have gained that would support them in researching a solution to a novel problem or consulting others about it became irrelevant in this paradigm. According to some advocates of kid coding, these sorts of skills and knowledge are some of the most important products of learning to code—part of what could be referred to as the cultural capital of computer literacy (Emihovich, 1990).

Finally, reasonable critics point out that the 1980s transfer studies on LOGO were, by today's standards, inadequate in their data collection and reporting. As Papert (1987) noted, transfer researchers of the 1980s tended to be trapped in the genres of experimental science. They were, after all, trained as experimental psychologists, not anthropologists; so they tended to treat the classroom context as a black box. Since their studies did not include systematic observation of classroom events, even if they had found evidence of successful transfer, they would not have been able to inform educators about how to reproduce it. If coding is conceived less as a specific intervention and more as a cultural building material (as Papert suggested it should be), then research needs to methodically study classroom culture if it is to appreciate the effective use of this building material in the learning and teaching of problem-solving. Recent research on young children learning to code has failed to do this (e.g., Dasgupta, Hale, Monroy-Hernández, & Mako Hill, 2016).

Research that addresses these fundamental flaws in the transfer studies of the 1980s will be of more than academic interest; because to successfully teach computational thinking for transfer beyond coding, the *why* of teaching coding (substantive rationality) must have greater influence over the *how* (functional rationality) than it presently has. To teach coding *well*, educators need a clear and long-range view of *why they are doing it* other than to prepare every child to potentially become a software developer. If new studies of the kind I will describe below fail once more to find evidence of far transfer, it would encourage the current advocates of kid coding to reformulate their movement on more sound foundations. In addition to being socially useful, the studies I imagine would provide an opportunity for learning scientists to demonstrate the worth of many of the methodological and conceptual tools they have developed since the 1980s.

7.3.2 Study Design

The literature of the Learning Sciences suggests that one should not expect cross-domain problem-solving from students who have been introduced to computational thinking as a simple add-on to their usual curriculum. A methodologically and pedagogically current set of studies on transfer of skill from coding would employ the paradigm of preparation for future learning (PFL) (Bransford & Schwartz, 1999) rather than sequestered problem solving (SQS). Rather than anticipating spontaneous far transfer from coding to a set of abstract problem-solving tasks, studies conducted in the PFL paradigm would evaluate the success of coding lessons and units on the basis of how they prepared learners for later lessons and units involving more ambitious problems—first set in a highly similar context, then in increasingly dissimilar ones.

Essential to this research design would be a sequence of age-appropriate tasks involving progressively distant domains of transfer, staged manageably for both students and teachers, and carefully observed as they are implemented. Depending on the age of the students involved and the length of the curriculum sequence, it is not without question for the sequence of transfer tasks to end with the very kind of solo, standardized, abstract paper-and-pencil tasks that typified the previous generation of transfer studies. Unlike the prior studies however, students' ability to transfer would be developed in a planful way, and classroom implementation would be carefully observed in a way typical of design-based research (Design-Based Research Collective, 2003). Rather than taking it as their sole purpose to demonstrate *whether* a particular lesson or unit was effective at enabling particular kinds of transfer *on average* (as in a typical classroom quasi-experiment), the new studies would also examine *for whom* it was effective, and *why*—respecting the fact that programming is as much an *experience* as it is a matter of abstract symbol manipulation (Sengupta et al., 2018).

What kinds of potentially general skills could these kinds of studies develop and assess? Two skills often mentioned by Wing and others as justifications for teaching coding are the same as those mentioned in the 1980s: the decomposition of complex problems into simpler ones, and debugging. One can imagine a months-long study designed using existing materials from Code.org or elsewhere, in which elementary school students would work in groups to decompose progressively complex problems, code solutions, and debug them until they work. Follow-up lessons in other areas of the school curriculum would be deliberately designed to afford opportunities to decompose problems and debug solutions in groups, with teachers drawing explicit parallels to the previous coding lessons as students work. Over time, students would be expected to exercise the same problem-solving strategies in other domains with greater independence. Following the preparation for future learning paradigm, researchers would evaluate students' success at transferring learning from each lesson both *into* and *out of* the next; and classroom observations would help examine the conditions and experiences that enabled or failed to enable transfer for particular students.

Beyond paper-and-pencil measures of problem-solving, other developments that have taken place in psychology since the 1980s might be of tremendous aid in this program of research. For instance, researchers could examine whether Dweck's measures of mastery orientation (Dweck, 2000) align with persistence at debugging in programming in other domains—which according to the claims of kid coding advocates, they should.

7.4 Conclusion

Today's kid coding movement has tremendous public relations and political momentum; but it is driven almost exclusively by corporate interests and functional rationality, and assumptions about transfer that learning scientists today consider naively optimistic. In this respect, its intellectual foundations are notably flabbier than those that undergirded the kid coding movement of the 1980s, which themselves failed to find decisive empirical support using the methods available in that day. Despite the tremendous growth of the computing industry since Papert's publication of *Mindstorms*, evidence for the relevance of coding for a broad swath of students is both as relevant and as absent now as ever, because it is still not possible for every child taught coding today to find use for programming in their future studies or the workforce. Now as in the 1980s, computational thinking has to translate into something other than programming skill in order to be socially valued in the long run.

Learning scientists now have at their disposal a much richer theoretical and methodological toolkit than was available in the 1980s to evaluate the broad relevance of computational thinking. The time is long past for us to step up and put this toolkit to use. Advocates of kid coding have been waiting a long time for a resurgence in public enthusiasm about computing science; but the current enthusiasm is fragile. Like the LOGO movement that came before it, the current computational thinking movement could fade quickly if parents and government officials get the sense that time spent coding in schools is not well spent. At some point, and possibly soon, empirical evidence will be demanded. Learning scientists should take this moment as an opportunity to demonstrate the worth of the theoretical and methodological tools that they have developed since the 1980s, and possibly put the kid coding movement on a stronger and more defensible intellectual foundation.

References

Aiken, R. M., Hughes, C. E., & Moshell, J. M. (1980). Computer science curriculum for high school students. Paper presented at the SIGCSE '80: The eleventh SIGCSE technical symposium on Computer science education, Minneapolis, Minnesota.

Alexander, P. A., & Murphy, P. K. (1999). Nurturing the seeds of transfer: A domain-specific perspective. *International Journal of Educational Research, 31*(7), 561–576.

Andreessen, M. (2011). Why Software Is Eating The World. The Wall Street Journal. Retrieved from https://www.wsj.com/articles/SB10001424053111903480904576512250915629460

Becker, H. J. (1984). Computers in Schools Today: Some Basic Considerations. *American Journal of Education, 93*(1),22–39.

Bransford, J. D., Brown, A. L., & Cocking, R. R. (2000). *How people learn: Brain, mind, experience, and school* (expanded ed.). Washington, DC: National Academy Press.

Bransford, J. D., & Schwartz, D. L. (1999). Rethinking transfer: A simple proposal with multiple implications. *Review of Research in Education, 24*, 61–100.

Bureau of Labour Statistics. (2017a). *Occupational employment statistics: Computer programmers*. Retrieved from https://www.bls.gov/oes/2017/may/oes151131.htm

Bureau of Labour Statistics. (2017b). *Occupational outlook handbook: Computer programmers*. Retrieved from https://www.bls.gov/ooh/computer-and-information-technology/computer-programmers.htm

Dasgupta, S., Hale, W., Monroy-Hernández, A., & Mako Hill, B. (2016). Remixing as a pathway to computational thinking. Paper presented at the 19th ACM Conference on Computer-Supported Cooperative Work & Social Computing.

Design-Based Research Collective. (2003). Design-based research: An emerging paradigm for educational inquiry. *Educational Researcher, 32*(1), 5–8.

Dweck, C. S. (2000). *Self-theories: Their role in motivation, personality and development*. Philadelphia: Psychology Press.

Emihovich, C. (1990). Technocentrism revisited: Computer literacy as cultural capital. *Theory Into Practice, 29*(4), 227–234.

Gick, M. L., & Holyoak, K. J. (1980). Analogical problem solving. *Cognitive Psychology, 12*, 306–355.

Greeno, J. (1998). The situativity of knowing, learning, and research. *American Psychologist, 53*(1), 5–26.

Grover, S., & Pea, R. (2013). Computational thinking in K–12: A review of the state of the field. *Educational Researcher, 42*(1), 38–43. https://doi.org/10.3102/0013189X12463051

Heath, N. (2016, October 7). If there's a tech skills shortage, why are so many computer graduates unemployed? *TechRepublic*. Retrieved from http://www.techrepublic.com/article/if-theres-a-tech-skills-shortage-why-are-so-many-computer-graduates-unemployed/

Lave, J., & Wenger, E. (1991). *Situated learning: Legitimate peripheral participation*. Cambridge: Cambridge University Press.

Macpherson, A. (2017, November 1). Tech sector thrilled about focus on coding; more computer courses are in works for elementary, high schools in province. *Star-Phoenix*, p. A4.

Mannheim, K. (1940). Man and Society in an Age of Reconstruction (E. Shils, Trans.). London: Routledge & Kegan Paul.

Mitterer, J. (2009). Personal communication.

Mitterer, J., & Rose-Krasnor, L. (1986). LOGO and the transfer of problem solving: An empirical test. *Alberta Journal of Educational Research, 32*(3), 176–194.

Papert, S. (1980). Mindstorms: Children, computers, and powerful ideas. New York: Basic Books.

Papert, S. (1987). Computer criticism vs. technocentric thinking. *Educational Researcher, 16*(5), 22–30.

Pea, R. D., Kurland, D. M., & Hawkins, J. (1985). Logo programming and the development of thinking skills. In M. Chen & W. Paisley (Eds.), *Children and microcomputers: Formative studies* (pp. 193–212). Beverley Hills, CA: Sage.

Rosoff, M. (2016). Here's how much each employee at a big tech company like Apple or Facebook is worth. Business Insider. Retrieved from http://www.businessinsider.com/revenue-per-employee-at-apple-facebook-google-others-2016-2?op=1

Sengupta, P., Dickes, A., & Farris, A. (2018). Toward a phenomenology of computational thinking in STEM education. In M. S. Khine (Ed.), *Computational thinking in the stem disciplines: Foundations and research highlights* (pp. 49–72). Cham: Springer.

Sengupta, P., Kinnebrew, J., Basu, S., Biswas, S., & Clark, G. (2013). Integrating computational thinking with K-12 science education using agent-based computation: A theoretical framework. *Education and Information Technologies, 18*(2), 351–380. https://doi.org/10.1007/s10639-012-9240-x

Silcoff, S. (2016, January 18). B.C. to add computer coding to school curriculum. *The Globe and Mail,* p. B1.

Singer, N. (2017, June 27). How Silicon Valley pushed coding into american classrooms. *The New York Times*. Retrieved from https://www.nytimes.com/2017/06/27/technology/education-partovi-computer-science-coding-apple-microsoft.html

Thibodeau, P. (2014). Displaced IT workers are being silenced. Computerworld. Retrieved from http://www.computerworld.com/article/2855642/displaced-it-workers-are-being-silenced.html

Thibodeau, P. (2016). IT layoffs at insurance firm are a 'never-ending funeral'. Computerworld. Retrieved from http://www.computerworld.com/article/3077302/it-careers/it-layoffs-at-insurance-firm-are-a-never-ending-funeral.html

Weintrop, D., Beheshti, E., Horn, M., Orton, K., Jona, K., Trouille, L., & Wilensky, U. (2016). Defining computational thinking for mathematics and science classrooms. *Journal of Science Education and Technology, 25*(1), 127–147. https://doi.org/10.1007/s10956-015-9581-5

Wing, J. M. (2006). Computational thinking. *Communications of the ACM, 49*(3), 33–35.

Wing, J. M. (Producer). (2010). Computational Thinking: What and Why? The Link. Retrieved from http://www.cs.cmu.edu/link/research-notebook-computational-thinking-what-and-why

Chapter 8
Beyond Isolated Competencies: Computational Literacy in an Elementary Science Classroom

Amanda C. Dickes and Amy Voss Farris

Abstract This work arises from the concern that investigations into children's computing have largely focused on learning to code as an isolated competency. This approach frames technology as a means to an end and unnecessarily narrows conceptual activity in the classroom to the (re)production of computational abstractions. Our approach is to argue for considering computational modeling and programming as part of a larger ensemble of STEM work in the elementary classroom, broadening and deepening what it means to code to include multiple forms and genres of representations. The distinction between focusing on computing as an isolated competency and our approach can be understood in light of diSessa's distinction between "material intelligence" and "literacies." DiSessa (2001) argued that while material intelligence can be understood as meaningful use of a technology, literacies are a lens through which we create, understand, and communicate with the world. It is our view that in elementary classrooms, computational modeling and programming can cease to exist merely as material intelligence and become a core component of scientific practice, particularly when activity is structured in certain ways.

This research was supported by the U.S. National Science Foundation (NSF CAREER Award #1150320), awarded to Pratim Sengupta. Feedback from two anonymous reviewers is gratefully acknowledged.

A. C. Dickes (✉)
Gulf of Maine Research Institute, Portland, ME, USA
e-mail: adickes@gmri.org

A. V. Farris
Department of Curriculum and Instruction, The Pennsylvania State University, University Park, PA, USA

P. Sengupta et al. (eds.), *Critical, Transdisciplinary and Embodied Approaches in STEM Education*, Advances in STEM Education,
https://doi.org/10.1007/978-3-030-29489-2_8

Keywords Computational literacy · Programming · Modeling · Elementary science · Computational thinking · Teacher professional learning

In the decades since their emergence in the late 1970s, personal computers have transformed nearly every aspect of human society and culture (diSessa, 2001; Maloy & LaRoche, 2014). A once-in-several-centuries innovation (Simon, 1983), computers have revolutionized disciplines such as science, engineering, and communication (diSessa, 2001; Woolf, 2010; Vee, 2013) and, more recently, have continued to diversify into such fields as art (Peppler & Kafai, 2005), architecture (Vee, 2013), and history (Maloy & LaRoche, 2014). Notable scholars have attributed the impact of computing and computers to what Papert (1980) has called their richness of material. Computers, as Papert notes, are "objects-to-think-with" (p. 11), providing a powerful frame through which users can represent and engage with the world.

Given this impact on culture and society, it is unsurprising that leading educational scholars have long argued for computational thinking, an analytic problem solving and design approach fundamental to computing (Wing, 2006), to be an essential focus of K12 curriculum (diSessa, 2001; Papert, 1980; Repenning, Basawapatna, & Klymkowsky, 2013; Sengupta, Kinnebrew, Basu, Biswas, & Clark, 2013; Weintrop et al., 2016; Wilensky, 1995; Wing, 2008). Computational thinking is included as an essential practice for science, technology, engineering, and math (STEM) education in the Next Generation Science Standards (NGSS, 2013). However, studies have shown that curricular integration of computational thinking and modeling is a complex and challenging endeavor (Basu et al., 2016; Lye & Koh, 2014; Barr & Stephenson, 2011; diSessa, 1991; Sherin, diSessa, & Hammer, 1993; Guzdial, 1994) which involves the introduction and adoption of new literacies (e.g., programming) to both teachers and students, alongside disciplinary ideas and practices that students already find challenging to understand (Sengupta et al., 2013).

In this chapter, we present one account of how computing was successfully integrated as part of a larger ensemble of STEM activity within a third-grade classroom. We report a study in which a third-grade teacher, in partnership with researchers, integrated an agent-based programming and modeling tool called ViMAP with her regular STEM curriculum over the course of an entire academic year. In-depth analysis of Emma's teaching, classroom discourse, and activity around ViMAP demonstrates how she positioned computing in ways that broadened and deepened computational experiences to include multiple forms and genres of representations beyond the (re)production of computational abstractions. Over time, Emma came to see coding as increasingly *valuable* and *teachable* (diSessa, 2001), and computing, in combination with other non-computational forms of activity, became an integral component of meaning construction, communication, and transformation in the STEM classroom. In turn, these supported students' learning in programming, mathematical reasoning, and scientific modeling in a *reflexive* (Sengupta et al., 2013; Kafai & Harel, 1991) manner. Emma's work is an example of how an elementary teacher without a background in computing began to support her students in transdisciplinary *computational literacy*.

8.1 Computational Literacy

Technology structures much of our daily lives, and shapes our actions and interactions with the world around us (Vee, 2013). Given technology's ubiquity in modern economies and workforces, scholars and educators have prioritized helping young people become computational thinkers who can expertly respond to an increasingly digital world (Barr & Stephenson, 2011). As a result, efforts to integrate *computational thinking* (Wing, 2006, 2008) into K12 STEM curricula have gained momentum (Grover & Pea, 2013a). However, as several scholars have argued (Sengupta, Dickes, & Farris, 2018; Lye & Koh, 2014; Grover & Pea, 2013b), current trends to integrate computing within K12 STEM overemphasize programming as the *key* to learning (e.g., Computer Science for All, 2016; ISTE, 2018; NRC, 2012), and, in particular, place a greater emphasis on learning the isolated technicalities of programming than on developing learners' understandings of themselves with relation to the technological world.

The present emphasis on developing "computational thinking skills" in computing (Computer Science for All, 2016) undermines Papert's (1980) original vision of computers as cognitive partners in learning and instead positions them as delivery mediums of isolated technical knowledge outside of the broader learning culture. This positioning unnecessarily narrows conceptual activity in the classroom to tool use and the (re)production of computational abstractions (Sengupta et al., 2018). Papert (1987) cautioned against technocentrism in educational computing, and his critique of classroom computing as "treatment" rather than "culture" (1987, p. 24), we believe, can be understood in light of diSessa's (2001) distinction between the roles that *material intelligences* and *literacies* play in knowledge production.

8.1.1 Pillars of Computational Literacy

DiSessa (2001) has argued that no computational technology is revolutionary unless it becomes transformed from a *material intelligence* into a *literacy*. DiSessa describes material intelligence as the deployment of skills and capabilities with computational technologies. In other words, material intelligence positions the computer as an instrument, a "thing in itself" (Papert, 1987, p. 24) which, if used intelligently, may deliver benefits to the user. Literacies, however, involve more than tool use. Literacies allow people to negotiate their world (Vee, 2013) through their impact on a wide variety of contexts, both mundane and profound (Holyoak, 1991; diSessa, 2001). DiSessa argues that while individuals greatly benefit from material intelligences, it is through literacies that knowledge is both influenced and generated.

Material intelligences transform into literacies when they become infrastructural to a society's communicative practices (diSessa, 2001). That is, there is widespread ability to compose and interpret with that technology, what diSessa terms "two-way" literacy. DiSessa (2001) and other scholars (Street, 1984; Vee, 2013) have argued that literacy of any form involves an interplay between *material*, *cognitive,* and *social* dimensions. Investigating each of these dimensions as well as their interactions in K12 settings can help us understand the nature of computational literacy and how to support it in K12 classrooms. The material dimension of literacy involves creating and modifying symbolic systems (e.g., coding) as well as physical computing (e.g., microcontrollers and 3D printers). Developing expertise along the material dimension in turn involves developing expertise along the cognitive dimension (e.g., learning to use programming commands and computational abstractions such as data structures). The social dimension is omnipresent. Learning—as well as manipulations and transformations of materials—occurs in specific social contexts where complex social forces of innovation, adoption, and interdependence transform material intelligences into literacies (diSessa, 2001; Street, 1984; Vee, 2013).

8.1.2 Supporting Literacy Through Heterogeneity of Coding Experiences

What does this mean for research on integrating computational thinking and modeling with science curricula in the elementary grades? We (Sengupta et al., 2018; Sengupta, Dickes & Farris, forthcoming) have argued that the experience of coding in STEM is inherently *heterogeneous*: Computing in STEM disciplines requires engagement with "multiple forms and genres of representations" beyond the reproduction of computational abstractions (Sengupta et al., 2018, p. 55). In the work of STEM, these representations might include, but are not limited to, mathematical representations and embodied and physical modeling experiences. For example, when students create computational models of the motion of an object, their computational work is interdependent with mathematical descriptions of the motion scenario they intend to model as well as their physical and/or embodied investigations of the phenomenon. Such heterogeneity in representation, we feel, broadens and deepens what is means to code in the K12 classroom by creating contexts for complex forms of experience which are grounded in the epistemic and representational work of science (Sengupta, Dickes & Farris, forthcoming; Farris, Dickes & Sengupta, 2019; Sengupta et al., 2013) and expanding the representational space within which teachers with no prior coding experience can integrate coding as part of their regular instruction (Dickes et al, 2019; Sengupta et al., 2018).

Our work also extends the argument that coding in the K12 STEM classroom is a *dialogue* that is emergent and unfolds within the production of knowledge in distributed social and material systems (Sengupta et al., forthcoming). Drawing on Bakhtin's dialogism, we argue that predominant views—in which children's coding artifacts are seen as the products of isolated material competency—stand to "lose insight into the dialogicality through which [coding artifacts] are shaped" and more importantly, "the dialogue those artifacts [continue to] shape or [are] about to shape" (Sengupta et al., forthcoming). In other words, designing a computational artifact should not merely culminate in the artifact itself; a dialogic view argues that the artifact, once designed, continues as part of an authentic and reflective dialogue grounded in the context of the use that it was designed for.

Unfortunately, modern K12 computing initiatives largely mirror the trouble with technocentrism that Papert (1987) critiqued: learners use commands and control structures to learn something about programming as a tool for abstraction and automation (Wing, 2008). However, *what is worth abstracting and automating* and for what ends are rarely questions learners are asked to grapple with. Instead, the focus is largely on mastery of "computational thinking" for problems and tasks that are invented by the designers of the system. There is little opportunity for students to use programming to represent or solve a problem or task that is not already a computational problem. A dialogic stance—in which students have opportunities to reflect in or on their own practices—is essential because it makes way for the learner to become enmeshed in the process of production towards their own goals (Sengupta et al., forthcoming).

Given diSessa's distinction between intelligence and literacy and Sengupta and colleagues' dialogic view, our goal in this paper is to demonstrate how reframing educational computing; not, as a dialogic and heterogeneous form of experience (Sengupta et al., 2018, forthcoming) can help to integrate diSessa's three pillars of computational literacy in the elementary STEM classroom. In this chapter, students use programming in ways that are scientifically relevant and also give them new ways to talk about and represent the world around them. That is, programming was used as a *literacy* through which students negotiated, thought about and represented the real world in ways that were deeply rooted in the social and material spheres of the class.

To this end, this chapter specifically investigates the following research questions:

1. How did students conceptual work with ViMAP progress from a material intelligence to a literacy during the course of instruction?
2. How did the classroom teacher facilitate this progression? Specifically, what instructional moves helped to situate work with ViMAP as part of a larger ensemble of STEM work in the classroom?

8.2 Method

8.2.1 The Programming Environment

We used ViMAP (Sengupta et al., 2015), an agent-based, visual programming language that uses the NetLogo modeling platform as its simulation engine (Wilensky, 1999). In ViMAP (Fig. 8.1), users construct programs using a drag-and-drop interface to control the behaviors of one or more computational objects. In terms of

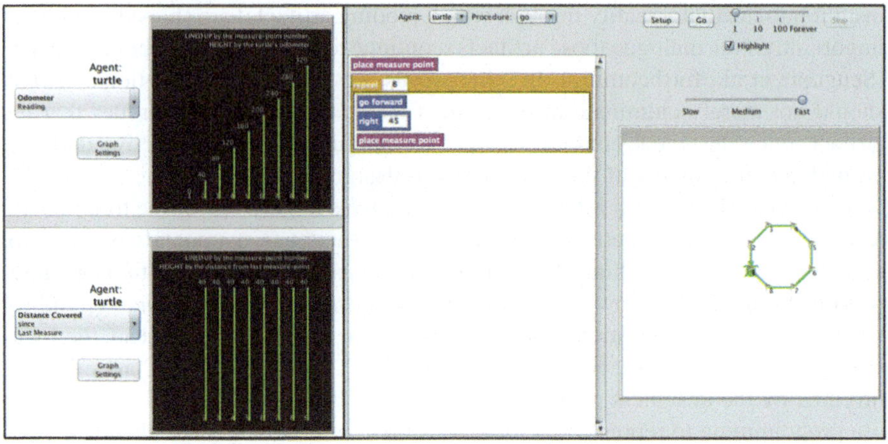

Fig. 8.1 ViMAP's measurement window and programming interface. Figure illustrates the program for generating a regular octagon, the enactment by the turtle agent, and graphical representations of length of each line segment (graph on lower left) and perimeter (graph on top left)

qualities of cognitive partners (Kim & Reeves, 2007), learning in ViMAP is both *flexible* and *open*. It carries both *general* and *specific* expertise by supporting student-driven investigations across the domains of mathematics, science, and computing through interaction with domain-specific and domain-general programming commands. ViMAP also has a graphing feature which allows learners to design mathematical measures based on periodic measurements of object-specific (e.g., speed, distance, energy) and aggregate-level variables (e.g., number of agents).

Figure 8.1 above depicts a prototypical program (and graph) for a regular octagon and is similar to the kinds of programs learners built early in the academic year. In the setup procedure (not shown), the learner has placed the command "pen down" and "set step size 40" in the construction window. These commands instruct the turtle agent to move forward a distance of 40 units each step, marking each step with a solid green line. In the "go" procedure (shown in Fig. 8.1), the learner has placed two "place measure point" commands (one outside the repeat block and one inside) plus a "go forward" and "right 45" within a "repeat 8" block. When the program is run, the shape shown on the right is drawn and bar graphs measuring perimeter (top) and length of side (bottom) are generated.

8.2.2 *Setting and Participants*

This study was conducted over the course of 7 months in a third-grade classroom. The school is a public charter school with a student composition that is 99% African-American and is located in a large metropolitan school district in the southeastern United States. Fifteen students—fourteen African-American and one Latino—participated in this study. The learning activities in this study were divided into two phases: Phase I (Geometry) and Phase II (Kinematics) and are described below in Table 8.1. Instruction during Phase I focused on shape drawing and the interpretation of graphs. Phase II focused on the invention and interpretation of mathematical measures and using ViMAP as a way to explain a real-life phenomenon involving motion (e.g., walking at a constant rate or two cars traveling at different rates for different periods of time). Researchers met weekly with the classroom teacher and iteratively co-designed the classroom activities.

The classroom teacher taught all lessons with the exception of initial shape drawing activities. During implementation, the classroom teacher often improvised instructional methods, connecting inscriptions and representational forms from other domains (e.g., mathematics) with the work students were doing on ViMAP. These improvisations were often made based on the classroom teacher's formal and informal assessments of student understanding of the material or in-the-moment responses to student ideas. Any adjustments to the activities often took the form of extending instructional time on a topic, and modifying the designed classroom materials to better meet mandated instructional goals. Our study focuses on how the teacher adapted and employed this approach as a way to integrate programming as one element of a larger ensemble of scientific activity within the classroom which in turn assisted in transforming code from an isolated technical competency to a literacy.

Table 8.1 Summary of learning activities during Phases I and II

Phase	Activity	Description
Phase I	Shape drawing	Students work in pairs writing rules and creating ViMAP programs for drawing squares, triangles, and circles
	Regular polygons	Students derive a formula for finding the exterior angle of regular polygons (# of sides/360) and use that formula to draw any regular shape in ViMAP
	Congruent shapes	Students program congruent shapes in ViMAP
	Perimeter	Students use ViMAP's graphing function to find the perimeter of geometric shapes. Students discuss how ViMAP graphs are "unfolded" polygons
Phase II	Leaving footprints	Students leave ink footprints on banner paper
	Generating measures	Students iteratively develop, apply, test, and refine a measurement of distance termed a "step size"
	Collecting step-size data	Students use the "step size" measurement convention to measure their personal step sizes
	Modeling step sizes in ViMAP	Students model their personal step sizes in ViMAP. Total distance graphs and predictions using ViMAP's grapher are generated and discussed
	Modeling motion as a process of continuous change	Students model motion scenarios in ViMAP and check the validity of those models using ViMAP's grapher and the total distance equation

8.2.3 Data and Analysis

Data for this work comes from informal interviews with the participants, video recordings of class activities and discussion, student artifacts (e.g., student representations; activity sheets; ViMAP models; and pre-, mid, and post-assessments), reflective interviews with the classroom teacher, and daily field notes. During instruction, the lead researcher and the classroom teacher conducted informal interviews during opportune moments while the students were engaged in single, pair, or small group work around modeling and representational activities. Classes were video recorded, and student-created artifacts (ViMAP models, written work) were also collected.

We present a qualitative, thematic analysis (Corbin & Strauss, 1990) of the data at different points during Phase I and II, with the goal of identifying how the classroom teacher's actions and interactions around ViMAP facilitated the integration of programming as an integral component of STEM practice in the classroom. Specifically, we investigate how the heterogeneity of activity supported by the classroom teacher positioned coding as *one* way (among others in the classroom) of knowing and representing real phenomena, thereby deepening students' mathematical and scientific work and allowing different dimensions of computational literacy to thrive in the classroom. We present this analysis in the form of three illustrative instructional episodes which make explicit how the classroom teacher's focus on heterogeneity of representation shifted students conceptual work with ViMAP from an intelligence (Episode 1) to a nascent literacy (Episode 3). Following diSessa (2001) and Berland (2016), these dimensions are summarized in Table 8.2. Each episode highlights the structure of activity, student conceptual work, and critical instructional moves by the classroom teacher's (possessive) which together supported the gradual take-up of code as a meaning making lens in the classroom.

Table 8.2 Dimensions of computational literacy

		Dimension	Definition	Description in practice
Literacy	Material intelligence	Material	Manipulation of external signs, symbols, depictions, and representations	Making things with computation
		Cognitive	What the human mind does in the presence of such external symbols and representations	The ability to think about problems, contexts, and the world in terms of what computation can and cannot do
		Social	The social context within which the material systems of computing are manipulated and adopted	The ability to identify and discuss the socially accepted ways of articulating meaning as well as understand what you and others need from computational artifacts

8.3 Findings

8.3.1 Episode I: Programming as a New Expression of Third Grade Mathematics

Phase I began by introducing students to the ViMAP programming and modeling language through shape drawing activities. Students worked in pairs to write rules for drawing squares, triangles, and circles (e.g., go forward 5 steps, turn right 90°, go forward 5 steps, etc.) and translated those rules into ViMAP programs (see Fig. 8.1 for an example of a similar program). These activities were designed by the researchers and were primarily taught by the first author. This does not mean, however, that Emma's role as the teacher was sidelined. During these early shape building tasks, Emma often worked with small groups to expand the representational space of classroom activity by connecting ViMAP's representations of geometric shapes (e.g., written code) to embodied actions outside of the computer. Emma's move to reframe activity within the computer as embodied actions outside of the computer helped students see the problem space from multiple perspectives, and as we discuss in Episode II, supported students in transferring programming work in ViMAP to a novel, isomorphic problem space.

Classroom discourse during early shape drawing activities primarily focused on understanding how manipulations of the ViMAP programming blocks (such as increasing or decreasing lengths of sides and changing angles) caused predictable changes in geometric shapes. It is worth noting that during these early activities, programming itself was the end product of knowledge, rather than a representational component of scientific exploration in the classroom. Put simply, programming was an *isolated competency* with instruction largely focused on small manipulations of simple programs. Emma initiated two key instructional moves that altered this trajectory towards programming becoming a competency that was integrated with other goals of the course.

During shape drawing tasks, Emma observed how control structures such as a repeat loop could be reframed as a representation of multiplicative reasoning. In an interaction with a student, Kendra, Emma asked Kendra to compare the number of commands, or rules, needed to make a square using the repeat loop versus the number of commands needed without a repeat loop. Kendra noticed that you needed eight rules to make a square without a repeat loop and only two with a repeat loop. Emma pointed to Kendra's repeat block, and asked Kendra to think about the number of repeat loops (4) in her program. Kendra realized her non-repeat and repeat programs for generating a square were equivalent, telling Emma that "four times two equals the eight commands [in the non-repeat program]." Emma challenged Kendra to think of another way to "make a square" and Kendra produced a ViMAP program using a repeat two with four commands in the repeat loop. What Emma termed Kendra's "three ways to make a square" are shown in Fig. 8.2.

Following her interaction with Kendra, Emma approached the lead author with several ideas for future work with ViMAP. In particular, she suggested enhancing

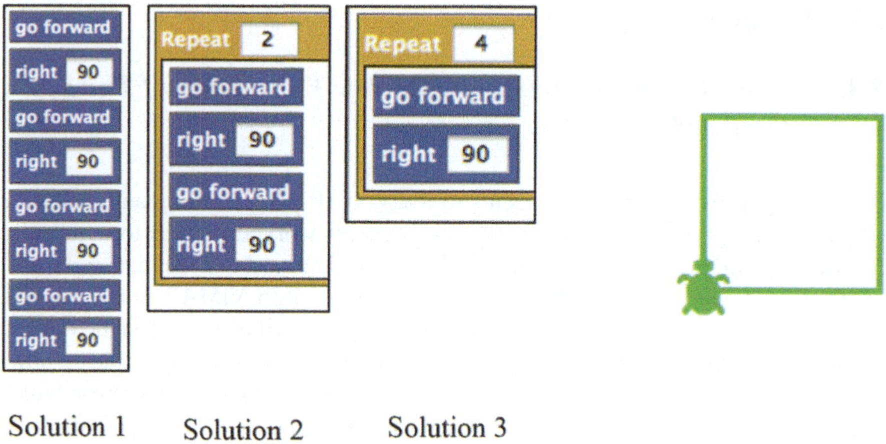

Fig. 8.2 Kendra's three solutions to "making a square"

mandated curricular goals in mathematics by using ViMAP to instantiate mathematical abstractions such as regular polygons, which in turn supported greater opportunities for students to engage with programming, mathematics, and later science in more authentic ways. Emma was a *key* architect of these new mathematics activities with ViMAP and, from that point forward, took over the lead instructional role from the first author.

We feel that this episode is critical to the development of computational literacy in the classroom for several reasons. First, it demonstrates how reframing programming as mathematics was a productive pedagogical lens for the classroom teacher. Emma, to use her words, saw "so much math" in ViMAP and used connections to mandated curricular goals in mathematics to more authentically integrate work with ViMAP within her instruction. In *Changing Minds* (2001), diSessa argues that the adoption of new literacies is dependent upon individuals, and particularly educators, finding those literacies valuable enough and useful enough to be worth the tremendous effort to teach to everyone. By reframing mathematical concepts and phenomena such as multiplicative reasoning and regular polygons as coding, programming became *valuable* and *useful* to Emma, and, because it was valuable, she designed opportunities to use programming as a problem-solving and representational tool in the classroom.

Emma's push to reframe classroom work with ViMAP as mathematics had another beneficial consequence. Integrating programming within third grade mathematics lessons shifted programming from an isolated competency to a nascent material intelligence within the class. Beyond just making shapes with programming, students were exploring and thinking through mathematical problems *with* programming. Critical to our research, Emma herself recognized the reflexivity (Kafai & Harel, 1991) between programming and math (and, in Phase II, programming and science) and took an active role in designing and teaching lessons which supported integrations with elementary content areas.

Emma's work to support more authentic integrations of programming by expanding the representational space of third grade mathematics and science to include both computational and non-computational representations continued into the next phase of student activity, Phase II Kinematics, which we discuss below.

8.3.2 *Episode II: Reasoning Across Different Instantiations of the Same Phenomenon*

Instruction during Phase II began with an investigation of animal tracks. Using the richly illustrated children's book "*Wild Tracks! A Guide to Nature's Footprints*" (Arnosky, 2008), Emma and her students discussed how animal footprints were data-laden. Among the ideas offered by students and privileged by Emma were animal tracks as histories of "where [the animal] started [moving] and where [the animal] stopped." Emma emphasized footprints as *measurable objects* and a credible account of the behavior of the footprint-leaver. In particular, Emma and her students hypothesized that footprints carried information on the rate ("how *fast* or *slow* it went") and distance traveled by an agent, a conjecture that students would devote all of Phase II investigating.

Following this discussion, Emma asked her students to create an embodied artifact: their own footprints inked onto a strip of banner paper. Guided by Emma, students invented a measurement convention, which they termed a "step-size" and used this measurement convention to measure and record their own step sizes which they then modeled in ViMAP. Additional work during Phase II focused on selection of "approximate" step-size values (Emma and the student's term for *typical* values) and repeating those values to derive a formula for calculating unknown total distances (Approximate Step Size x Number of Steps = Total Distance). This formula was later applied to an isomorphic problem space, constant speed, and translated into its more recognizable form: Rate x Time = Distance. Emma co-designed Phase II activities with the lead author and taught all lessons. Her instruction emphasized developing normative criteria for what counted as "good" measures and models of motion, and meaningfully framing coding as a language of science through the design of measures and multiple forms of modeling, such as those described above.

Case 1 In the first case, a student, Angelo, has used ViMAP to find how far he could walk if he took 20 steps of approximately 15 units in length. Prior to the interaction between Angelo and the researcher in Table 8.3, Emma had given the class the challenge of finding their total distance walked at 10 steps, 15 steps, and 20 steps. She had purposely selected numbers "too big to add up one by one" to "force" students to use programming to solve the problem. In a conversation with the lead author, Emma mentioned she wanted students to use ViMAP to solve the problem because she wanted her students to understand that ViMAP had the power to model movement they (her students) had not actually walked. Prior kinematics however lessons were entirely focused on using ViMAP to re-represent embodied actions,

Table 8.3 Angelo uses ViMAP to "Know"

	Utterance	Line
Researcher	How far did you walk after taking 15 steps?	1
Angelo	300 distance	2
Researcher	That's exactly right	3
Angelo	So, if somebody bet that I won't make it farther than 100 *I know* that I will	4
Researcher	That's right. That's how a formula for approximate distance can help you	5
	If someone said "I bet Angelo would only walk 150 inches in 15 steps," but	6
	you knew what your approximate step size was, could you prove them	7
	wrong?	8
Angelo	Yes	9
Researcher	How?	10
Angelo	I could look at my graph	11
Researcher	Or you could do what?	12
Angelo	I could use a calculator. Fifteen times 20 equals 300	13

this lesson designed by Emma was the first to model what Emma called "unknown distances." Table 8.3 illustrates one student's interpretation of his 20-step ViMAP model. The student, Angelo, interprets his ViMAP model as a means to "win a bet." Angelo comments in line 4 that if someone bet him that he would only travel less than or equal to 100 units of distance, he would *know* that they were wrong based on his ViMAP program. Angelo's ViMAP program is shown in Fig. 8.3.

We feel this first case demonstrates how Emma's focus on including multiple, heterogeneous forms of representation and models was crucial in supporting students representational and epistemic work in science. During Phase II, students computational work to discretize motion as a process of continuous change was interdependent with their embodied actions (footprints) and mathematical descriptions of the phenomena (steps size x number of steps = total distance traveled): students used their bodies to discretize motion into literal "steps," reconstructed that motion by repeating typical step-size values to define motion algebraically, and finally programmed motion to represent movement computationally. In the case above, Angelo's program demonstrates that he is able to mathematically summarize discrete values to program and model continuous patterns of change. Additionally, he describes that ViMAP is how he "know[s]." Epistemologically, this is a significant move. We believe that Angelo's explanation of "betting" and "knowing" is his intuitive way of explaining how ViMAP, as a programming and modeling tool, has epistemic and representational power that can help him find *unknown* solutions (to borrow Emma's language) to real problems.

Case 2 Our second illustrative case demonstrates how ViMAP, as a representational tool, had become *infrastructural* to classroom STEM work. Near the end of Phase II, students could easily solve "step-size" problems, prompting Emma to extend students' work with step sizes into isomorphic problem spaces of rate, distance, and time, what Emma termed a "real world problem." Extending student

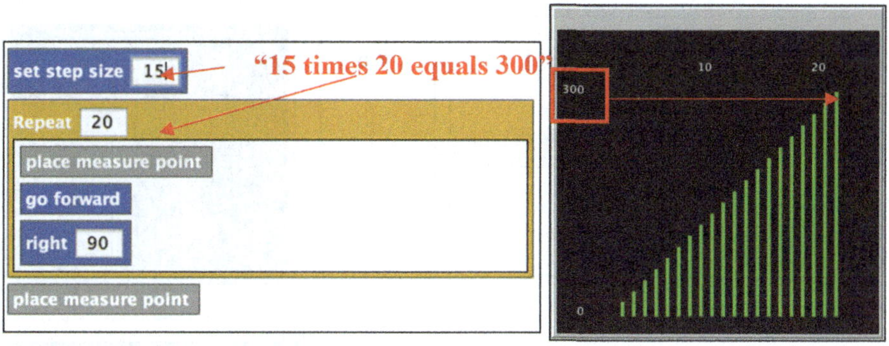

Fig. 8.3 Angelo's ViMAP model

step-size work in ViMAP to more general problems of rate and distance was accomplished, in large part, by Emma's careful attention to the heterogeneity of the representational infrastructure and design of productive ways to harness those different elements to make both science and computing intuitive and authentic for students.

Emma began explorations of rate by returning to embodied enactments of speed and distance. She asked her students to consider "who walked faster," a student who walked five steps with a step size of 10 or a student who walked five steps with a step size of 15? Two students volunteered to embody the fictitious students in Emma's hypothetical scenario and walked five steps each (one 10 and 15) at the head of the classroom. Before this demonstration, students had hypothesized that the student with a step size of 15 was "faster," but they had difficultly explaining in mathematical terms why that might be the case. During the demonstration, students noticed that the student with a step size of 15 traveled further in the same amount steps, and, with Emma's help, defined "speed" as distance traveled in one step and "faster" as *more* distance traveled in one step.

In order to extend these ideas to canonical rate, distance and time problems, Emma asked her students if "cars (or other things that move) took steps." Students responded that while cars did not "walk" as they themselves could, they did move at a set "speed," which, after the demonstration at the front of the class, students could now formalize as distance per step. Emma then introduced her "real world" problem, in which students were asked to determine which of two cars, Car 1 or Car 2, traveled further during a 4-h period of time. Car 1 traveled at a speed of 45 mph for 3 h. Car 2, on the other hand, traveled at a speed of 35 mph for 4 h. Emma presented the problem to students and told them to "ViMAP it." A sample student program and solution to the two-car problem is presented below in Fig. 8.4.

Students' solutions to the two-car problem demonstrate how programming, as part of an ensemble of representational and epistemic work in STEM, helped students represent and talk about the real world. It is worth commenting, however, that *time* was a difficult construct for students.

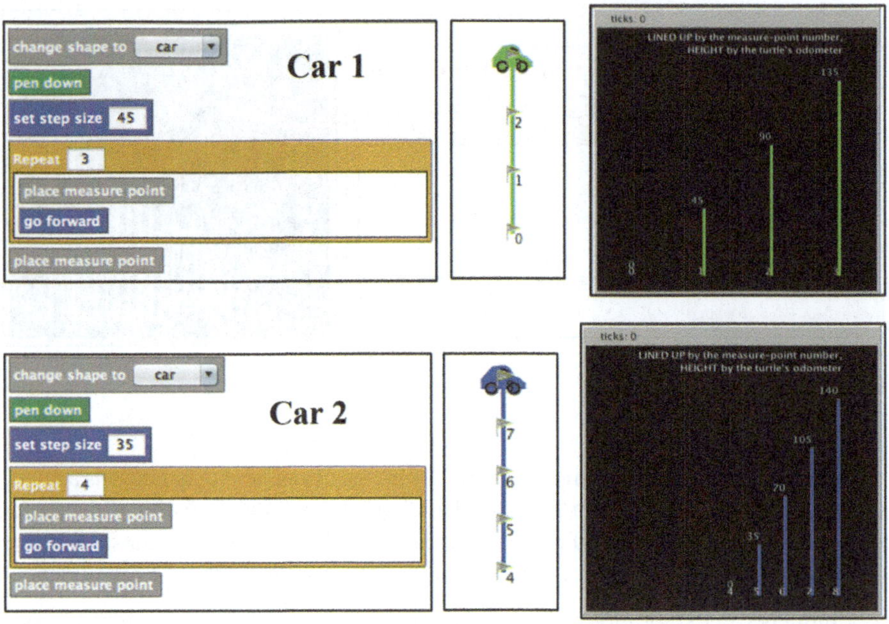

Fig. 8.4 A student's solution to the two-car problem using ViMAP

Students proved proficient in solving the two-car problem with ViMAP and other rate problems designed by Emma, but experienced difficulty *formally* re-representing "number of steps" as *time* in the standard rate equation even though they intuitively used repeat loops to represent time in their ViMAP programs. We do not view this as necessarily problematic and hypothesize that the ease with which students solved rate problems using ViMAP was, in large part, attributable to the work Emma had done deconstructing, representing, re-constructing, and finally re-representing motion in the classroom. Students had deeply experienced motion as a phenomenon—both inside and outside of the computer—and the forms of knowledge they had acquired made it possible to adapt those schemes to novel situations (Vergnaud, 2009).

In light of diSessa's distinctions between material intelligences and literacies, we also feel this case shows how work with ViMAP had become infrastructural to the class' communicative practices in science. In the two-car problem, Emma's use of the phrase "ViMAP it" as an instructional directive suggests how integrated programming and ViMAP had become to how students reasoned through scientific phenomena. As diSessa (2001) and Vee (2013) argue, the critical distinction between a material intelligence and a literacy is the positioning of material technologies as *central* to a community's social and communicative practices. This is particularly evident in the fact that the phrase "ViMAP it" would carry little meaning to those outside of the classroom community.

8.3.3 Episode III: Considering Audience and Communicative Function of Models

Students ended their investigations of motion with a final activity that presented students with three equivalent models of motion, each designed to privilege the elements—code, enactment, and grapher—of ViMAP's representational space. Emma's goal with this task was to encourage students to think about how ViMAP communicated information about motion. Students organized themselves into three groups, and each group was given one of the models. While in their groups, Emma asked her students to summarize the motion (i.e., identify the rate and total distance traveled and provide evidence for how you know) as well as consider how a novice might engage with the different representational elements.

In each model, the turtle agent traveled a total distance of 120. In the first model, the turtle made eight right turns of 110° and did not include a "place measure point" command. Thus, no graphs of motion were generated by the program. In the second model, the code represented motion as a straight line marked with measure flags, but the students were only asked to interpret the code and the agent enactment. The third model was equivalent to the second model, but the students in Group 3 were also asked to interpret ViMAP's graphs of motion. The different models provided to student groups are shown in Fig. 8.5.

The discussion that occurred within and across groups during this activity suggests that students had developed proficiency reasoning across ViMAP's different representational forms, but that they had also established community norms on the quality of ViMAP models. All groups, regardless of the model they had been given, successfully determined the total distance traveled by the turtle agent. Group 1 (model 1) mentioned that the "turns didn't matter," they simply added "8 fifteen times" to find the correct total distance. Group 2 (model 2) used a method similar to Group 1, but noted that there was a second solution to the problem. They pointed out

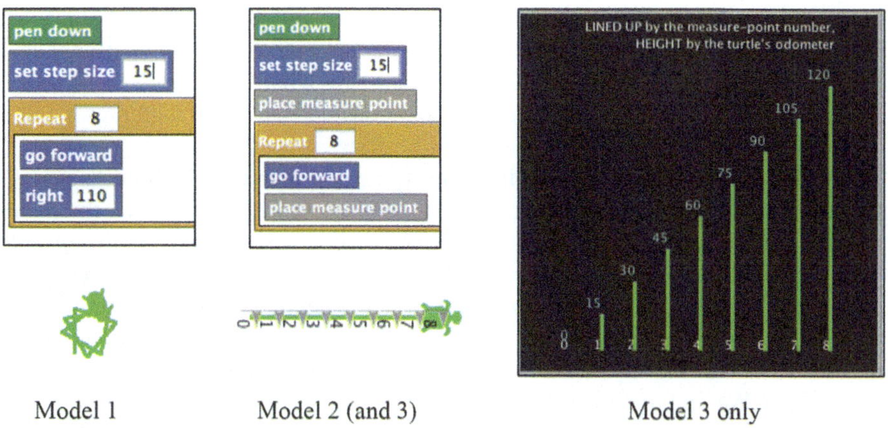

Fig. 8.5 Three models of motion

that you could "count the flags" to know how many steps the turtle had gone. Group 3 (model 3) felt they had the "easiest" model because it included a graph. They pointed to the last column in the graph as the solution and mentioned that nothing else, including the code, was needed to solve the problem. When Emma asked students to evaluate which of the three models was the "best," students agreed that Group 3's model was "the best" because you could quickly and easily solve for both speed (length of first bar in graph) and total distance by interpreting the graph. Graphs, then, were seen as *the* indicator of the mathematical quality of models of motion and were valued by students for their communicative power.

When asked to consider how individuals unfamiliar with ViMAP might interact with each model of motion, students reasoned that novices would have difficulty interpreting the ViMAP enactment as well as the actual ViMAP code. One student, Martin, explained to the class that he felt that a novice unfamiliar with ViMAP as a modeling tool would be unable to interpret the turtle enactment as it related to the phenomena, remarking that if novices "look at the [enactment and code] it wouldn't be understandable." He underscored the communicative quality of canonical bar graphs, remarking that "if [novices] look at the graph, they will *know* [the solution to the problem]," mirroring Angelo's phrasing from earlier in the year.

This episode highlights an important development in students' epistemological work in science. The development of normative criteria around the mathematical quality of models makes it clear that computing had become a critical component of the learning culture in science, and that within this culture students had come to value and assess the different ways ViMAP helped you talk about and show motion phenomenon. By the end of phase II, students had developed fluency across the different representational components of ViMAP. This is evident in their ability to easily discern multiple solutions to the same problem (using repeat, measure flags and graphs to find total distance) and ignore facts unrelated to the problem under investigation (turn angles). Moreover, this episode demonstrates that students, guided by Emma, had begun to consider the social context within which the representational systems of computing are manipulated and adopted (diSessa, 2001) by identifying and discussing socially accepted ways of articulating meaning (graphs) as well as understanding how others might interact with these computational artifacts (Berland, 2016). In the process of recognizing what was "hard" for novices, Emma and her students settled upon what they considered a universal representation of meaning—graphs.

DiSessa (2001) argues that for literacies to be considered as such the meaning-production practices of a community must involve interactions between the material, cognitive, and social dimensions described earlier in this chapter. In the episode described above, the material elements of programming (ViMAP's code, enactment and graphs) were each evaluated based on what they could or could not show (cognitive dimension) based on socially defined and accepted ways of articulating meaning within both the classroom and broader communities of practice (social dimension). We feel that over the course of Phase II, Emma's focus on representational heterogeneity—i.e., developing and integrating diverse forms of modeling—

allowed for complex social forces of innovation and adoption to transform work with ViMAP from isolated technical work to an integral component of representational and epistemic work in STEM, and thus a literacy.

8.4 Discussion

In this chapter, we have demonstrated how the integration of computing as part of an ensemble of STEM work in the elementary classroom involves careful consideration of the complex interplay among materials (both computational and non-computational), cognition, and classroom culture. We argue that this integration supports a transition from computing as isolated material intelligences to part of the broader culture of learning within the elementary STEM classroom. So, why should we endeavor to support integration of computing as *literacy* rather than *isolated material intelligences* in the STEM classroom?

Our prior work has argued that reducing code to the (re)production of symbolic forms leaves out several important elements of the *experience* of coding such as "seeing the same code from different perspectives, designing code for others, and talking about and interpreting code" (Sengupta et al., forthcoming). In most research on educational computing, the experience of code is often limited to the code created by students, leaving teachers and students with few opportunities to experience code beyond the ways designed by educational researchers. As Heidegger (1977) cautions, neither technology (or code), nor how it *enframes* (p. 19) the world around us, is the issue here, but rather our relationship to technology as the only frame through which we experience the world. The world can show or reveal itself to us in different ways, and attention to different ways of showing and knowing can help us recognize that technology is itself one of these ways, but, as Stenner (1998) argues, it is only one.

Reproduction of symbolic forms (material intelligences) falls into the enframing trap by limiting our interaction with objects to only questions of what those objects can and cannot do for us, echoing Papert's (1988) critique of an "information centered" (p. 5) approach to educational computing. Literacies, on the other hand, interact with objects (materials) within cultural contexts, bringing multiple perspectives into contact with one another and thereby expanding, rather than constricting, the ways in which we perceive and interact with reality. In our work, we have reimagined classroom computing as a fundamentally heterogeneous activity. That is, the meaning of computation, in Heideggerian terms, is not necessarily "technological"; instead it becomes reframed through complementary forms of modeling, such as embodied and physical modeling. This distribution of heterogeneous forms for sense-making creates contexts in which students' representational and epistemic work is transformative for both the teacher and the students in terms of the development of their epistemologies about the relationships among reality, scientific representation, and programming as ways of making sense of and explaining the world.

References

Arnosky, J. (2008). *Wild tracks! A guide to Nature's footprints*. New York: Sterling Children's Books.

Barr, V., & Stephenson, C. (2011). Bringing computational thinking to K-12: What is involved and what is the role of the computer science education community? *ACM Inroads, 2*(1), 48–54.

Basu, S., Biswas, G., Sengupta, P., Dickes, A., Kinnebrew, J. S., & Clark, D. (2016). Identifying middle school students' challenges in computational thinking-based science learning. *Research and Practice in Technology Enhanced Learning, 11*(1), 13.

Berland, M. (2016). Making, tinkering, and computational literacy. *Makeology: Makers as Learners, 2*, 196–205.

Corbin, J., & Strauss, A. (1990). Grounded theory research: Procedures, canons and evaluative criteria. *Zeitschrift für Soziologie, 19*(6), 418–427.

Dickes, A. C., Kamarainen, A., Metcalf, S. J., Gün-Yildiz, S., Brennan, K., Grotzer, T., & Dede, C. (2019). Scaffolding ecosystems science practice by blending immersive environments and computational modeling. *British Journal of Educational Technology*.

diSessa, A. A. (1991). An overview of boxer. *The Journal of Mathematical Behavior, 10*(1), 3–15.

diSessa, A. A. (2001). *Changing minds: Computers, learning, and literacy*. Cambridge, MA: MIT Press.

Farris, A. V., Dickes, A. C., & Sengupta, P. (2019). Learning to Interpret Measurement and Motion in Fourth Grade Computational Modeling. *Science & Education*, 1–30.

Grover, S., & Pea, R. (2013a). Computational thinking in K–12: A review of the state of the field. *Educational Researcher, 42*(1), 38–43.

Grover, S., & Pea, R. (2013b, March). Using a discourse-intensive pedagogy and android's app inventor for introducing computational concepts to middle school students. In *Proceeding of the 44th ACM technical symposium on computer science education* (pp. 723–728). New York: ACM.

Guzdial, M. (1994). Software-realized scaffolding to facilitate programming for science learning. *Interactive Learning Environments, 4*(1), 001–044.

Heidegger, M. (1977). The question concerning technology. In *The question concerning technology and other essays* (W. Lovitt, Trans.). New York: Harper & Row.

Holyoak, K. J. (1991). Symbolic connectionism: Toward third-generation theories of expertise. In K. A. Ericsson & J. Smith (Eds.), *Toward a general theory of expertise: Prospects and Limits* (pp. 301–335). Cambridge: Cambridge University Press.

International Society for Technology in Education (ISTE). (2018). *National educational technology standards for teachers*.

Kafai, Y., & Harel, I. (1991). Learning through design and teaching: Exploring social and collaborative aspects of constructionism. In I. Hare & S. Papert (Eds.), *Constructionism* (pp. 85–106). Norwood, NJ: Ablex.

Kim, B., & Reeves, T. C. (2007). Reframing research on learning with technology: In search of the meaning of cognitive tools. *Instructional Science, 35*(3), 207–256.

Lye, S. Y., & Koh, J. H. L. (2014). Review on teaching and learning of computational thinking through programming: What is next for K-12? *Computers in Human Behavior, 41*, 51–61.

Maloy, R. W., & LaRoche, I. S. (2014). *We, the students and teachers: Teaching democratically in the history and social studies classroom*. Albany, NY: SUNY Press.

National Research Council. (2012). *A framework for K-12 science education: Practices, crosscutting concepts, and core ideas*. Washington, DC: National Academies Press.

NGSS Lead States. (2013). *Next generation science standards: For states, by states*. Washington, DC: National Academies Press.

Papert, S. (1980). *Mindstorms: Children, computers, and powerful ideas*. New York: Basic Books.

Papert, S. (1987). Information technology and education: Computer criticism vs. technocentric thinking. *Educational Researcher, 16*(1), 22–30.
Papert, S. (1988). A critique of technocentrism in thinking about the school of the future. In *Children in the information age* (pp. 3–18). Cambridge, MA: MIT.
Peppler, K., & Kafai, Y. (2005). Creative coding: Programming for personal expression. Retrieved August, 30(2008), 314.
Repenning, A., Basawapatna, A., & Klymkowsky, M. (2013, September). Making educational games that work in the classroom: A new approach for integrating STEM simulations. In *Games Innovation Conference (IGIC), 2013 IEEE International* (pp. 228–235). Piscataway, NJ: IEEE.
Sengupta, P., Dickes, A., & Farris, A. V. (2018). Toward a phenomenology of computational thinking in STEM. In M. S. Khine (Ed.), *Computational thinking in STEM: Research highlights* (pp. 49–72). New York: Springer.
Sengupta, P., Dickes, A.C., & Farris, A.V. Voicing Code in STEM: A Dialogical Imagination. Cambridge, MA: MIT Press. Forthcoming 2020.
Sengupta, P., Dickes, A., Farris, A. V., Karan, A., Martin, D., & Wright, M. (2015). Programming in K-12 science classrooms. *Communications of the ACM, 58*(11), 33–35.
Sengupta, P., Kinnebrew, J. S., Basu, S., Biswas, G., & Clark, D. (2013). Integrating computational thinking with K-12 science education using agent-based computation: A theoretical framework. *Education and Information Technologies, 18*(2), 351–380.
Sherin, B., diSessa, A. A., & Hammer, D. (1993). Dynaturtle revisited: Learning physics through collaborative design of a computer model. *Interactive Learning Environments, 3*(2), 91–118.
Simon, H. A. (1983). Why should machines learn?. In Machine learning (pp. 25-37). Morgan Kaufmann.
Stenner, P. (1998). Heidegger and the subject: Questioning concerning psychology. *Theory and Psychology, 8*(1), 59–77.
Street, B. V. (1984). *Literacy in theory and practice* (Vol. 9). Cambridge: Cambridge University Press.
Vee, A. (2013). Understanding computer programming as a literacy. *Literacy in Composition Studies, 1*(2), 42–64.
Vergnaud, G. (2009). The theory of conceptual fields. *Human Development, 52*(2), 83–94.
Weintrop, D., Beheshti, E., Horn, M., Orton, K., Jona, K., Trouille, L., & Wilensky, U. (2016). Defining computational thinking for mathematics and science classrooms. *Journal of Science Education and Technology, 25*(1), 127–147.
Whitehouse.gov. (2016). *Computer science for all.*
Wilensky, U. (1995). Paradox, programming, and learning probability: A case study in a connected mathematics framework. *The Journal of Mathematical Behavior, 14*(2), 253.
Wilensky, U. (1999). {NetLogo}.
Wing, J. M. (2006). Computational thinking. *Communications of the ACM, 49*(3), 33–35.
Wing, J. M. (2008). Computational thinking and thinking about computing. *Philosophical Transactions of the Royal Society of London, A: Mathematical, Physical and Engineering Sciences, 366*(1881), 3717–3725.
Woolf, B. P. (2010). Building intelligent interactive tutors: Student-centered strategies for revolutionizing e-learning. Morgan Kaufmann.

Chapter 9
Development of a CDIO Framework for Elementary Computational Thinking

Stephanie Hladik, Laleh Behjat, and Anders Nygren

Abstract Computational thinking involves using computer science concepts to design systems, find solutions to problems, and understand human behavior. Recent reports recommend the addition of computational thinking into K-6 education, and computational thinking has been taught in different learning environments, including formal and informal environments, and has also been taught alongside engineering design thinking. The CDIO Initiative includes a Conceive-Design-Implement-Design process, which can be adapted for use in guiding computational thinking activities at the elementary (K-6) level. This chapter draws on a phenomenological approach to computational thinking to provide the justification and method for adapting the C-D-I-O design process for teaching K-6 computational thinking. It also describes the design requirements and methods for creating four scaffolded computational thinking activities, which are discussed in detail. For each design, the connection with computational thinking concepts and the proposed framework are provided. This chapter also includes a discussion of the scaffolding techniques between the activities, and how the fading of scaffolding was used to improve learning and confidence. By drawing upon a phenomenological approach to computational thinking, we can make rooms for different ways of knowing and representation in computational thinking education to better connect with students' lived experiences.

Keywords Computational thinking · Coding · Programming · K-6 · Elementary · Engineering design · Framework · CDIO · Experience · Phenomenology

S. Hladik (✉)
Werklund School of Education, University of Calgary, Calgary, AB, Canada
e-mail: skhladik@ucalgary.ca

L. Behjat · A. Nygren
Schulich School of Engineering, University of Calgary, Calgary, AB, Canada

P. Sengupta et al. (eds.), *Critical, Transdisciplinary and Embodied Approaches in STEM Education*, Advances in STEM Education,
https://doi.org/10.1007/978-3-030-29489-2_9

9.1 Introduction

Computational thinking was first defined by Jeannette Wing in 2006 as a form of thinking that involves "solving problems, designing systems, and understanding human behavior, by drawing on the concepts fundamental to computer science" (Wing, 2006, p. 33). As research into the inclusion of computational thinking and coding in K-12 classrooms has grown, teachers and researchers have found ways to include computational thinking in formal learning environments by utilizing computational thinking's links to science and math (Sengupta et al., 2015; Weintrop et al., 2016; Wilkerson-Jerde, Wagh, & Wilensky, 2015). These researchers have also argued that the contextualization of computational thinking in the form of curricular activities requires careful attention (Sengupta et al., 2015; Sengupta, Kinnebrew, Basu, Biswas, & Clark, 2013; Weintrop et al., 2016; Wilkerson-Jerde et al., 2015). The link between computational thinking and design thinking (Lee et al., 2011; Sengupta et al., 2013; Weintrop et al., 2016) allows for established design thinking frameworks to provide structure and support for integrating computational thinking activities into elementary classrooms. Additionally, by drawing upon a phenomenological approach to computational thinking (Sengupta, Dickes, & Farris, 2018), we can make rooms for different ways of knowing and representation in computational thinking education to better connect with students' lived experiences.

This chapter proposes a new framework for computational thinking in K-6 through modification of the existing Conceive-Design-Implement-Operate design process in the CDIO post-secondary engineering education framework (Crawley, Malmqvist, Östlund, & Brodeur, 2014). The goals of the modified framework are to support the development of computational thinking activities appropriate for K-6 classrooms and encourage implementation of these activities by elementary teachers. In this chapter, we present a pedagogical approach for appropriating the Conceive-Design-Implement-Operate (CDIO) design process for teaching computational thinking in K-6 classrooms. To illustrate our design approach, we present descriptions of four activities developed using this framework, including their design requirements and connections to the modified framework, as well as discussion of relevant scaffolds.

9.2 Background

9.2.1 Computational Thinking in K-12

Brennan and Resnick (2012) argued that computational thinking should be understood in terms of concepts, practices, and perspectives. It is also noteworthy that as Sengupta et al. (2013) argued, abstractions and concepts in computational thinking are only evident in the form of contextualized representations of code. Along similar lines, Brennan and Resnick (2012) also argued that *concepts* in computational thinking are evident in how programmers represent sequences, loops, parallelism, events, conditionals, operators, and data in their code. Computational thinking also involves using problem-solving *practices* such as being incremental and iterative, and abstracting and modularizing (Brennan & Resnick, 2012). Finally, computational thinking *perspectives* highlight how computational thinkers connect to the world around them, by expressing themselves, connecting with new audiences, and questioning the technological world (Brennan & Resnick, 2012).

Computational thinking has been taught in both formal and informal learning settings. Informal spaces such as after school clubs (Kafai, Fields, & Burke, 2008), public spaces (Sengupta & Shanahan, 2017), and museums (Horn, Leong, & Block, 2012) can be designed to support learning activities for computational thinking. In the context of K-12 classrooms, researchers have also argued that the contextualization of computational thinking in the form of curricular activities in Science and Math requires careful attention so that epistemic and representational practices associated with computational thinking can be reflexively integrated with the disciplinary learning (Sengupta et al. 2013, 2015; Weintrop et al., 2016; Wilkerson-Jerde et al., 2015). Some recommendations have been made by Lye and Koh (2014) regarding teaching computational thinking in K-12, such as designing authentic learning activities to which the students can relate their prior experiences, creating prompts for reflection, and including scaffolding techniques to reduce frustration, such as highlighting important features, managing carefully the degrees of freedom of the activity, and demonstrating the activity (Lye & Koh, 2014).

As Sengupta et al. (2013) argued, if we want children to develop computational thinking, we need to provide students with opportunities to create authentic computational representations, because representational work and computational thinking go hand-in-hand. It is therefore no surprise that the emphasis on integration of CT with STEM classrooms has resulted in the development of both new technologies and curricular activities. Weintrop et al. (2016) created various lesson plans and activities which tied to high school physics, biology, chemistry, and mathematics. For example, students investigated how different video games used the laws of physics, or used an interactive simulation to understand how properties of a gas (such as volume, temperature, and pressure) affect its behavior and interactions. Wilkerson-Jerde designed the DeltaTick simulation interface for NetLogo (Wilensky, 1999) using which middle and high school students can model animal populations in ecosystems to better understand natural selection and population dynamics (Wilkerson-Jerde et al., 2015). Sengupta and colleagues developed specialized programming languages—ViMAP (Sengupta & Farris, 2012; Sengupta et al., 2015) and CTSiM (Sengupta et al., 2013)—for elementary students to introduce them to computational modeling of kinematics and ecology.

However, integrating computational thinking with science and math curricular topics through computational modeling is only one way to integrate computational thinking into elementary classrooms. The work presented in this chapter is grounded in a somewhat different perspective anchored in engineering education. We present a pedagogical framing of computational thinking in K-12 classrooms as experiences that encompass multiple disciplinary forms in the elementary classroom without necessarily being committed to only mathematics and science curricular topics. Similar to Sengupta et al. (2018), we adopt a phenomenological approach which foregrounds the richness and complexity of learning experiences over cognitivist views of computing and computational thinking. At the heart of this approach, as Sengupta et al. (2018) have argued, is the emphasis on interpretive and intersubjective experiences that learners and teachers can engage in using multiple forms of computing (e.g., material, virtual, and embodied modeling), rather than only focusing on the use of virtual symbolic abstractions (e.g., use of specific data structures in SCRATCH programs) as the signature of learning. As Sengupta et al. (2018) argued, such a phenomenological approach can challenge a technocentric (Papert, 1987) view of learning computing, in which assessments of learning are focused on the technological "products" of learning (e.g., computer programs), rather than focusing on the richness of learning experiences.

A notable example of such a pedagogical approach can be found in the work of Danish (2014), who argued that research on educational computing need to move beyond the capabilities of the programming language and consider aspects such as embodiment and representational and participatory activities. Danish (2014) used activity theory to design a series of interrelated activities to support student understanding of honeybees, including representing bees in drawings or sculptures, acting out (embodying) the behavior of these bees, and exploring the BeeSign simulation to understand hive behaviors. In this way, he progressed away from a technocentric approach which reduces classroom activities (and discussion of those activities) to the use of the programming language or simulation. Instead, his work mirrored the call for a phenomenological approach to computing (Sengupta et al., 2018) by including representational work and embodied learning as a way to connect with students' lived experiences (see also Lee & Wilkerson, 2018). In this way, computational thinking can become transdisciplinary, taking place across different disciplinary contexts, and incorporating different ways of knowing.

To summarize, in order to challenge a technocentric approach to computational thinking education, we must consider embodiment, narratives, and artistic representations as ways to engage students and teachers in computing. Merleau-Ponty (1962) defined "sense experience" as "that vital communication with the world which makes it present as a familiar setting of our life" (Merleau-Ponty, 1962, p. 61). Challenging a technocentric approach to computational thinking can lead to new sense experiences (Merleau-Ponty, 1962) of computational thinking by locating it in familiar settings and interactions in the lives of both students and teachers. As well, by doing so coding and computational thinking can extend beyond the computer and become part of the lived world of the students. Our goal is to provide a pedagogical approach for elementary classrooms that leverages these insights, grounded in the CDIO framework, which we explain next.

9.2.2 *CDIO Initiative*

The Worldwide CDIO Initiative is a framework developed by educators at the Massachusetts Institute for Technology to guide post-secondary engineering education (CDIO, 2017c; Crawley et al., 2014). This framework has a vision for engineering education that includes teaching technical skills alongside professional skills such as communication and teamwork and featuring active and experiential learning through design-built-test projects for students at all levels (CDIO, 2017b). The CDIO Syllabus was created which includes a breakdown of attributes and skills that are desirable in engineers, and this syllabus can be used to develop teaching materials and assessments (Crawley, Lucas, Malmqvist, & Brodeur, 2011; Crawley et al., 2014). For example, some of the main topics of the syllabus include having a strong engineering knowledge base, understanding ethics and equity in engineering, cultivating good communication and teamwork skills, and using the Conceive-Design-Implement-Operate design process when completing engineering projects.

9.2.2.1 Conceive-Design-Implement-Operate

Sections 4.3 to 4.6 of the CDIO Syllabus detail the attributes of the Conceive-Design-Implement-Operate design process (CDIO, 2017a). This design process has been used in a variety of design-build-test projects at both the post-secondary and K-12 levels. Table 9.1 shows the breakdown of this design process into the second and third levels of detail (CDIO, 2017a). A fourth level of detail specifying the topics for each third level category can be found online in the full CDIO Syllabus (CDIO, 2017a).

The Conceive-Design-Implement-Operate design process can be understood through the example of designing and releasing a smartphone app for food delivery. The first step in the design process is to *conceive* whatever is being designed and built. The engineers should understand the opportunities and needs of their client and consider what are their requirements for the app, and set goals. In this stage the engineer should also consider what the necessary functions of the app are (such as connecting to GPS to determine a location, or access to the speaker to call the food delivery person) and what the overall architecture and layout of the app would be. The conceive step also includes basic project management with respect to project cost, performance, and schedule. The *design* stage is where the engineers will use the design process to come up with an initial design for the app, consider alternatives and risks with those designs, complete prototypes and mock-ups of the app, and settle on a final design. These designs take into account not only technical knowledge (how to write the code), but also creative thinking and utilization of prior work in the field, such as looking at existing apps or services. The engineers should also consider safety of the app (does it store credit card numbers?), aesthetics (a nice

Table 9.1 Sections 4.3.X to 4.6.X of the CDIO syllabus

4.3—Conceiving, Systems, and Engineering Management	4.4—Designing	4.5—Implementing	4.6—Operating
4.3.1—Understanding needs and setting goals 4.3.2—Defining function, concept and architecture 4.3.3—System engineering, modeling and interfaces 4.3.4—Development Project Management	4.4.1—The design process 4.4.2—The design process phrasing and approaches 4.4.3—Utilization of knowledge in design 4.4.4—Disciplinary design 4.4.5—Multidisciplinary design 4.4.6—Design for sustainability, safety, aesthetics, operability and other objectives	4.5.1—Designing a sustainable implementation process 4.5.2—Hardware manufacturing process 4.5.3—Software implementing process 4.5.4—Hardware software integration 4.5.5—Test, verification, validation and certification 4.5.6—Implementation management	4.6.1—Designing and optimizing sustainable and safe operations 4.6.2—Training and operations 4.6.3—Supporting the system life cycle 4.6.4—System improvement and evolution 4.6.5—Disposal and life-end issues 4.6.6—Operations management

color scheme and intuitive layout), and any environmental impacts. Next, the engineers will *implement* the app by working through the software implementation process of breaking down the code into functions and modules, designing algorithms, and ensuring each part of the app's code works together. This step is where they would test their app on multiple devices and operating systems, and validate that the app's performance meets the customer's standards. Finally, the *operating* step is where engineers must consider what training is necessary for customers to use the app, how it will be updated (automatically? Downloaded from an app store?), and brainstorm possible improvements or updates.

9.2.2.2 Application of CDIO to K-12 Education

The CDIO Syllabus has been used outside of post-secondary engineering programs, including at the K-12 level. Eleven-year-old students in Sweden took part in an egg-drop challenge which highlighted design-built-test skills (Traff, Wedel, Gustafsson, & Malmqvist, 2007). The C-D-I-O design process was used to create and implement activities to meet learning outcomes around electricity at the grade 5 level (Marasco & Behjat, 2013). Finally, the CDIO Syllabus has also been used to train K-12 teachers. Post-secondary students in a B.Sc in Science and Technology Education program used the CDIO approach to balance engineering fundamentals with pedagogy and teaching practice (Verner, 2015).

9.3 Designing a CDIO-Based Approach for Computational Thinking in Elementary Classrooms

9.3.1 Bridging Epistemology and Practice

As outlined in previous sections, this work is epistemologically committed to phenomenological approaches for computational thinking. Sengupta et al. (2018) center their argument around the idea of *sense experience* (Merleau-Ponty, 1962). Merleau-Ponty argues that "sense experience… invests the quality with vital value, grasping it first in its meaning for us, for that heavy mass which is our body, whence it comes about that it always involves a reference to the body" (Merleau-Ponty, 1962, p. 61). He describes sense experience as a way in which our thoughts and understandings of objects and concepts can change due to their interactions with our senses. Sengupta et al.'s (2018) concern is that because recent approaches to teaching computational thinking and coding in K-12 education have centered around a particular programming language, application, or platform, such approaches can become disconnected from learners' sense experiences, thus creating a gap between their lived experiences in the world and what they believe coding to be. In order to explore students' sense experiences with coding, we should consider aspects of materiality and subjectivity to be important parts of computing.

Additionally, Sengupta et al. (2018) highlight "the importance of grounding computational thinking in representational and epistemic practices that are central to *knowing* and *doing* in science and, more broadly, in STEM education" (p. 51). They critique commonly held decontextualized notions of abstraction in computational thinking and argue that contextualization is key to a phenomenological approach to computational thinking. Professional computer scientists typically create and apply abstractions to a particular design goal or context in professional practice (Schmidt, 2006). What does this mean for young learners new to computing? We believe that engaging in *authentic* computational work demands that their use of computational abstractions and representations be grounded in rich disciplinary and representational work, that in turn is valued as part of their everyday classroom activities. In the elementary classroom, such activities could include writing a story or creating a piece of digital artwork. Even outside of STEM-specific activities, students can engage in CT-practices of modeling, decomposition, and verification that are key to problem-solving in scientific, engineering, and mathematical disciplines (Weintrop et al., 2016; Sengupta et al., 2013).

In the context of STEM education, a phenomenological approach to computing and computational thinking draws upon the *science as practice* perspective (Lehrer, 2009; Pickering, 1995). Pickering (1995) describes the "mangle of practice" as the ways in which scientists investigate by using theories and instruments, on one hand, to study a natural world that they expect to perform in a certain way. This is a dynamical interaction between material (instruments and the natural world) and human agency (Lehrer, 2009; Pickering, 1995), and something that learners in computing must deal with as they theorize and refine their models. They must manage this uncertainty as best they can, by making subjective decisions based on their experiences. The mangle of materiality, agency, and uncertainty is not something to be eliminated from computational thinking education, but instead is something that should be considered a fundamental aspect of the learning experience in computing.

In order to incorporate these phenomenological elements of sense experience, contextual work, and the mangle of science and practice into computational thinking, we should consider a framework from engineering education which centers these aspects in practice.

9.3.2 CDIO: A Practice-Based Approach for Engineering Education

By using the Conceive-Design-Implement-Operate framework from the CDIO Initiative, we are able to make explicit our commitments to a phenomenological approach to computational thinking. One of the biggest strengths of the C-D-I-O framework is that it widens the scope of a computational thinking activity beyond a simple task that invites learners to write code to solve a problem. In the conceive step, learners must set their own goals and requirements and develop some level of project management. By asking learners to specify their own goals and requirements, they will come to understand the mangle of practice as they begin to work through the activity—their needs and goals will affect the work they try to do, which in turn may cause them to reflect on and change their original assumptions. This mangle will also be affected by both the knowledge and tools they choose to use, both of which play key roles in the C-D-I-O process.

Additionally, by specifically asking learners to set their own goals within an open-ended, creative project, they are able to align computational thinking directly with their interests. This alignment helps learners connect computational thinking with their own personal sense experiences of the world; as illustrated in the activities later in this chapter, learners may create a narrative of a scene in their lives, or create a piece of artwork that they find aesthetically pleasing. Their computational work is then also contextually situated in their lives. Rather than computational abstractions, loops and conditional statements become key aspects of their narratives or dances, grounded in their personal and embodied experiences. Again, these personal connections are only possible because the C-D-I-O framework provides space for them to conceive their own "problems to solve" in the form of puzzles, stories, or dance.

The iterative nature of the C-D-I-O design process helps learners to understand the dynamical relationships between their goals, the tools they use, and what they are trying to create. By moving beyond design-implement activities that have them solving pre-specified problems, learners are able to contextualize computational thinking in sense experiences that are meaningful to them. For these reasons, the CDIO framework is an excellent place to start for the creation of a modified framework for computational thinking that is epistemologically committed to a phenomenological approach.

9.3.3 Creating the Modified Framework

Creating the modified CDIO framework required an in-depth look at the Conceive, Design, Implement, and Operate steps of the design process. To do so, a consultation of the online CDIO Syllabus (CDIO, 2017a) was required to break down each C-D-I-O step into its specific topics. Next, various sources were consulted to determine a traditional teaching approach for how to solve programming problems. Each source has similar steps, and a 7-step approach for solving programming problems from Cornell University (and used in an introduction to programming course on the online platform Coursera) was chosen as it encapsulated all of the steps and was intended for an audience with little to no programming experience, which is in line with the intended users of this framework: teachers and students who may not have any prior experience with computational thinking (Hilton & Bracy, 2015). This 7-step approach was mapped against the C-D-I-O steps, which led to the discovery that each of the 7 steps fits into the Design step of the C-D-I-O process. That result made it clear that in order to create open-ended problems based on the entire Conceive-Design-Implement-Operate design process, it would be necessary to move beyond simply solving programming problems to creative assignments with open-ended questions. Other resources to guide the creation of the framework were various computational thinking syllabi from organizations and countries which include it in their elementary curriculum, which led to the inclusion of topics such as digital literacy and digital citizenship.

In order to be applicable to the scope of an elementary school project, some topics in the original C-D-I-O design process were removed, such as disposal issues, detailed project management, and system engineering. As well, since this framework is intended for computational thinking activities that do not require any digital technology, topics related to hardware were also stripped, though it should be noted that those topics may be applicable to some activities that do make use of digital technology and hardware such as robotics kits. Finally, the technical language used in the CDIO Syllabus was not appropriate for K-6 students or teachers without an engineering background, and therefore terminology was simplified.

9.3.4 A Phenomenological Approach for Designing CDIO-Based Learning Activities

We used the Modified CDIO Framework for K-6 Computational Thinking to design four computational thinking activities. Our underlying commitment to a phenomenological approach was evident in the ways in which we advocated the use of digital technologies, our choice of the programming language, our emphasis on cross-curricular application, and on attending to the importance of discourse among students, embodied reasoning, and aesthetic experiences (Sengupta et al., 2018).

Different schools may have different levels of access to technology; one school may have a single computer lab for the entire student population, where another may have a cart of laptops for each grade, or even each class. Therefore, it was important to design some activities which did not require the use of digital technology such as laptops or tablets. As well, having device-less computational thinking activities expands the context of computational thinking beyond programming on a laptop and into various daily activities, such as dancing or cooking a meal. By including computational thinking activities which do not solely rely on using programming languages, we move away from a technocentric approach to computational thinking. This distinction will also help teachers connect computational thinking with students' sense experiences (Merleau-Ponty, 1962), reframing familiar settings and activities as sites in which computational thinking can be explored and understood, and in turn making computational thinking more familiar and accessible to them. Finally, we believe that untethering computational thinking from programming languages also makes it more likely for teachers to connect it with other subjects in the school curriculum.

The second design requirement was the selection of appropriate programming languages. In order to be applicable to students of all ages, as well as provide adequate scaffolding between activities, block-based and text-based programming languages were used in two different activities. One activity uses Scratch, MIT's block-based programming language that is well-known in the K-12 environment (Sáez-López, Román-González, & Vázquez-Cano, 2016; Vaca Cárdenas et al., 2015). Scratch is known to be a "low floor, high ceiling" programming language, which means that though little knowledge is needed to create a basic programming, it is possible to create technically advanced projects as well (Papert, 1980). Finally, block-based programming languages remove syntax constraints which may be frustrating for students who are just learning how to read and write (Lye & Koh, 2014). The second activity uses the processing programming language, which is an open-source text language based on Java (Processing, 2017). Processing was chosen as it has the tools to create visual output of shapes and colors using few lines of code. As well, it can provide a good stepping stone for teachers and students who want to move on to Android or Arduino development.

It has been shown that students may be losing interest in STEM subjects as early as grade 5 (Arnot, James, Gray, Rudduck, & Duveen, 1998; Bussiere, Cartwright, & Knighton, 2004). Creating opportunities to learn and use computational thinking outside of the context of traditional STEM subjects may increase student interest in the subject, especially if it ties to subjects such as sports, fine arts, or digital media (Guzdial, 2009; Marasco, 2013). Working within arts-related subjects can provide students with more opportunities to explore ideas of aesthetics in their work (Azevedo, 2018; Farris & Sengupta, 2016). As well, cross-curricular connections with other mandatory subjects including English Language Arts and social studies provide the potential for teachers to teach multiple topics at the same time, and emphasize the connections between computational thinking and other subjects. This saves time for teachers, as they do not have to fit a new, non-mandatory subject into their already packed curricula.

9.4 Illustrative Examples of CDIO Activities for Supporting Computational Thinking

The proposed Modified CDIO Framework for K-6 Computational Thinking can be found in Table 9.2. The left column of the table contains the third level of detail of the Conceive-Design-Implement-Operate design process from the CDIO Syllabus. The right column contains the new proposed framework. Specific topics and attributes of K-6 computational thinking projects for each attribute of the C-D-I-O design process are listed, as well as some questions that can guide the completion of the project. These topics are directly related to the fourth level of detail of the C-D-I-O design process from the CDIO Syllabus; some of the topics are directly included, while others are changed to be made more specific to computational thinking.

It can be seen from this framework that computational thinking projects for K-6 students can contain all four of the Conceive-Design-Implement-Operate design steps. As well, the framework not only includes the "coding" part of the activity, but also puts emphasis on knowing exactly what the code needs to do, designing the code, and what happens when the code is finished (such as considering whether others need training to view the final product, or acknowledging the role of digital citizenship and safety when posting online). The next section of this chapter will detail the connections between this proposed modified CDIO framework and the four developed computational thinking activities.

9.4.1 Programming Puzzles

Programming Puzzles is an activity that introduces students to basic computational thinking concepts of sequences and conditional statements, and the practices of being incremental and iterative and debugging. Both Programming Puzzles and the second activity are activities which do not use electronic devices, as it has been shown that conducting programming and modeling activities outside the computer can greatly facilitate the adoption of programming (Sengupta et al., 2015). Programming Puzzles begins with a discussion about computers, hardware vs. software, and how different robots or electronic devices may have different "instructions" in their programming that help them accomplish different tasks according to their design. The hands-on activity begins with the group being split into pairs, and each pair is given a set of colored paper squares (white, green, red, and blue) and an animal figurine. The students will use the colored squares to create a maze, which

Table 9.2 Proposed modified CDIO framework for K-6 computational thinking

CDIO Syllabus 2.0 (Engineering)	Computational thinking/programming for K-6 (Proposed)
4.3—Conceiving	
4.3.1—Understanding needs and setting goals	–Define the problem to be solved –Determine what the program/algorithm needs to do –Performance metric/rubric (how will we know if it worked?) –Societal context
4.3.2—Defining function, concept, and architecture	–What functions are needed –Determine what technology to use
4.3.3—System engineering, modeling, and interfaces	
4.3.4—Development project management	–Consider schedules, time limits –Allocate resources, both human and technological –Consider risks and alternatives
4.4—Designing	
4.4.1—The design process	–Requirements for each element or component derived from system level goals and requirements –Alternatives in design –The initial design –Life cycle consideration in design –Experimental prototypes and test articles in design development –Appropriate optimization in the presence of constraints –Iteration until convergence –The final design –Accommodation of changing requirements
4.4.2—The design process phrasing and approaches	
4.4.3—Utilization of knowledge in design	–Use technical and scientific knowledge –Different types of thinking, including problem solving, inquiry, creative thinking, critical thinking –Consider standardization –Using prior work (reusing code)
4.4.4—Disciplinary design	–Consider the appropriate programing language –Model the task/program (i.e., pretend to be the robot, step through the program)
4.4.5—Multidisciplinary design	–Interactions between disciplines
4.4.6—Design for sustainability, safety, aesthetics, operability and other objectives	–Reliability –Consider sustainability, safety –Digital literacy/citizenship –Code readability, is it easy to understand? (aesthetics)
4.5—Implementing	
4.5.1—Designing a sustainable implementation process	

(continued)

Table 9.2 (continued)

CDIO Syllabus 2.0 (Engineering)	Computational thinking/programming for K-6 (Proposed)
4.5.2—Hardware manufacturing process	
4.5.3—Software implementation process	–The breakdown of high-level components into module designs (including algorithms and data structures) –Algorithms (data structures, control flow, data flow) –The programming language and paradigms –The low-level design (coding) –The system build
4.5.4—Hardware software integration	
4.5.5—Test, verification, validation and certification	–Does it work according to the design? –Does it solve the problem?
4.5.6—Implementation management	–Consider process improvements: Was there a better way to do the programming?
4.6—Operating	
4.6.1—Designing and optimizing sustainable and safe operations	–Sustainable, safe, secure operation –Digital literacy/citizenship
4.6.2—Training and operations	–Training to use final product
4.6.3—Supporting the system life cycle	
4.6.4—System improvement and evolution	–Consider ways to improve the product
4.6.5—Disposal and life-end issues	
4.6.6—Operations management	

will be a pattern of squares starting on green and ending on red, with some white and blue squares in the middle. Squares must be aligned properly such that there is a clear path, and diagonal movement is not allowed. Next, the students will need to write the instructions for the animal figurine to reach the end. However, they are choosing from a limited set of commands: forwards, backwards, left, right, and “special,” where the “special” instruction is for use on the blue squares only: the animal will make its noise. Once the pair has their maze and instructions complete, they scramble the maze and their instructions to another pair. Those students have to try and recreate the other pair’s maze using only their written instructions. An example of a programming puzzle and its instructions is included in Fig. 9.1.

This activity highlights the importance of correct instructions (through testing and modeling) and their clear communication though words or pictures. Programming Puzzles is cross-curricular with mathematics concepts such as pattern creation. Programming Puzzles can also be completed in a life-sized manner, with larger squares and using students to walk through the maze rather than animal figurines, which gives it cross-curricular ties with physical education and embodied ways of knowing. Previous research has shown that even young learners can understand complex systems and agent-based modeling through explanations and activities which make use of the child’s embodied actions (Danish, 2014; Papert, 1980;

Fig. 9.1 Programming puzzle example

Sengupta et al., 2018), just as Papert (1980) described young children *thinking like/with the turtle* when creating LOGO programs.

While Programming Puzzles is a simple activity, it uses all four of the C-D-I-O steps. Students need to *conceive* what the maze will look like, understand restrictions on movement, commands, and materials, and know the purpose of the algorithm that they will be writing. Next, they *design* the maze by considering different layouts and using creativity, working through a small activity, modeling the maze using their figurine, and choosing a way to write their instructions (top to bottom, left to right, arrows, or words). In the *implement* stage, they break the maze into parts and write the code, testing as they go to ensure it is correct. Finally, they *operate* the maze by seeing if others can accurately follow their code, if any training is required (if they used alternate commands), and if the maze or commands can be improved.

9.4.2 Teach a Robot to Dance

Teach a Robot to Dance is an activity that requires no digital technology other than an internet connection to watch a video. It adds the concepts of program control in the form of loops and events to the sequences and conditional statements from Programming Puzzles. The basis of this activity is that a choreographed dance can be a great example of computational thinking: it has a series of steps executed in order, which may include repeated actions or events, such as starting a new part of the dance when the song reaches the chorus. In this way, the dance becomes an enactment of thought through embodied movement (Danish, 2014; Hwang & Roth, 2011; Papert, 1980). To highlight the connections to computational thinking, the activity begins with the viewing of a popular song with a specific dance, such as La Macarena or the chicken dance. The students then have to write code, or instructions, of how to complete the dance. This part of the activity will emphasize specific instructions; "put hand out" does not tell the dancer exactly what to do, while "put right hand out at shoulder height, palm facing up" does. Another option for this part of the activity would be to have pre-made blocks with the necessary instructions which would then need to be put into the correct order for the dance. To connect with the future Scratch Stories activity, these blocks can be color coded to match Scratch's blocks (i.e., blue for movement, orange for events, etc.). Example blocks used in grade 1–3 classrooms are shown in Fig. 9.2.

Once students have had practice in writing specific, detailed instructions for a dance, they are separated into groups, and each group will create their own dance to a song. They can either write down their instructions using words, or for younger students, pre-made dance move blocks can be used to create the dance. They will have time to practice their dance and refine their instructions, and should be using multiple computational thinking practices such as being incremental and iterative in breaking the song down and practicing different parts together, testing as they go, and even reusing or remixing dance steps from an official or fan-made video. Finally, once their dance is finished, groups will switch dances to see if they can follow another group's instructions to accurately perform their dance in a dance party setting. This activity ties into cross-curricular learning outcomes in physical

Fig. 9.2 "Teach a Robot to Dance" cards for pre-reader students

education and music/dance, two popular hobbies for grade 5 students (Marasco, 2013), which can help those students connect computational thinking with their interests. Like *Programming Puzzles*, this activity is another example of using embodied modeling and programming to introduce programming to young students, which has been shown to deepen understanding of programming and modeling concepts (Danish, 2014; Farris, Dickes, & Sengupta, 2016; Hwang & Roth, 2011; Papert, 1980; Sengupta et al., 2015).

In Teach a Robot to Dance, students must *conceive* the factors that will affect their dance, such as the number of people in the group, the song to use, its length, and if the dance moves are appropriate for all students (i.e., if a student is on crutches). They must also decide how to tell if their instructions are "good enough," or how many mistakes in the final presentation are still acceptable. Next, they *design* the dance by brainstorming dance moves, pulling in material from videos online or other sources, and determining the exact wording that would make their desired dance moves unambiguous and easy to follow. While *implementing* the dance, it would be broken into sections (i.e., verse vs. chorus) and into each line of the song, figuring out the dance moves for each line before putting it all together and making sure the dance moves flow into each other, and that the timing makes sense. They can also check with other students outside of their group to make sure their written words accurately reflect what they are doing. Finally, they consider *operating* their dance by ensuring each dancer knows when to start or how long to do certain moves, brainstorming how the dance can be changed or improved, or considering digital citizenship, privacy, and intellectual property if they were to put up a video of their dance online.

Teach a Robot to Dance was the favorite activity of most of the teachers who implemented these activities in their classrooms. The overall atmosphere was energetic and even slightly chaotic: students ran from dance to dance, trying them out even as they requested a new song. For grade 1–3 classrooms, I printed out some common dance moves on cards (as shown in Fig. 9.2), and also provided blank cards for students to write their own dance moves. One aspect of the activity that was particularly interesting was the way in which students used and modified these cards. They followed popular trends (such as "dabbing"), but also discussed with their partners and other students the best way to verbalize or draw the particular dance move they wanted others to do. Additionally, some students even modified the blocks to include elements of parallelism that were not present in the original activity. If students wanted others to jump and spin at the same time (a popular combination), they used different representations to get that meaning across. For example, some students used a blank card and wrote "at same time" and put it above those two dance moves. Others simply stacked the cards they wanted dancers to execute simultaneously. At that point, we had not even touched upon the idea of parallelism in the class discussions, but the students' lived experiences and desires pushed them to find novel ways to represent parallelism in order to create the dance they were imagining.

9.4.3 Scratch Stories

Scratch Stories is the first developed activity which uses computers to write code. In this case, MIT's Scratch was chosen as the programming language as it is block-based, which makes it a good starting point for younger students while also acting as a stepping stone to more complicated text-based programming languages. The use of Scratch to introduce young students to computational thinking and programming is well documented (Brennan & Resnick, 2012; Kafai et al., 2008; Lye & Koh, 2014; Meerbaum-Salant, Armoni, & Ben-Ari, 2013; Sáez-López et al., 2016; Vaca Cárdenas et al., 2015).

Before beginning this activity, students should be introduced to Scratch, either through direct instruction or pre-made videos. These tutorials should cover the basics of the Scratch environment, sprite movement, control using loops and events, and conditional statements and control. Scratch Stories is a cross-curricular activity in which students use computational thinking to write a story that has a beginning, middle, and end. Teachers can set additional requirements and constraints, such as a maximum number of characters, minimum number of scenes, or a central topic or theme. Scratch Stories allows learners to create their own narratives and connect to personal experiences or imaginary worlds through computing. In Fig. 9.3, a student explains their program to me, showing how their code choices affect the movement of the characters and the overall story.

Scratch Stories provides complete coverage of the computational thinking concepts, as students will use sequences, loops, events, parallelism, conditionals, operators, and data. They will also need to use the computational thinking practices to break down their story into pieces, work incrementally, and test and debug their story. Finally, these stories provide a medium through which students can express

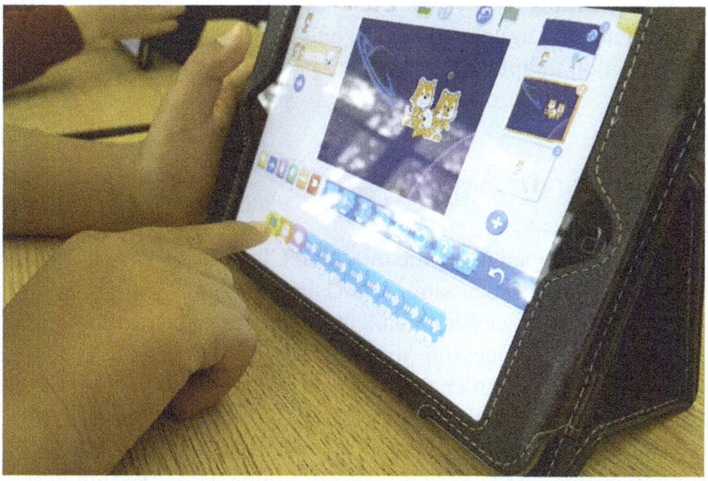

Fig. 9.3 Student Scratch Story example

themselves, connect with others, and question the role of technology in storytelling and other aspects of their lives. Scratch Stories can also be modified for younger or pre-reader students by using the app Scratch Jr. rather than the online version of Scratch. Scratch Jr. has many of the same features of Scratch, including backgrounds, characters, and programming blocks for movement, looks, sounds, and control including loops. The selection of blocks is more limited, and meanings for each block are conveyed through pictures rather than written words.

Scratch Stories provides a good environment for practicing the C-D-I-O steps of the new K-6 framework for computational thinking in a larger-scale project. First, students need to *conceive* what the story will be about: what is the setting, what will the characters need to be able to do? If the teacher has not given specific goals, the students can set some for themselves, whether it is a time limit (i.e., story need to be at least 1 min long), or simple enjoyment of the final product. In the *design* step, students should break down their story into scenes, and those scenes into actions for each character. They can model each scene from the point of view of each character; what do they need to do, in what order? Students can also consider different plot developments based on the blocks available, and use creativity to use new blocks they have not seen in the tutorials. This step also encourages students to reuse code, either from their previous Scratch projects or online Scratch projects from other Scratchers. The *implement* step of the process is where everything comes together: code for each scene, scene changes, costume changes, and timing. Testing and debugging will be necessary to make sure everything lines up properly (e.g., that characters wait their turns to talk in a conversation). As well, the implement step is where students can check to make sure their story can be "reset" properly. Since Scratch sprites do not return to their starting positions automatically when the code finishes running, that is something the students will need to build into their stories. Finally, students will *operate* their stories, and the stories of others, by ensuring that others can figure out how to make the story actions take place (by clicking the flag, clicking a sprite, hitting a key, etc.) and thinking of ways their stories can be improved or extended. Digital citizenship and privacy also play a role here, as students have the option to upload their project publicly to the Scratch website.

9.4.4 Processing Art

The final activity developed using the modified CDIO framework for K-6 computational thinking is called Processing Art. Using the open-source, Java-based programming language of processing, students will create interactive digital artwork. This is made simple in Processing thanks to its built-in functions for shapes and color. Processing has been shown to be a good language for creating simple simulations and models which can then be "hacked" by beginner programmers, such as creating a simulation which depicts flocking of birds and then allowing individuals to modify certain values in the code, changing their behavior (Sengupta & Shanahan, 2017). By creating artwork, learners create personal representations of computing and consider aesthetic and emotional ways of knowing (Farris & Sengupta, 2016; Wickman, 2017) that can then be translated into code.

Similar to the Scratch Stories activity, students should first be given an introduction to the Processing programming language, with a set of video tutorials showing commands for basic shapes and colors and use of the draw() function to make the artwork interactive. For example, a shape could change colors when it is clicked, or another shape could follow the mouse on the screen. This interactivity is mainly due to the nature of the draw() function, as it is run in a loop once the setup of the program is complete. Values in the loop can be updated on each iteration to change color shades or positions of shapes. As well, conditional statements can be incorporated by checking to see if the mouse has been clicked or a key has been pressed. To help facilitate the jump from block-based programming to text-based programming, students are encouraged to look at the various tutorials, examples, and reference sheets on the Processing website. While it is important to note that a text-based programming language may be too difficult for many elementary school children, it was chosen to give context to how the skills they have developed in prior activities apply to a more traditional programming language. As well, in the setting of a professional development workshop for teachers, it gives teachers practice with a tool that may be more complicated than those used in class, to help strengthen their confidence in their skills with Scratch and the previous activities, as well as connect to their previous notions of what coding is.

Making a piece of art using Processing requires the use of the C-D-I-O design steps. Students will *conceive* what their art will look like by getting a feel for the different shape and color functions available. They should also think about how the art will be "graded," or what their goals are: to use a certain number of different shapes? To have at least one interactive component? In this activity, it is also important that students think about the time limit for the activity; if they try to make a very complicated image as their first attempt, they may run out of time. Next, they need to *design* their artwork by planning each component. Grid paper is especially useful here for students to understand the grid coordinate system they will need to position their shapes properly and the size constraints of the computer screen. Modeling can help them note how each shape will overlap those drawn previously, and therefore plan the order of shapes to be drawn. The design step is also where they can make use of the various reference materials for Processing to get inspiration and reuse and remix code from other sources. The students can *implement* their artwork by coding one shape at time, running the code to make sure the shapes are in the right positions and are overlapping according to the design. They can compare their coded artwork with their original graph paper drawing to make sure the positions and orientations match. Finally, when they *operate* their artwork they should consider how it will be shared: online? As a printed copy? They will also need to make sure other users know how to make any interactive components in the artwork "work." Finally, they can brainstorm ways to improve their artwork, by using different shapes and colors, or by incorporating different interactive elements. One example of a work of Processing Art can be seen in Fig. 9.4. It is based on the popular children's book "The Very Hungry Caterpillar" by Eric Carle (1969), in which a hungry caterpillar

Fig. 9.4 Processing Art example from "The Very Hungry Caterpillar"

eats its way through many different foods. In the interactive processing art, when the screen is clicked, a hole appears in the orange (signifying that the caterpillar has eaten it, in line with the illustrations in the book).

9.5 Progression of Activities as Fading of Scaffolds: A Reflective Summary

One of Lye and Koh's (2014) most important recommendations for teaching computational thinking and coding in K-12 was to use scaffolding (Wood, Bruner, & Ross, 1976) in order to reduce frustration and promote learning and success. Scaffolding was defined by Wood et al. (1976) as having a teacher (or knowledgeable peer) assist novice students to complete tasks and solve problems that they would be unable to complete on their own, without assistance. Learners would require three different types of support in these activities, which are adapted from the ideas Reid, Zhang, and Chen (2003) proposed for support for scientific study. Interpretive support helps learners interpret and understand domain knowledge (such as computational thinking concepts), experimental support helps leaners set up and understand experiments (or in this case, design and carry out coding activities), and reflective support helps learners to reflect on their learning process (which may be led by the instructor in a class setting) (Reid et al., 2003).

Each computational thinking activity was broken down into small steps to make the overall task manageable, and the instructor provided support to the novice learners in order to facilitate their completion of the activities. For example, in *Scratch Stories*, the researcher shows a set of tutorials to teach the basics of Scratch, which get progressively harder and introduce more computational thinking concepts. The first tutorial may cover basics such as selecting a background and having one character move, the second tutorial would move to having two characters talking together with proper timing, and the final tutorial may introduce conditional statements to have characters perform certain actions when they bump into each other. Each of the activities also include demonstrations by the teacher/researcher for the activities, which often include either going through a worked example together as a class, or including tutorial videos which cover the necessary skills to create the project alongside completed examples which can be examined and modified by students. Finally, in both the program tutorials or project explanations, the instructor highlights critical features of the program and how each individual programming statement affects the program as a whole. For example, since novice programmers may have difficulty understanding the relationships between different commands (Lye & Koh, 2014), they may cause multiple characters to speak at the same time in their story. One of the tutorials explicitly addresses this issue by showing both the problem and the solution (inserting "wait" blocks between conversation lines, and writing out the conversation on paper first).

Scaffolding also takes place in terms of the computational thinking concepts used in each of the activities. The sequence of the computational thinking activities uses progressive "fading" of scaffolding (Basu, Sengupta, & Biswas, 2015; Guzdial, 1994; McNeill, Lizotte, Krajcik, & Marx, 2006; Sherin, Reiser, & Edelson, 2004; Stone, 1998) to gradually reduce the scaffolds provided to students in order to help them become independent in carrying out the tasks. For example, the coverage of computational thinking concepts seen in *Programming Puzzles* and *Teach a Robot to Dance* is more limited than the activities which use text-based or block-based programming languages, which is to be expected as they are more simple activities intended to provide an introduction to computational thinking. These initial concepts of sequences, loops, events, and conditionals are built upon in the following two activities. These concepts are not re-explained in the context of the new programming languages; rather, the direct instruction on these basic computational thinking concepts "fades" as students begin to use their conceptual understanding of sequences, loops events, and conditionals on their own, without prompting from the instructor, in order to complete the more complex tasks in *Scratch Stories* and *Processing Art*. They can also build upon that knowledge as they dive into new computational thinking concepts, such as parallelism and data. *Scratch Stories* covers all of the computational thinking concepts, practices, and perspectives, while *Processing Art* uses everything except parallelism.

Finally, the progression from activities without a digital device to those that use block- or text-based programming languages was specifically done in order to separate the computational thinking concepts from specific programming languages. For example, if students can understand the concept of conditional statements as they apply in everyday life (e.g., *if* it is raining, *then* I will need an umbrella), they can solidify that conceptual understanding and then have an easier time applying it in different programming languages, such as making a character say something when bumped in Scratch or having a shape change to a different color when it is clicked in Processing. This process can help students focus less on the particular syntax of a concept and think more about when and how it should be used in overall program design. As well, the movement from block-based to text-based programming languages is another example of the "fading" of scaffolding evident in these activities. Learners are originally presented with the block-based programming language to scaffold their learning and assist them in making the transition between basic understandings of computational thinking concepts and applying them to programing languages on the computer. However, once they have applied that knowledge and become comfortable with block-based programming, that scaffold fades and they are presented with the text-based programming language of Processing.

9.6 Discussion: CDIO and Computational Thinking Beyond Technocentrism

Papert believed that "children can use to learn computers in a masterful way, and that learning to use computers can change the way they learn everything else" (Papert, 1980, p. 8). At the heart of his constructionist perspective was the idea that the versatility and malleability of computational platforms such as LOGO could help learners build deep expertise across within and across disciplines such as science, mathematics, and languages. However, in the years following the uptake of LOGO and other educational computing technologies, Papert (1987) warned against the fallacy of technocentrism—the fallacy of referring all questions about technology to the technology itself. Sengupta et al. (2018) recently reminded us of this argument, warning that a technocentric approach to computational thinking and programming reduces human experience to technological production and ignores other aspects of the lived experiences of students. Our work presented here further advances this argument by illustrating a pedagogical approach based on the CDIO framework in ways that highlight phenomenological perspectives for integrating computational thinking in the elementary classroom.

The application of the Conceive-Design-Implement-Operate design process to elementary computational thinking makes space for the design of new computational thinking activities which challenge a technocentric approach to computational thinking and computing in the classroom. Moving beyond a focus on computing itself, learners can understand computational thinking through embodiment, narratives, and aesthetics. In the activities described in this chapter, coding becomes puzzles, dancing, stories, and art. Coding is framed as transdisciplinary, unconstrained by a focus on computational thinking concepts or a necessary connection to STEM. The framework can facilitate the design of transdisciplinary activities that incorporate aspects of the learners' lived experiences of coding, including their sense experiences. In educational settings, technical frameworks can sometimes be used as a way to constrain students' learning—they are expected to meet each requirement as if they are checking off a box. A technical framework used in this way will inhibit learners' possible experiences of computing. We can lose sight of the intersubjectivity that is inherent in complex and interpretive experiences that play central roles both in computing and in STEM. The proposed modified CDIO framework in this chapter is intended as a way to expand the ideas of what computational thinking activities could look like, going beyond the simple design and implementation of code to consider these broader and deeper forms of experiences as central to learning. The fading of scaffolding through the discussed activities can facilitate deeper understanding and help students and teachers transition from basic computational thinking concepts in device-less activities to using all computational thinking concepts in block-based, and eventually, text-based programming. However, this work expands beyond the computational languages themselves to include representational work that is meaningful to the learners, and thus coding exists in spaces outside of the confines of the technology. Future work will continue to leverage a phenomenological approach to computing in order to design and implement new, experiential coding opportunities for learners.

References

Arnot, M., James, M., Gray, J., Rudduck, J., & Duveen, G. (1998). *Recent research on gender and educational performance*. London: The Stationery Office.

Azevedo, F. S. (2018). An inquiry into the structure of situational interests. *Science Education, 102*(1), 108–127.

Basu, S., Sengupta, P., & Biswas, G. (2015). A scaffolding framework to support learning of emergent phenomena using multi-agent-based simulation environments. *Research in Science Education, 45*(2), 293–324. https://doi.org/10.1007/s11165-014-9424-z

Brennan, K., & Resnick, M. (2012). New frameworks for studying and assessing the development of computational thinking. *Annual American Educational Research Association Meeting*, Vancouver, BC, Canada, (2) (pp. 1–25). Retrieved from http://web.media.mit.edu/~kbrennan/files/Brennan_Resnick_AERA2012_CT.pdf.

Bussiere, P., Cartwright, F., & Knighton, T. (2004). *The performance of Canada's youth in Mathematics, Reading, Science and problem solving: 2003 first findings for Canadians aged 15*. Ottawa.

Carle, E. (1969). *The very hungry caterpillar*. Cleveland, NY: World Publishing Company.

CDIO. (2017a). *CDIO syllabus 2.0—Worldwide CDIO initiative*. Retrieved from http://cdio.org/benefits-cdio/cdio-syllabus/cdio-syllabus-topical-form

CDIO. (2017b). *CDIO vision—Worldwide CDIO initiative*. Retrieved from http://cdio.org/cdio-vision.

CDIO. (2017c). *Member schools—Worldwide CDIO initiative*. Retrieved from http://cdio.org/cdio-collaborators/school-profiles.

Crawley, E. F., Lucas, W. A., Malmqvist, J., & Brodeur, D. R. (2011). The CDIO syllabus v2.0 an updated statement of goals for engineering education. In *Proceedings of the 7th International CDIO conference*. Copenhagen: Technical University of Denmark.

Crawley, E. F., Malmqvist, J., Östlund, S., & Brodeur, D. R. (2014). *Rethinking engineering education* (2nd ed.). Berlin: Springer. https://doi.org/10.1007/978-3-319-05561-9

Danish, J. A. (2014). Applying an activity theory lens to designing instruction for learning about the structure, behavior, and function of a honeybee system. *Journal of the Learning Sciences, 23*(2), 100–148.

Farris, A. V., Dickes, A. C., & Sengupta, P. (2016). Development of disciplined interpretation using computational modeling in the elementary science classroom methods. In *Proceedings of the 12th International conference of the learning sciences*.

Farris, A. V., & Sengupta, P. (2016). Democratizing children's computation: Learning computational science as aesthetic experience. *Educational Theory, 66*(1–2), 279–296.

Guzdial, M. (1994). Software-realized scaffolding to facilitate programming for science learning. *Interactive Learning Environments, 4*(1), 1–44. https://doi.org/10.1080/1049482940040101

Guzdial, M. (2009). Teaching computing to everyone. *Communications of the ACM, 52*(5), 31–33. https://doi.org/10.1145/1506409.1506420

Hilton, A., & Bracy, A. (2015). *All of programming*. Ithaca, NY: Cornell University.

Horn, M., Leong, Z. A., & Block, F. (2012). Of BATs and APEs: An interactive tabletop game for natural history museums. In *Proceedings of the 2012 ACM annual conference on human factors in computing systems* (pp. 2059–2068). https://doi.org/10.1145/2207676.2208355

Hwang, S. W., & Roth, W. M. (2011). The (embodied) performance of physics concepts in lectures. *Research in Science Education, 41*(4), 461–477. https://doi.org/10.1007/s11165-010-9175-4

Kafai, Y. B., Fields, D., & Burke, Q. (2008). Entering the clubhouse: Case studies of young programmers joining the scratch community. *Journal of Organizational and End User Computing (JOEUC), 22*, 21–35. https://doi.org/10.4018/joeuc.2010101906

Lee, I., Martin, F., Denner, J., Coulter, B., Allan, W., Erickson, J., … Werner, L. (2011). Computational thinking for youth in practice. *ACM Inroads, 2*(1), 32–37.

Lee, V. R., & Wilkerson, M. (2018). *Data use by middle and secondary students in the digital age: A status report and future prospects*. Washington, DC.

Lehrer, R. (2009). Designing to develop disciplinary dispositions: Modeling natural systems. *American Psychologist, 64*(8), 759.

Lye, S. Y., & Koh, J. H. L. (2014). Review on teaching and learning of computational thinking through programming: What is next for K-12? *Computers in Human Behavior, 41*, 51–61. https://doi.org/10.1016/j.chb.2014.09.012

Marasco, E. (2013). *Development of a CDIO-based creative cross-disciplinary curriculum and associated projects for elementary engineering education*. Calgary, AB: University of Calgary.

Marasco, E., & Behjat, L. (2013). Integrating creativity into elementary electrical engineering education using CDIO and project-based learning. In *2013 IEEE International Conference on Microelectronic Systems Education, MSE 2013* (pp. 44–47). https://doi.org/10.1109/MSE.2013.6566701

McNeill, K. L., Lizotte, D. J., Krajcik, J., & Marx, R. W. (2006). Supporting students' construction of scientific explanations by fading scaffolds in instructional materials. *Journal of the Learning Sciences, 15*(2), 153–191. https://doi.org/10.1207/s15327809jls1502_1

Meerbaum-Salant, O., Armoni, M., & Ben-Ari, M. (2013). Learning computer science concepts with scratch. *Computer Science Education, 233*(03), 239–264. https://doi.org/10.1080/08993408.2013.832022

Merleau-Ponty, M. (1962). *Phenomenology of perception*. New York: Routledge.

Papert, S. (1980). *Mindstorms: Children, computers, and powerful ideas*. New York, NY: Basic Books.

Papert, S. (1987). Computer criticism vs. technocentric thinking. *Educational Researcher, 16*(1), 22–30. https://doi.org/10.2307/1174251

Pickering, A. (1995). *The mangle of practice: Time, agency, and science*. Chicago, IL: University of Chicago Press.

Processing. (2017). Processing.org.

Reid, D. J., Zhang, J., & Chen, Q. (2003). Supporting scientific discovery learning in a simulation environment. *Journal of Computer Assisted Learning, 19*(1), 9–20.

Sáez-López, J.-M., Román-González, M., & Vázquez-Cano, E. (2016). Visual programming languages integrated across the curriculum in elementary school: A two year case study using "scratch" in five schools. *Computers & Education, 97*, 129. https://doi.org/10.1016/j.compedu.2016.03.003

Schmidt, D. C. (2006). Model-driven engineering. *Computer, 39*(2), 25–31.

Sengupta, P., & Farris, A. V. (2012). Learning Kinematics in elementary grades using agent-based Computational modeling: A Visual programming-based approach. In *IDC'12 Proceedings of the 11th International conference on Interaction Design and Children*. (pp. 78–87).

Sengupta, P., Dickes, A. C., & Farris, A. V. (2018). Toward a phenomenology of computational thinking in STEM education. In *Computational thinking in STEM: Foundations and research highlights*, (January) (pp. 49–72). Retrieved from http://doi.org/arXiv:1801.09258v1.

Sengupta, P., Dickes, A. C., Farris, A. V., Karan, A., Martin, D., & Wright, M. (2015). Programming in K-12 science classrooms. *Communications of the ACM, 58*(11), 33–35. https://doi.org/10.1145/2822517

Sengupta, P., Kinnebrew, J. S., Basu, S., Biswas, G., & Clark, D. (2013). Integrating computational thinking with K-12 science education using agent-based computation: A theoretical framework. *Education and Information Technologies, 18*(2), 351–380. https://doi.org/10.1007/s10639-012-9240-x

Sengupta, P., & Shanahan, M.-C. (2017). Boundary play and pivots in public computation: New directions in STEM education. *International Journal of Engineering Education, 33*(3), 1124–1134.

Sherin, B., Reiser, B. J., & Edelson, D. (2004). Scaffolding analysis: Extending the scaffolding metaphor to learning artifacts. *Journal of the Learning Sciences, 13*(3), 387–421. https://doi.org/10.1207/s15327809jls1303_5

Stone, C. A. (1998). The metaphor of scaffolding: Its utility for the field of learning disabilities. *Journal of Learning Disabilities, 31*(4), 344–364.

Traff, P.-A., Wedel, M. K., Gustafsson, G., & Malmqvist, J. (2007). To rescue eggs; a design-build-test experience for children. In *Proceedings of the 3rd International CDIO conference*. Cambridge.

Vaca Cárdenas, L., Bertacchini, F., Tavernise, A., Gabriele, L., Pantano, P., Valenti, A., … Bilotta, E. (2015). *Coding with Scratch: The design of an educational setting for Elementary pre-service teachers*, (September) (pp. 1171–1177). Retrieved from http://doi.org/978-1-4799-8706-1.

Verner, I. M. (2015). Technology teacher education and outreach using the CDIO approach. In *Proceedings of the 11th International CDIO conference*. Chengdu, Sichuan, China.

Weintrop, D., Beheshti, E., Horn, M., Orton, K., Jona, K., Trouille, L., & Wilensky, U. (2016). Defining computational thinking for mathematics and science classrooms. *Journal of Science Education and Technology, 25*(1), 127–147. https://doi.org/10.1007/s10956-015-9581-5

Wickman, P.-O. (2017). Back to the drawing board: Examining the philosophical foundations of educational research on aesthetics and emotions. In A. Bellocchi, C. Quigley, & K. Otrel-Cass (Eds.), *Exploring emotions, aesthetics and wellbeing in science education research* (Vol. 13, pp. 9–37). Dordrecht: Springer. https://doi.org/10.1007/978-3-319-43353-0

Wilensky, U. (1999). *NetLogo (and NetLogo user manual). Center for Connected Learning and Computer-Based Modeling*. Evanston, IL: Northwestern University.

Wilkerson-Jerde, M., Wagh, A., & Wilensky, U. (2015). Balancing curricular and pedagogical needs in computational construction kits: Lessons from the DeltaTick project. *Science Education, 99*(3), 465–499. https://doi.org/10.1002/sce.21157

Wing, J. M. (2006). Computational thinking. *Communications of the ACM, 49*(3), 33–35. https://doi.org/10.1145/1118178.1118215

Wood, D., Bruner, J. S., & Ross, G. (1976). The role of tutoring in problem solving. *Journal of Child Psychology and Psychiatry, 17*(2), 89–100. https://doi.org/10.1111/j.1469-7610.1976.tb00381.x

Chapter 10
Playfully Coding Science: Views from Preservice Science Teacher Education

Pratim Sengupta, Beaumie Kim, and Marie-Claire Shanahan

Abstract There is now a growing body of research focused on integrating computational thinking and modeling in teacher education, ranging from studies that investigate preservice teachers' perceptions of computational thinking to those that evaluate the efficacy of computational tools that can support such integration. Our work extends this literature by investigating how preservice science teachers can be introduced to computational thinking and modeling by playfully designing computer simulations and games for modeling kinematics and ecological interdependence. Adopting a phenomenological research agenda, we focus on how preservice science teachers experience coding and computational modeling as pedagogical experiences for science education. In doing so, our goal is to contribute to an epistemological, rather than an instrumental, understanding of computational thinking and modeling in the context of preservice science teacher education.

Keywords Computational thinking · Play · Phenomenology · Science education · Modeling · Teacher education

10.1 In.troduction

Over the past several years, computational thinking (Wing, 2006) has emerged as one of the centerpieces in K-12 STEM education. Computational thinking has been commonly positioned by scholars as involving analytical skills that draws on concepts and practices from computer science, as well as a more fundamental ability that can be used by and useful for all people (Wing, 2006). Computational thinking in Wing's (2006) words involves "... breaking down a difficult problem into more familiar ones

P. Sengupta (✉) · B. Kim · M.-C. Shanahan
Werklund School of Education, University of Calgary, Calgary, AB, Canada
e-mail: Pratim.sengupta@ucalgary.ca

P. Sengupta et al. (eds.), *Critical, Transdisciplinary and Embodied Approaches in STEM Education*, Advances in STEM Education,
https://doi.org/10.1007/978-3-030-29489-2_10

that we can solve (problem decomposition), using a set of rules to find solutions (algorithms), and using abstractions to generalize those solutions to similar problems" (p. 33). In the context of science and STEM education, computational thinking must be thought of contextually in light of disciplinary practices, which involves developing epistemic and representational practices such as thinking algorithmically, use of data structures and other relevant forms of computational abstractions, and designing and creating computational artifacts such as programs and simulations (Sengupta, Kinnebrew, Basu, Biswas, & Clark, 2013; Weintrop et al., 2016).

There is now a growing body of research focused on integrating computational thinking and modeling in teacher education, ranging from studies that investigate preservice teachers' perceptions of computational thinking (Bower & Falkner, 2015; Sands, Yadav, & Good, 2018) to those that evaluate the efficacy of computational tools that can support such integration (Kalogiannakis & Papadakis, 2017). Our work seeks to extend this body of work by investigating how preservice science teachers can be introduced to computational thinking and modeling by playfully designing computer simulations and games for modeling kinematics and ecological interdependence.

However, following Sengupta, Dickes & Farris (2018), rather than adopting a technocentric (Papert, 1987) approach, where the primary (and often the sole) emphasis remains on evaluating computational artifacts generated by participants to assess how they have used and applied computational abstractions, our approach is phenomenological in nature. In a phenomenological approach participants' *sense experience* becomes objects of inquiry (Sengupta, Dickes & Farris, 2018). Merleau-Ponty (1962) defined sense experience as a dynamic and dialectical form of experience: "that vital communication with the world which makes it present as a familiar setting of our life" (Merleau-Ponty, 1962, p. 61). The emphasis on understanding participants' sense experience must necessarily go beyond the sphere of givenness—i.e., the world as it is already known—and reveal participants' originary sense-making (Merleau-Ponty, 1962; McMahon, 2017). In the present context, this means that our focus is not on assessing the efficacy of our pedagogical approach in terms of helping preservice teachers apply computational abstractions re-contextualized as computational models of scientific phenomena. Instead, we are interested in preservice science teachers' originary sense-making of coding and computational modeling as pedagogical experiences for doing and learning science. In doing so, our goal is to contribute to an epistemological, rather than an instrumental, understanding of computational thinking and modeling in the context of preservice science teacher education.

10.2 Research Question

Specifically, we ask the following research question: how do preservice science teachers view computing and coding as pedagogical experiences in the context of doing science through playfully engaging in computational modeling and game design?

10.3 Background

10.3.1 Productive Uncertainty and Play in Science Education

Studies of scientists at work reveal an image that is far from being one of certitude, despite the latter being the more commonly represented image of science in public education (Duschl, 2008). For example, Pickering (1995) illustrated how scientific advancement necessitates a deep entanglement of theories and materiality, and of conceptual and representational work, thereby rendering a far more nuanced character than what is commonly represented in the public imagination. Ochs, Gonzales, and Jacoby (1996) highlighted the central role of interpretive work in creating scientific knowledge, and illustrated how this interpretive uncertainty is also tied to the representational infrastructure. This is echoed by Daston and Galison (2007), who pointed out that as representational technologies evolve and new ones emerge in order to support scientific advancement, their use, on the other hand, often results in new forms of uncertainty and interpretive work.

There is a growing recognition among science educators that an essential aspect of the teachers' work is developing a more nuanced view of scientific uncertainty and supporting students in such nuanced scientific inquiries. Aikenhead (2003) identified that grappling with the feeling of "playing in the subculture of science" both as insider and outsider is essential to humanistic pedagogies. Manz and Suárez (2018) proposed some strategies that can promote such pedagogies, such as beginning with complex phenomena, iterating on investigations, and leveraging variability in students' ways of conducting investigations. Similarly, Farris, Dickes, and Sengupta (2019) argued that when teachers pay attention to students' errors and uncertainties during the process of designing computational models in science classrooms, they can support students to deepen their engagement with scientific practices.

We believe that positioning computational modeling as playful engagement with science and computing can also support preservice teachers' engagement in such experiences that value, rather than ignore, interpretive uncertainties. Playful learning environments can greatly facilitate learning of complex topics across a range of STEM disciplines (Berland & Lee, 2011; Sengupta, Krinks, & Clark, 2015; Kim & Ho, 2018; Sengupta & Shanahan, 2017). The notion of play challenges the expectations of disciplinary rigidity that often keep newcomers from participating in scientific inquiry (Sengupta & Shanahan, 2017). Playful engagement with virtual learning environments can help learners reshape the learning activities even within a structured setting, such that the activities are both personally meaningful and relevant to the disciplinary context of learning (Farris & Sengupta, 2016; Kim & Ho, 2018). As Kim and Ho (2018) and Sengupta and Shanahan (2017) pointed out, central to positioning personal meaningfulness alongside disciplinary relevance is the harnessing, rather than discarding, of possibilities that emerge from interpretive flexibilities and uncertainties. Participants' dilemmas and uncertainties, we therefore believe, can become resources in playful engagement with STEM disciplines.

10.3.2 Computational Thinking and Modeling in Teacher Education

Several scholars have argued for democratizing computation by integrating computing with other disciplines and existing courses (such as science and math) that all children are required to take, rather than trying to create room for computer science as a new curricular domain (Sengupta et al., 2015; Wilensky, Brady, & Horn, 2014). We posit that the same arguments must be extended for preservice teachers, especially given that many of them may not have prior experience in computational modeling and programming. Rather than learning computer programming as a separate discipline, as Yadav, Stephenson and Hong (2017) also argued, we believe that in their science-methods courses, preservice teachers could be introduced to computational thinking through computational models. Such experiences, we believe, can help them deepen their (future) students' engagement with conceptual and representational practices that are central to the development of both scientific and computational expertise in a reflexive manner (Sengupta et al., 2013).

A growing body of literature advocates engaging preservice and in-service science teachers in computational thinking and modeling through positioning them as creators of computational models and artifacts. For example, Wilkerson et al. (2016) found that when preservice teachers are provided opportunities for constructing simulations in science, they are able to engage in practices that are central to scientific modeling, such as model evaluation and revision, in ways that are deeply connected to key conceptual ideas relevant to the phenomenon being modeled. Leonard et al. (2018) also found that culturally and contextually embedded game design activities and robotics can support teachers in developing dispositions central to computational thinking. Our work seeks to contribute to this literature by offering insights into how preservice teachers frame (and re-frame) code and coding from a pedagogical perspective in the context of scientific modeling, as part of their teacher preparation coursework.

10.4 Our Pedagogical Approach: Integrating Playfulness and Mathematization to Support Preservice Science Teachers' Computational Work

Our work is premised on the position that engaging preservice teachers in computational modeling can be supported by emphasizing playfulness in their interactions with computational artifacts. This is particularly important given that many preservice science teachers may not have prior experience with programming (e.g., Yadav et al., 2017). Positioning computational and scientific work as play would allow teachers, regardless of their prior background, to interact with computing in their *regimes of competence* (diSessa, 2001). diSessa (2001) reminded us "that resources for learning don't always look just like the product of learning" (p. 84) from the perspective afforded by the regime of competence. This means that learners (broadly speaking, and including preservice teachers in this case) may begin from a place that may not be initially recognizable as the putative discipline to be learnt. Instead, the regime of competence—e.g., practices and hobbies that the learners may already be interested in outside the discipline—may offer learners productive resources, using which they can develop a deep and meaningful relationship with the discipline (Azevedo, 2018).

Our previous work with in-service science teachers offers some insights into what such regimes of competence might look like (for science teachers). For example, we found that framing programming as "mathematizing" in the science classroom can serve as a productive pedagogical approach for integrating programming in the K-12 science classroom (Sengupta et al., 2013, 2015; Sengupta, Brown, Rushton, & Shanahan, 2018; Dickes, Sengupta, Farris, & Basu, 2016; Farris, Dickes & Sengupta, 2019). In this approach, programming is used in the context of creating computational models of scientific phenomena through designing discrete mathematical representations of units of change, for representing change over time. Similarly, Sands et al. (2018) found that all teachers in their study—both primary and secondary—viewed mathematical work in the classroom as a form of computational thinking. This is also not surprising given the heavy emphasis on mathematical work within K-12 science curricula, as we found in our work with K-12 teachers both in the USA and Canada (Sengupta, Brown, Rushton, & Shanahan, 2018; Farris, Dickes & Sengupta, 2019).

Such a reframing of computing as *mathematizing* is in line with a phenomenological approach in which computational thinking is viewed not as merely a set of prescribed performances of technological and symbolic dexterity, rather as more complex and heterogeneous forms of experience (Sengupta, Dickes & Farris, 2018). While at first glance, it might seem counterintuitive to claim that an emphasis on mathematizing may position teachers and preservice teachers in regimes of competence, Farris, Dickes and Sengupta (2019) found that when teachers and students amplify (rather than ignore) the interpretive dilemmas and uncertainties involved in (and inherent in) mathematizing code for designing scientific models, their work can become progressively more creative and at the same time, computationally more intensive.

With this in mind, we designed a set of learning activities in which preservice teachers were presented with two computational simulations designed in the NetLogo Web platform: Lunar Lander (Fig. 10.1), and Bird-Butterfly-Flower Ecosystem (Fig. 10.2). In both the simulations, the activities involved not only manipulating parameters and variables that controlled the simulation, but also modifying the underlying NetLogo code. Modeling in NetLogo involves instantiating the individual elements of a system - the "agents" - and simulating their interactions using NetLogo code. A particular affordance of NetLogo code is that it has been shown to be effective in supporting science learners and newcomers to computing engage deeply with computational modeling, by drawing upon their intuitive and embodied knowledge (Wilensky & Reisman, 2006). In addition, agent-based modeling and programming is fundamentally aligned with mathematization because the

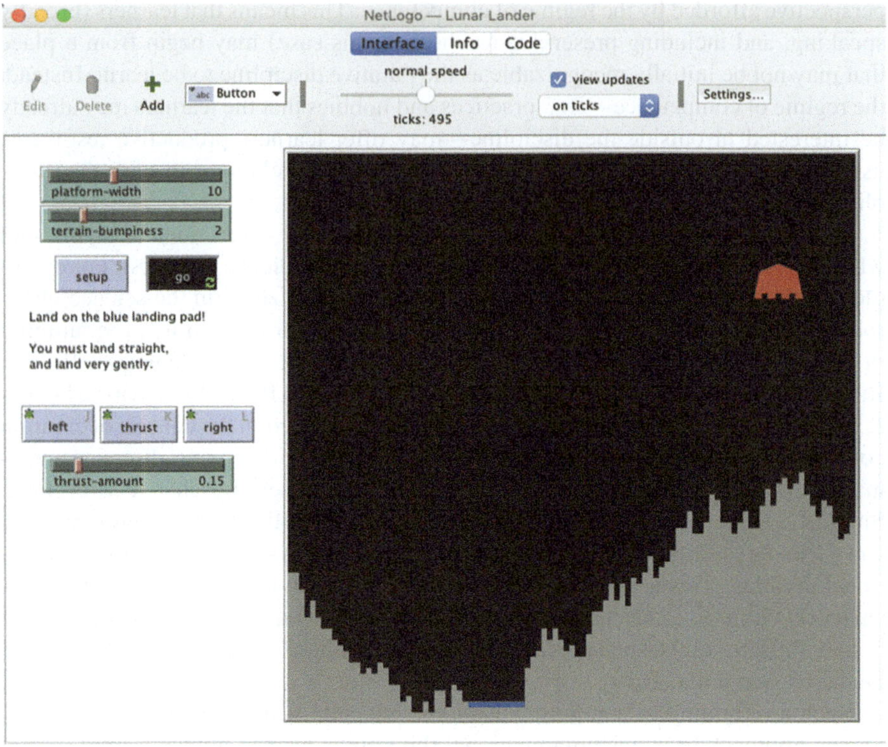

Fig. 10.1 A screenshot of the Lunar Lander NetLogo simulation (Game)

activity of programming the behavior of agents requires the learners to define an event using discrete mathematical measures (Sengupta et al. 2015).

In the Lunar Lander simulation, the goal of the "player" was to land the spaceship safely on the lunar surface. This meant that the ship had to land at very low speed, and vertically. Controlling the trajectory of the ship involved adding thrust (sudden bursts of acceleration) along any of the four directions: top, down, left, or right. Based on the classic premise of early moon landing games (e.g., Atari's 1979 Arcade Game *Lunar Lander*), we used a modified version of the Lunar Lander simulation in the NetLogo Models Library (Wilensky, 1999). The NetLogo simulation is designed as a game and we modified it such that it would be very difficult for players to land successfully. We did so by modifying the underlying NetLogo code so that the length of the landing strip on the lunar surface was nearly equal to the width of the ship. That is, in trying to land, the ship would inevitably crash due to hitting rocks adjacent to the landing strip. The framing of the simulation as a game and their activity as game hacking, we posited, would encourage our participants to dig deeper into the code.

The Bird-Butterfly-Flower simulation (Dickes & Sengupta, 2013) modeled predator–prey dynamics in an ecosystem of flowers, butterflies, and birds. The overarch-

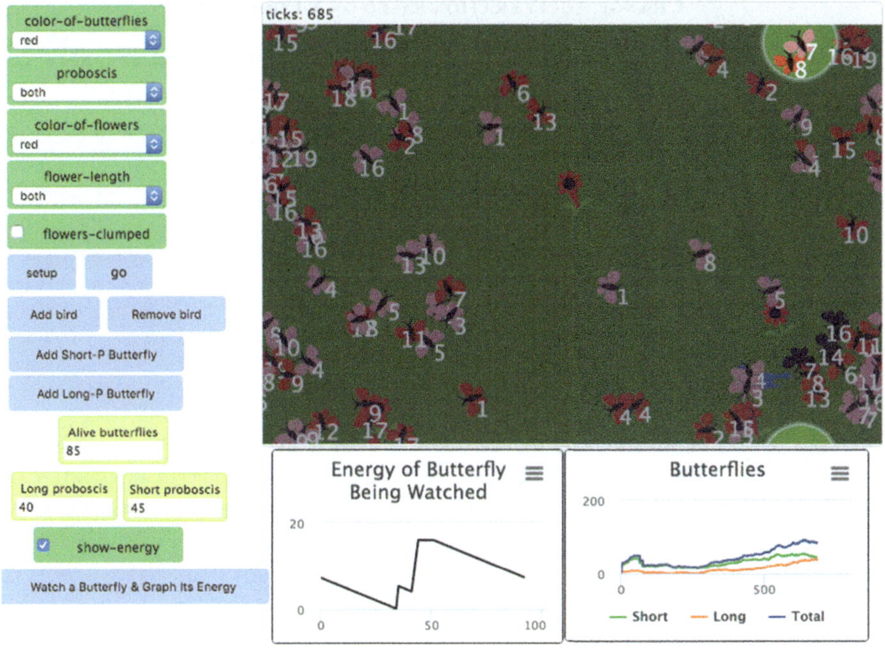

Fig. 10.2 A screenshot of the Bird-Butterfly-Flower simulation

ing goal of this activity was to identify the model parameters that resulted in a thriving butterfly population, manipulating variables such as proboscis length, flower length, flower location, color of flower, color of butterfly, and speed of predator movement. In addition, similar to the Lunar Lander simulation, participants also had the opportunity to significantly expand the scope of the simulation by introducing additional predators, altering structure–function balance by changing the morphological characteristics of the agents, etc.

In order to encourage and scaffold preservice teachers to “hack” and “debug” the NetLogo code, we commented the code heavily and provided a printed programming guide (see Figs. 10.3 and 10.4). The comments within the NetLogo code explained how each line of code affected the simulation. The printed programming guide explained how relevant code segments represented the science concepts and mathematical relationships, as well as how to change relevant code segments in order to change different elements of the simulation (e.g., changing the gravity on the moon, the size of the landing strip, acceleration and deceleration due to the thrusters). In class, we also encouraged the students to dig into the code as part of the “playful” experience and answered their questions to help them along their journey.

Change Acceleration based on Y coordinate

```
[ setxy (xcor + xvel) (ycor + yvel) set size 1 stamp set size 10]
;; exert the force of gravity
set yvel yvel - 0.001
;; detect crashes and insufficiently soft landings
if [pcolor] of patch-at 0 -2 != black ;or pcolor = yellow
```

The goal here is to tie the ship's speed or acceleration to its y coordinates.

The line of code that we are looking for here is:

set yvel yvel - 0.001

This can be changed to:

set yvel (yvel - 0.001) - (ycor * 0.0001)

What does this change do? Every step, the new y-velocity, which is specified by the "variable" **y-vel**, is equal to the old y-velocity minus the y coordinate multiplied by 0.0001. This number was based on playing around with the code; it is quite likely that you may want to try out other options for better outcomes.

Fig. 10.3 Excerpt from the Lunar Lander Activity Guide

How Do Butterflies Avoid Birds?

Here's the relevant code:

```
to avoid-birds
 if count birds in-radius 3 > 0
 [
 set heading towards one-of birds
 rt 180
 fd 0.4
]
end
```

Each butterfly looks for birds within a radius of 3 units.

If they see a bird, they first face the bird (set heading towards one-of birds), then turn away from the birds (rt 180), and them move away from them by a distance of 0.4 units (fd 0,4).

You can change the angle of turning and/or the distance by which the butterfly moves away from the bird. What do you think will happen as a result of your change to the overall population?

Fig. 10.4 Excerpt from the Bird Butterfly Flower Simulation Activity Guide

10.5 Method

10.5.1 Setting and Participants

We conducted our study in a secondary science education methods course in a Canadian research-intensive university. The third author of the paper was the instructor of the course, and all three authors collaboratively designed the computational artifacts and programming guides used in the study. All three authors were present during Day 1, whereas the first two authors were present during Day 2. There were 27 students in the class, a majority of whom participated in two intensive sessions on computational modeling. All the students were enrolled in a two-year, after-degree B.Ed program with a specialization in secondary science. All students held a Bachelors or Masters degree (or beyond) in a scientific or engineering discipline, and several of them had professional careers in their previous specializations.

10.5.2 Data and Analysis

The data for this study comes from two days of classroom activities, totalling to approximately 6 h of class time. During Day 1, participants worked on the Lunar Lander simulation and game, while on Day 2 they worked on Bird-Butterfly-Flower simulation. During each day, the participants were introduced to key elements of the underlying code of the relevant simulation or game by one of the researchers. Participants worked in groups of two or three throughout the duration of the study. The researchers also visited each group to work with them as needed both on their computational modeling and their pedagogical focus. They also led class discussions in which participants shared reflections on their experiences of the classroom activities.

Overall, the activities were framed as playful pedagogical exercises. That is, the participants were first asked to take on the perspective of their future students by engaging with the computational models and games following the programming and activity guide. They were specifically encouraged to discuss with their partners the challenges and successes that would emerge in their interactions. They were then asked to redesign the simulation or game in order to deepen their students' engagement and/or to address the challenges they experienced initially.

We collected three forms of data: (a) students' written memos on their experiences relevant to the course, (b) computational artifacts designed by students, (c) video and audio recordings of classroom conversations and interviews with the participants. These interviews were conducted while the researchers were working with the participants in small groups. During the interviews, we asked the participants to explain their challenges and how they were planning on addressing them, as well as how and why they would further modify the underlying NetLogo code.

We conducted thematic analysis using the check coding method (Miles & Huberman, 1994) to analyze the interview data, classroom conversations, and student artifacts. The class discussions provided us with initial insights into emergent *themes*, i.e., views that were shared by several groups of participants. We then analyzed interviews with specific students and small groups in order to develop a more detailed understanding of the participants' experiences referred to during the class discussions. We iteratively compared our emergent observations as evident from both these data sets, and identified three salient themes that were experienced broadly by the class. All the authors collaboratively discussed and identified these themes.

10.6 Findings

Our analysis identified three themes along which preservice teachers' viewed code and coding as pedagogical objects and experiences for deepening scientific inquiry.

10.6.1 Theme 1: Interacting with Code Can Deepen Conceptual Engagement in Science

In order to understand preservice teachers' understandings of how coding can support scientific inquiry, in the first illustrative case, we focus primarily on Adela and Jerry, who participated in all the modeling activities reported in this chapter. Adela has a Master's degree in biology, and Jerry had recently completed a doctorate in astrophysics. Neither of them identified themselves as coders or had taken any computer science course, although both of them had some prior experience with programming. In the excerpt below, we illustrate how Adela's and Jerry's experiences with playful coding shaped how they saw coding in their (future) science classrooms as a pedagogical tool.

In Excerpt 1, Adela and Jerry are explaining to Pratim one of the changes they were thinking of making to the NetLogo code in order to make it easier to land the spaceship. They were discussing the possibility of changing the code so that as the ship used the thruster, it would also become lighter. Jerry immediately recognized this as senior level undergraduate physics, because this uses the notion of differential mass. Adela, on the other hand, was not convinced that it would be difficult for students to understand the idea, because for her, the notion of differential mass, in this context, involved "the concept of burning something a.. and losing mass with it" (Turn 3). Jerry argued that the students would still not be familiar with the formal mathematics involved, but Adela argued that "as a teacher and you would provide them the code you wouldn't expect them to come up with it but you could have them to explore the results of it" (Turn 6).

Excerpt 1: Lunar Lander

Turn 1: Adela**:** What we were thinking of is the laws of thermodynamics, so conservation of mass, so as you're using your thruster your ship will get lighter

Turn 2: Jerry**:** That's a very crazy 4 year physics, with differential mass, yeah I think that's well beyond

Turn 3: Adela**:** The concept of burning something a.. and losing mass with it?

Turn 4: Jerry**:** Yeah you can talk about it but mathematically speaking they will not have [the understanding]

Turn 5: Pratim: Yeah but I think what she's saying is that this would actually make that understandable

Turn 6: Adela**:** Right so as a teacher and you would provide them the code, you wouldn't expect them to come up with it but you could have them to explore the results of it

Turn 7: Dorothy)**:** If you set up the code then they can manipulate it

Turn 8: Adela**:** Yeah yeah exactly

This excerpt is insightful in two senses. First, Adela's explanation here is indicative of the framing of code itself as a pedagogical object—i.e., as an object that can be manipulated by the students for deepening their conceptual engagement with science. This is also supported by another preservice teacher in the class (Dorothy), who agreed that code can be "set up" in such a way so that it can be manipulated by the students (Turn 7). Second, as Pratim interpreted and re-articulated Adela's comments (Turn 5), another important implication is that through interacting with the code, secondary students, by using their intuitive understanding of "burning something" and "losing mass," can begin to explore conceptual issues that are typically reserved for upper undergraduate physics.

10.6.2 Theme 2: Coding for Scaffolding as a Form of Productive Uncertainty

The idea of "setting up" the code, as Dorothy put it in Excerpt 1 (Turn 7)—i.e., designing code in order to pedagogically support particular forms of student interactions—deserves further unpacking, because it can bring to light a form of productive uncertainty experienced by the participants during their own playful engagement with the simulation and the code. We found that several participants realized the need to modify the code not only for deepening students' conceptual engagement with science, but also to scaffold their experience of play. In the process, they experienced dilemmas regarding whether scaffolding their students would help or limit their scientific inquiry. We see this dilemma as a form of productive uncertainty that in turn deepened their own understanding of the relationship between coding and pedagogical design in the science classroom. The following excerpt (Excerpt 2), which reports a conversation between Marie-Claire and two participants, Ronnie and Negin, provides a rich illustration:

Excerpt 2

Turn 1: Ronnie: So like if it tell you the cause of death then you can adjust to your play

Turn 2: Marie-Claire: Right, so if it tells you more specific feedback

Turn 3: Ronnie: Right exactly - it took us a few tries to realize that we were going too fast

Turn 4: Marie-Claire: So it wasn't just about position whereas it might be different for someone else who is slightly off position

Turn 5: Ronnie: Yes exactly

Turn 6: Marie-Claire: Cool, yeah that makes sense -Did you manage to land?

Turn 7: Negin: Once

Turn 8: Marie-Claire: So if you wanted to change it to add feedback- would you give categories of feedback like - "you died due to speed, You died due to" and sort of have

Turn 9: Ronnie: Yeah, what is the alternative?

Turn 10: Marie-Claire: I'm not sure. How would you envision that feedback?

Turn 11: Ronnie: Directly that you died due to the velocity, was too fast, the speed was too fast

Turn 12: Negin: That's very straightforward though- you could maybe show this guy falling, and he falls off the track you could show him falling off the track and rolling over or blowing up if it was too fast.

Turn 13: Negin: Or maybe some warning

Turn 14: Marie-Claire: -Beep beep beep beep

Turn 15: Ronnie: Yeah some warning that you are going to fast - so before you die you have the chance to save yourself

Turn 16: Marie-Claire: Or abort mission - eject eject - Haha - but I assume you do know what you want to change

Turn 17: Ronnie: Yep, here it was very vague - I was like what are we doing, why did we die?

Turn 18: Negin: Yep and then I increased the platform here, and I still died, so I was like there is something going on here

Turn 19: Marie-Claire: Right - so it's not alignment, and it's not

Turn 20: Ronnie: But there is a good thing in not giving any feedback for learning because then it makes it more inquiry - when we did figure it out that it was due to velocity - there is a good aspect of not having any warning

Turn 21: Marie-Claire: Yeah - I was going to ask that - do you think there is something - is there something for students to learn in that process of figuring out why - why did I die

Turn 22: Ronnie: Yeah, because we eventually figured it out - because it was too fast right because we were concerned because it landed right perfectly on the blue line - so we eliminated that

Turn 23: Marie-Claire: Other elements

Turn 24: Ronnie: Yeah, so it's not totally bad that there is no - but at the same time I don't know….

Turn 25: Marie-Claire: I wonder if it depends on what you want them to get out of it

Turn 26: Ronnie: Yeah - maybe you can have the option you can play it without any feedback - and first play it like that and figure out why you died - and then you can switch it to a different mode

At the beginning of the excerpt, the participants explained to Marie-Claire that they were thinking about changing the code so that the player (or student) would get a visual feedback that would specify and explain the cause of the crash of the spaceship in the Lunar Lander simulation. Ronnie mentioned that it took her and Negin (her partner) a few attempts to figure out that they were crashing because the speed of the spaceship was too high (Turn 3). Negin explained that they arrived at higher speed as being the cause of the crash only after they made alterations to the code to "increase the landing"—i.e., to flatten the landing surface (see Fig. 10.5a, b), so that it would be easier for the ship to land, and then they realized that doing so still did not solve their problem (Turn 18). At this point, Ronnie pointed out an advantage of not receiving feedback from the system (simulation) earlier regarding their crashes: it made them inquire more deeply into the issue. They started thinking about multiple factors that could be responsible for the crash (Turn 20) in a systematic manner: "Yeah, because we eventually figured it out - because it was too fast right because we were concerned because it landed right perfectly on the blue line - so we eliminated that" (Turn 22). Ronnie further commented that not receiving feedback can be helpful for learning, but at the same time, she was unsure (Turn 24).

We find this uncertainty to be productive along two dimensions. First, from a pedagogical perspective, the lack of feedback, as both Ronnie and Negin realized through their own experiences, could also be an opportunity for students to dig deeper into the code as well as conceptual issues in physics. Second, from an epistemological

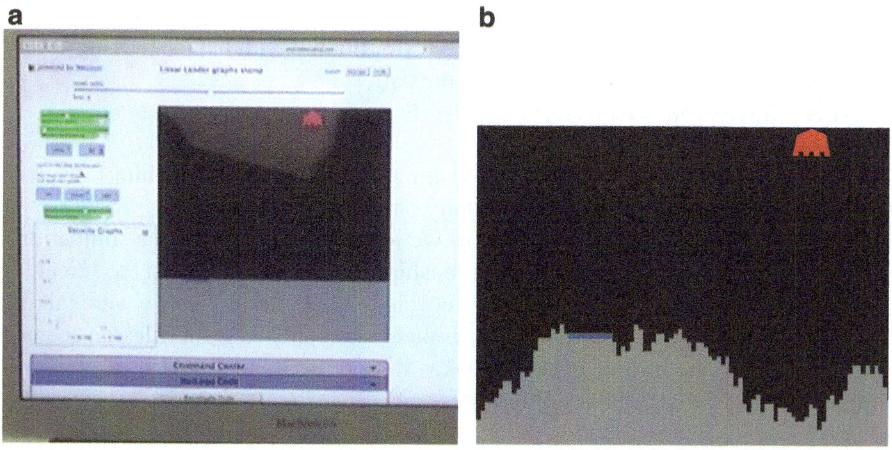

Fig. 10.5 (**a**) (left): Ronnie and Negin's altered simulation showing a flat lunar landing surface; (**b**) (right): The original simulation as provided to the students had a rocky terrain as the lunar landing surface

perspective, for both Ronnie and Negin, in their roles as preservice teachers, this allowed them to view code and coding as pedagogical objects that can deepen their students' scientific inquiry. In this view, coding was not merely a skill to be learnt, rather a means to further student engagement and scientific inquiry.

10.6.3 Theme 3: Coding to Deepen Students' Personal and Playful Engagement with Science

The following excerpt (Excerpt 3) reports a classroom-wide discussion led by Beaumie and Pratim, in which they asked the class to discuss the changes to the code and the simulation that they had considered in order to engage their future students, especially those who may not be interested in the topic. The context of this conversation was the Bird-Butterfly-Flower simulation, which, along with the printed programming guide, provided students opportunities to manipulate the variables in the simulation and the underlying code.

Excerpt 3

Turn 1: Beaumie: So I was wondering actually some of you talked about how, because it was a game and you can think about different colours, there is a different entry point for kids - is there anything that you would change or do something with this other simulation that would provide a different entry point for kids who are not interested in biology?

Turn 2: Adela: One thing after you said lunar lander is made to be able to pick a single butterfly because it would be too messy otherwise and have the trail of the butterfly so you could see its "behaviour"

Turn 3: Pratim: See the graph

Turn 4: Class: That's just energy

Turn 5: Adela: Right but I mean to track [motioning with hands tracking]

Turn 6: Pratim: Oh you want to see the path

Turn 7: Mel: You want to see the story of a specific butterfly

Turn 8: Adela: Yeah

Turn 9: Mel: You can build a relationship with the butterfly [laughing]

Turn 10: Adela: No, no, no I am totally on that

Turn 11: Pratim: That is exactly how we wrote the paper about the simulation - about how the students can build a relationship with the butterflies on the screen

Turn 12: Mel: Even if you had the lifecycle of the butterfly - so we were talking about having caterpillars and certain birds that would only eat butterflies - it seemed the lifecycle of one bird or butterfly makes it more personal… and some students might want to follow the story of it more

[…]

Turn 15: Pratim: Did any of you feel that this is more simulation and lunar lander is more of a game?

Turn 16: Emily: I felt this was more of a simulation because there is no way to win it

Turn 17: Pratim: And do you think that has implications for learning

Turn 18: Emily: Not necessarily- because I made up my own challenges like, can I fill the screen with butterflies or how many birds does it take to get rid of all the butterflies? So I made up my own goals and played with the different settings that way, but students who might not be able to create their own goals may be bored very easily.

The conversation that ensued consists of two distinct parts. In the first part (Turns 1–12), the conversation focused on altering the code and the simulation in order to help students develop a personal connection with the phenomena being modeled. Here, Adela pointed out (to Pratim) that based on her experience with the Lunar Lander simulation, where the focus was on the behavior of a single computational agent, she realized that focusing on the behavior of a single butterfly in the second simulation would make it easier for students to understand what is going on in the model (Turn 2). When Pratim inquires whether Adela wanted to see the path of the butterfly (Turn 6), Adela's partner, Mel, clarified that the goal was not to merely observe the path of the butterfly and graph its energy change over time (the simulation already allowed participants to do that); instead, their goal was to see the "story" (Turn 7) of the butterfly, i.e., its life cycle. Doing so, according to Mel, would enable the students to "build a relationship" with the butterfly (Turn 9). As Mel further elaborated, poignantly, in Turn 12: "Even if you had the lifecycle of the butterfly - so we were talking about having caterpillars and certain birds that would only eat butterflies - it seemed the lifecycle of one bird or butterfly makes it more personal… and some students might want to follow the story of it more."

In the second part (Turns 15–18), the conversation focused on how coding and other interactions with the simulation were also playful, even though unlike the Lunar Lander game, the Bird-Butterfly-Flower simulation wasn't initially framed as a game. This was evident in the words of Emily, (Turns 16 and 18), who explained that even though there was no way to "win" the simulation, she still made up her own "challenges" such as "can I fill the screen with butterflies or how many birds does it take to get rid of all the butterflies?" (Turn 18). Emily further implied that because she was able to create new goals on her own in order to interact with the simulation (this involved her altering the underlying code along with her partner, as well as changing the variables on the simulation's graphical interface), she was engaged in the activity, noting that students who might not be able to "create their own goals may get bored very easily."

There are two insights to be gained from this conversation. First, it is well established in the educational computing literature that an important affordance of agent-based models is the ability of the learners to easily take on the perspectives of the computational agents in the models (Levy & Wilensky, 2008). Adela and Mel's comments (Turns 1–13) echo this finding. This is important because they believed that this could be pedagogically important, as it would allow their students to "build a relationship" with the scientific phenomenon. At the same time, the variations in their approaches indicate how they use their regimes of competence as productive resources in creating such relationships. Second, Emily's comments (Turns 15–18) illustrate that even coding and working with computer simulations could be framed as play, which in turn could encourage students to take risks that might take them even deeper within the discipline.

10.7 Summary and Discussion

This chapter makes two contributions. Along one dimension, we present a pedagogical approach for integrating computational modeling in preservice science teacher education by emphasizing preservice teachers' interpretive dilemmas, flexibility, and uncertainties in playfully interacting with the code and computational models. Along another dimension, we also illustrate how code, coding, and computational models get reframed by preservice teachers as pedagogical objects and experiences for doing and teaching science. The analysis presented here focuses on the participants' conversations about both code and pedagogy, a key feature being their deeply intertwined nature.

For example, in Theme 1, we saw how participants, through their own interactions with altering the underlying code of the Lunar Lander model, realized that they could make modifications to the code in order to facilitate their (future) students' engagement with key scientific concepts relevant to understanding the phenomenon represented by the game. In Theme 2, we saw how participants also came to a similar realization—that they could alter the NetLogo code in order to facilitate their students' engagement with scientific concepts—but also experienced a form of productive uncertainty. Upon reflecting on the value of their own scaffolded experience of coding, they wondered whether scaffolding their students might also prevent them from the form of deep explorations that the participants themselves experienced. And finally, in Theme 3, we saw how participants experienced playfulness even when they were presented with a simulation rather than a digital game, realizing the value of being able to set their own goals in their exploration of both the simulation and the underlying code. Across these themes, the frame of mathematization is present throughout—as our participants' engagement with the underlying code often involved altering underlying mathematical parameters and units of measurement (e.g., altering the speed and acceleration of the Lunar Lander, and rate of reproduction in the Bird-Butterfly-Flower simulation). The framing of mathematization is significant given recent findings from several studies that teachers do indeed view coding as mathematization in their classrooms (Sands et al., 2018; Sengupta et al., 2015; Farris, Dickes, & Sengupta, 2019).

Furthermore, as evident in Theme 3, participants also wanted to make alterations to the code so that their students could get opportunities to follow the narrative of an individual agent as a means to help them develop a deep understanding of the complexity of ecological interdependence. Herein lies an often noted affordance of agent-based modeling—that it provides opportunities for learners to take on perspectives of the computational agents, and even draw upon their own embodied and intuitive knowledge to make sense of the computational representations (Wilensky & Resnick, 1999; Levy & Wilensky, 2008; Farris & Sengupta, 2014). This is in no way an insignificant insight: as Keller (1984) noted, it was thinking like the agent (e.g., a chromosome) enabled the Nobel Laureate Barbara McClintock to make significant advances in her research on human genetic structures.

Overall, as stated in the beginning of this paper, our goal here is to present an epistemological perspective of how preservice science teachers view computational models and code in science education. Each of the three themes we have identified here are examples of epistemological stances in the sense that they reveal how the participants connected their experiences of computational modeling and coding in science with how they would support their future students' scientific inquiry. In this sense, our work also advances a phenomenological agenda in educational computing (Sengupta, Dickes, & Farris, 2018), as our emphasis is on identifying preservice science teachers' sense experiences (Merleau-Ponty, 1962) of coding and computational modeling—i.e., their framing and reframing of what code, coding, and computational modeling can become and can look like in their own imagined futures in science classrooms. The themes we have identified in our analysis, we believe, offer some useful resources that preservice teachers already bring to the table, and a pedagogical approach to build on these resources. We hope our work will inspire more scholarship on phenomenologically grounded, epistemological investigations of computing in K-12 teacher education.

References

Aikenhead, G. S. (2003, August). Review of research on humanistic perspectives in science curricula. In *4th Conference of the European Science Education Research Association (ESERA), Research and the Quality of Science Education. Noordwijkerhout, The Netherlands (August 19–23)*.

Azevedo, F. S. (2018). An inquiry into the structure of situational interests. *Science Education, 102*(1), 108–127.

Berland, M., & Lee, V. R. (2011). Collaborative strategic board games as a site for distributed computational thinking. *International Journal of Game-Based Learning (IJGBL), 1*(2), 65–81.

Bower, M., & Falkner, K. (2015, January). Computational thinking, the notional machine, preservice teachers, and research opportunities. In *Proceedings of the 17th Australasian Computing Education Conference (ACE 2015)* (Vol. 27, p. 30).

Daston, L., & Galison, P. (2007). *Objectivity*. Brooklyn, NY: Zone Books.

Dickes, A., Sengupta, P., Farris, A. V., & Basu, S. (2016). Development of mechanistic reasoning and multi-level explanations in 3rd grade biology using multi-agent based models. *Science Education, 100*(4), 734–776.

Dickes, A. C., & Sengupta, P. (2013). Learning natural selection in 4th grade with multi-agent based computational models. *Research in Science Education, 43*(3), 921–953.

DiSessa, A. A. (2001). *Changing minds: Computers, learning, and literacy*. Cambridge, MA: MIT Press.

Duschl, R. (2008). Science education in three-part harmony: Balancing conceptual, epistemic, and social learning goals. *Review of Research in Education, 32*(1), 268–291.

Farris, A. V., Dickes, A. C., & Sengupta, P. (2019). Learning to interpret measurement and motion in fourth grade computational modeling. *Science & Education*. https://doi.org/10.1007/s11191-019-00069-7

Farris, A. V., & Sengupta, P. (2014). Perspectival computational thinking for learning physics: A case study of collaborative agent-based modeling. In *Proceedings of the 12th international conference of the learning sciences* (pp. 1102–1107). ICLS.

Farris, A. V., & Sengupta, P. (2016). Democratizing children's computation: Learning computational science as aesthetic experience. *Educational Theory, 66*(1–2), 279–296.

Kalogiannakis, M., & Papadakis, S. (2017, August). Pre-service kindergarten teachers acceptance of "ScratchJr" as a tool for learning and teaching computational thinking and science education. In *Proceedings of the 12th Conference of the European Science Education Research Association (ESERA), Research, practice and collaboration in science education* (pp. 21–25). Dublin: Dublin City University and the University of Limerick.

Keller, E. F. (1984). *A feeling for the organism, 10th anniversary edition: The life and work of Barbara McClintock*. New York: Macmillan.

Kim, B., & Ho, W. (2018). Emergent social practices of Singapore students: The role of laughter and humour in educational gameplay. *International Journal of Child-Computer Interaction, 16*, 85–99.

Levy, S. T., & Wilensky, U. (2008). Inventing a "mid-level" to make ends meet: Reasoning between the levels of complexity. *Cognition and Instruction, 26*(1), 1–47.

Leonard, J., Mitchell, M., Barnes-Johnson, J., Unertl, A., Outka-Hill, J., Robinson, R., & Hester-Croff, C. (2018). Preparing teachers to engage rural students in computational thinking through robotics, game design, and culturally responsive teaching. *Journal of Teacher Education, 69*(4), 386–408.

Manz, E., & Suárez, E. (2018). Supporting teachers to negotiate uncertainty for science, students, and teaching. *Science Education, 102*, 1–25. https://doi.org/10.1002/sce.21343

McMahon, L. (2017). Phenomenology as first-order perception: Speech, vision, and reflection in Merleau-Ponty. In K. Jacobson & J. Russon (Eds.), *Perception and its development in Merleau-Ponty's philosophy* (pp. 308–334). Toronto: University of Toronto Press.

Merleau-Ponty, M. (1962). *Phenomenology of perception*. New York: Routledge.

Miles, M. B., & Huberman, A. M. (1994). *Qualitative data analysis: An expanded sourcebook*. Thousand Oaks: Sage.

Ochs, E., Gonzales, P., & Jacoby, S. (1996). "When I come down I'm in the domain state": Grammar and graphic representation in the interpretive activity of physicists. *Studies in Interactional Sociolinguistics, 13*, 328–369.

Papert, S. (1987). Information technology and education: Computer criticism vs. technocentric thinking. *Educational Researcher, 16*(1), 22–30.

Pickering, A. (1995). *The mangle of practice: Time, agency, and science*. Chicago, IL: University of Chicago Press.

Sands, P., Yadav, A., & Good, J. (2018). Computational thinking in K-12: In-service teacher perceptions of computational thinking. In *Computational thinking in the STEM disciplines* (pp. 151–164). Cham: Springer.

Sengupta, P., Brown, B., Rushton, K., & Shanahan, M. C. (2018). Reframing coding as "Mathematization" in the K12 classroom: Views from teacher professional learning. *Alberta Science Education Journal, 45*(2), 28–36.

Sengupta, P., Dickes, A. C., Farris, A. V., Karan, A., Martin, K., & Wright, M. (2015). Programming in K12 science classrooms. *Communications of the ACM, 58*(1), 33–35.

Sengupta, P., Krinks, K. D., & Clark, D. B. (2015). Learning to deflect: Conceptual change in physics during digital game play. *Journal of the Learning Sciences, 24*(4), 638–674.

Sengupta, P., Kinnebrew, J., Basu, S., Biswas, G., & Clark, D. (2013). Integrating computational thinking with K12 science education using agent-based modeling: A theoretical framework. *Education and Information Technologies, 18*, 351–380.

Sengupta, P., & Shanahan, M. C. (2017). Boundary play and pivots in public computation: New directions in STEM education. *International Journal of Engineering Education, 33*(3), 1124–1134.

Sengupta, P., Dickes, A., & Farris, A. (2018). Toward a phenomenology of computational thinking in STEM education. In *Computational thinking in the STEM disciplines: Foundations and research highlights* (pp. 49–72). Cham: Springer International Publishing.

Weintrop, D., Beheshti, E., Horn, M., Orton, K., Jona, K., Trouille, L., & Wilensky, U. (2016). Defining computational thinking for mathematics and science classrooms. *Journal of Science Education and Technology, 25*(1), 127–147.

Wilensky, U. (1999). *NetLogo*. Evanston: Center for Connected Learning and Computer-Based Modeling, Northwestern University. Retrieved from http://ccl.northwestern.edu/netlogo/

Wilensky, U., & Reisman, K. (2006). Thinking like a wolf, a sheep, or a firefly: Learning biology through constructing and testing computational theories—an embodied modeling approach. *Cognition and instruction, 24*(2), 171–209.

Wilensky, U., & Resnick, M. (1999). Thinking in levels: A dynamic systems approach to making sense of the world. *Journal of Science Education and Technology, 8*(1), 3–19.

Wilensky, U., Brady, C. E., & Horn, M. S. (2014). Fostering computational literacy in science classrooms. *Communications of the ACM, 57*(8), 24–28.

Wilkerson, M. H., Andrews, C., Shaban, Y., Laina, V., & Gravel, B. E. (2016). What's the technology for? Teacher attention and pedagogical goals in a modeling-focused professional development workshop. *Journal of Science Teacher Education, 27*(1), 11–33.

Wing, J. M. (2006). Computational thinking. *Communications of the ACM, 49*(3), 33–35.

Yadav, A., Stephenson, C., & Hong, H. (2017). Computational thinking for teacher education. *Communications of the ACM, 60*(4), 55–62.

Part III
Bodies, Hegemony and Decolonization in STEM Education (Theme 2)

Chapter 11
Rethinking Bodies of Learners Through STEM Education

Miwa Aoki Takeuchi and Shima Dadkhahfard

Abstract In the context of public discourse, STEM education is often coupled with its utilitarian value for economic growth and productivity. Under such discourse, learners are reduced at best to human capital, focusing on the production of economic value. Sen (World Development 25(12), 1959–1961, 1997) contrasted human capital with what he termed as human capability, which is "the ability of human beings to lead lives they have reason to value and to enhance the substantive choices they have" (p. 1959). What images of STEM education can we visualize if we place human capabilities at the center? Rather than treating learners as human capital or disembodied entities, we attempt to shed light on learner bodies. Drawing from the integral theoretical perspective of sociocultural theory with queer theory and critical race theory, we conceptualize learner bodies as the locus of negotiating the norm, emotions, and desires, and view them as fundamentally cultural and historical. Utilizing the counter-storytelling practices framed by critical race theory, we introduce the stories of two learners, May and Karim. May's story tells us how the informal mathematics knowledge she embodied came to be subjugated through formal school curriculum and pedagogy. Karim's story illustrates how his body queered normative mathematical representation and that facilitated a shift in his positional identity and participation in mathematics learning. The stories of learners with a fuller account of their cultural and historical bodies can help interrogate the underlying assumptions surrounding the current mathematics education. Reconceptualizing learner bodies prompts us to examine how we can mobilize the traditional boundaries of STEM education.

Keywords Learner bodies · Human capital and human capabilities · Sociocultural theory · Critical epistemologies (queer theory and critical race theory) · Equity in mathematics/STEM education

---

M. A. Takeuchi (✉) · S. Dadkhahfard
University of Calgary, Calgary, AB, Canada
e-mail: miwa.takeuchi@ucalgary.ca

P. Sengupta et al. (eds.), *Critical, Transdisciplinary and Embodied Approaches in STEM Education*, Advances in STEM Education,
https://doi.org/10.1007/978-3-030-29489-2_11

In the context of public discourse, Science, Technology, Engineering and Mathematics (STEM) education is often coupled with its utilitarian value for economic growth and productivity (e.g., Committee on Science, Engineering and Public Policy, 2007; Council of Canadian Academies, 2015; Science Technology and Innovation Council, 2009, 2015). Historically, in the United States, STEM education was proposed in response to the rising threat for national security during and after the Sputnik-spurred education reforms (Bybee, 2010). Such discourses linking scientific innovation with national security were prominent in driving the development of a curriculum document in the United States known as *A Nation at Risk* (National Commission on Excellence in Education, 1983). This document characterized the strong link between STEM education and national security, until the discourse surrounding STEM education changed in the post-Sputnik era. Broadly speaking, one of the leading narratives surrounding STEM education in the post-Sputnik era is filling workforce demands in the globalized and dynamically changing market (as represented in Committee on Science, Engineering and Public Policy, 2007). Standardized STEM education practices prevalent in the United States have also been critiqued as they prioritize economic competition and neoliberalism through the medium of education (Hoeg & Bencze, 2017; Strong et al., 2016).

Internationally, the leading narratives that shape STEM education in the United States are not necessarily shared with other countries. Shanahan, Burke, and Francis (2016) maintain that STEM education in a Canadian context can be best viewed as a boundary object with a definition that is not fixed or monolithic but partially shared and situationally defined among various stakeholders. Still, one of the most influential and conspicuous discourses around STEM education is filling STEM-related workforce demands and boosting economy as observed in recent policy-related documents in Canada (Council of Canadian Academies, 2015; Science Technology and Innovation Council, 2009, 2015). In another country such as Japan, which is the context of one of the stories introduced in the following section, STEM education has begun to be discussed in relation to promoting students' employability in the STEM fields (especially in computer sciences) (Japanese Cabinet, 2016). Under the dominant discourse around STEM education, learners are reduced at best to human capital, in which the aspect of producing economic value is overemphasized.

In lieu of human capital, Amartya Sen emphasized the theoretical construct of human capabilities. According to Sen (1997), human capital focuses on "skill and knowledge as well as effort—in augmenting production possibilities" (p. 1959); in contrast, the concept of human capabilities underlines the significance of "the ability of human beings to lead lives they have reason to value and to enhance the substantive choices they have" (p. 1959). Sen does not deny the fact that economic growth can lead to the expansion of human freedom to choose the kind of lives they want to live. However, as is often the case, the argument around human capital tends to concentrate on productivity and economic growth and does not extend the discussion to "why economic growth is sought in the first place" (Sen, 1997, p. 1960). Alternatively, Sen claims the centrality of individual freedom as the force and means of social development as well as end of development.

Sen (1999) draws on a childhood recollection to demonstrate his motivation to highlight human capabilities. As a child, Sen observed a man, Kader Mia, a Muslim daily worker, get knifed and killed in a largely Hindu area. Kader had to travel to the area hostile to Muslims in search of work to financially support his family. Being impacted by this encounter with Kader, Sen searched for ways that centralize human capabilities in economic and societal development, with his strong belief that "human beings are not merely means of production" (p. 296). Including the layer of human capabilities highlights "the instrumental role of capability expansion in bringing about social change (going well beyond economic change)" (p. 296). In light of human capabilities, freedom is conceptualized as the means and end of development. The means and end of STEM education, therefore, exceed beyond growing human capital, which mainly concentrates on the production of commodity and economic value, and should include fostering agency among students for social and political development along with the well-being and freedom of people.

11.1 Theoretical Framework: Integrating Perspectives from Sociocultural Theory, Queer Theory, and Critical Race Theory to Conceptualize Learner Bodies

What images of STEM education can we visualize if we place human capabilities at the center? Rather than treating learners as human capital or disembodied entities, the lens of human capabilities allows our focus to be on the cultural and historical nature of learner bodies, which hold the stories of who they are and who they are becoming. This lens sheds light on the expansion of freedom of bodies for all the learners through STEM education.

In the domain of mathematics education, learner bodies came to light in relation to the complex, embodied mathematical thinking (e.g., Abrahamson & Sánchez-García, 2016; de Freitas & Sinclair, 2013; Hwang & Roth, 2011; Lee, 2015; Ma, 2017; Nemirovsky, Tierney, & Wright, 1998; Radford, 2009). These studies on learner bodies draw from various epistemologies and ontologies: the ecological dynamics (as seen in Abrahamson & Sánchez-García, 2016), sensuous cognition (as seen in Radford, 2009), material phenomenology (as seen in Hwang & Roth, 2011), new materialism (as seen in de Freitas & Sinclair, 2013), and distributed cognition (as seen in Ma, 2017). Others (e.g., Lee, 2015; Nemirovsky et al., 1998) designed the interaction among the body, mathematics learning, and technology and made the interaction explicit. We seek to further advance this line of research by shedding light on how certain bodies are forced to be *hidden* in the public space of learning, how the mobilities of certain bodides can be restricted or liberated, and how such negotiation of bodies interact with the stories and histories of the learner. As detailed in the following sections, we believe that insights from queer theory and critical race theory will lend a hand in this endeavor.

In conceptualizing learner bodies as cultural, historical, and political, we integrate insights from sociocultural theory, queer theory, and critical race theory. Sociocultural theory emphasizes the cultural and historical nature of our bodies in learning, as the human mind is conceptualized as extending beyond the skin and is mediated by cultural and symbolic tools (Vygotsky, 1978; Wertsch, 1998). Philosophically, one of the central mandates of sociocultural theory is to overcome the Cartesian dualism, which is well represented by Il'enkov (1977) as "*thought lacking a body* and *a body lacking thought*" (p. 19). Instead, sociocultural theory conceptualizes the "*thinking body*" (p. 18) of living being. From this perspective, seemingly biological functions such as sleeping patterns are afforded and constrained within the cultural practices that we engage in. Rogoff (2003) succinctly summarizes this perspective by stating that humans are "biologically cultural" (p. 63). From this perspective, learner bodies are fundamentally cultural and historical, as they are shaped through the cultural practices, for example, as demonstrated in Saxe and Esmonde's (2005) trace of the change in Oksapmin bodily counting system over time.

11.1.1 Insights from Queer Theory

Sociocultural theories, however, have not explicitly addressed power, access, and privilege associated with learning (except for recent works advancing this area, Esmonde & Booker, 2016; Gutiérrez & Jurow, 2016; Nasir & Bang, 2012). In order to fully conceptualize cultural and historical bodies of learners, we incorporate perspectives from critical race theory and queer theory building further on our earlier work (Esmonde, Brodie, Dookie, & Takeuchi, 2009). Queer theory considers the norms, emotions and desires associated with performing certain bodies and also the costs of not-performing the normalized body (Butler, 1993; Foucault, 1980; Moraga & Anzaldúa, 1981). "Queer" in queer theory functions as "a marker representing interpretive work that refuses what Halley has called 'the heterosexual bribe'—that is, the cultural rewards afforded those whose public performances of self are constrained within that narrow band of behaviours considered proper to a heterosexual identity" (Sumara & Davis, 1999, p. 192). Queer theory thus meets pedagogy and curriculum by broadening what counts as knowledge: "not just knowledge about sexuality, but knowledge about how forms of desire are inextricable from processes of perception, cognition, and interpretation" (Sumara & Davis, 1999, p. 192). Taking the epistemology of queer theory thus amounts to questioning what counts as knowledge in STEM or mathematics education.

Queer theory also helps us see how power penetrates into our body. Foucault (1980) maintains that power "reaches into the very grain of individuals, touches their bodies and inserts itself into their actions and attitudes, their discourses, learning processes and everyday lives" (p. 39). Power is thus exercised "within the social body, rather than from above it" (p. 39). As such, power reaches into the way learners use or do not use their bodies and how their bodies are positioned, mobilized, or constrained through the configuration of classrooms and schools.

Power is exercised over non-normative bodies through materialization and abjection of bodies. Butler (1993) elucidates tacit exclusion of unintelligible bodies with the concept of abject. Butler explains how the boundary created to generate certain subjects simultaneously excludes *other* bodies or *unintelligible* bodies. According to Butler, abject beings are "those who are not yet 'subjects,' but who form the constitutive outside to the domain of the subject. The abject designates here precisely those 'unlivable' and 'uninhabitable' zones of social life which are nevertheless densely populated by those who do not enjoy the status of the subject, but whose living under the sign of the 'unlivable' is required to circumscribe the domain of the subject" (p. xiii). This tacit way of exclusion works to oppress the outside of materialized norm and materialized bodies.

Such exclusion can manifest in our daily encounters–how we move our bodies and how we feel about our bodies in social spaces. Adding the lens of racialized bodies into queer theory, Ahmed (2006) discusses how certain bodies are extended or not extended, and become more or less mobile in social spaces:

> For the bodies that are not extended by the skin of the social, bodily movement is not so easy. Such bodies are stopped, where the stopping is an action that creates its own impression. Who are you? Why are you here? What are you doing? (p. 139)

In these moments, racialized bodies can feel the loss of place and become estranged. While acknowledging the emotional stress and social and physical pressure experienced by the estranged bodies, Ahmed also describes the possibility of such bodily encounters for *queering* space by disturbing the order of things and the normative ways of living.

11.1.2 Insights from Critical Race Theory

Critical race theory has its roots in critical legal studies which examine the way law encodes cultural and racialized norms and it assumes that racism is not a series of isolated acts but is rather endemic and systemic (Ladson-Billings & Tate, 1995). Critical race theory legitimatizes racialized individuals' bodies and voices by using stories as a vehicle (Dixson & Rousseau, 2006; Ladson-Billings & Tate, 1995). Counter-storytelling informed by critical race theory offers space to challenge dominant, deficit narratives (Solórzano & Yosso, 2002). Counter-storytelling theorized through critical race theory challenges the deficit master narrative that overemphasizes and individualizes the deficits of non-dominant students and eventually forces non-dominant students to assimilate into the mainstream (Fernández, 2002; Ladson-Billings & Tate, 1995; Solórzano & Yosso, 2002). We will elaborate on how we integrated insights from critical race theory into our methodology in the following section. Integration of queer theory and critical race theory allows intersectional storytelling with STEM education (as seen in Leyva, 2016). We hope to bring forth the richness of mathematical knowledge those learners, bodies embodied and the agency that they exercise to queer the norm.

11.2 Methodology

The overarching methodology for this chapter is framed as critical race methodology centralizing counter-storytelling (Solórzano & Yosso, 2002). Solórzano and Yosso define the counter-story as "a method of telling the stories of those people whose experiences are not often told (i.e., those on the margins of society). The counter-story is also a tool for exposing, analyzing, and challenging the majoritarian stories of racial privilege" (p. 32). There are several types of counter-stories: personal and autobiographical narratives, a third person voice of other people's stories or narratives, and composited stories and narratives drawing on various forms of data (Solórzano & Yosso, 2002). Counter-storytelling locates a learner within broader sociocultural contexts and therefore should include political and societal contexts surrounding a learner when describing a story (Fernández, 2002).

In our counter-storytelling, we decided to use a third person voice to introduce the stories of two learners. A third person voice narrative allowed us to weave our reflexivity—our reflection on the relationship with participants—into these stories, while vividly depicting portraits of these learners (Langer, 2016). By utilizing the medium of counter-stories, our hope is to provide fuller and concrete pictures of who those learners are and who they are becoming in the space and within the norms of school mathematics learning.

Our storytelling focuses on the negotiation of learner bodies as informed by sociocultural theory and queer theory and also motivated by the counter-storytelling practices framed by critical race theory. The creation of counter-stories was guided by our theoretical sensitivity that surfaced the subtleties of meaning and significance to the particular segments of data (Solórzano & Yosso, 2002). In our analysis, we selected the stories that are illustrative to depict particular slices of students' experiences of learning mathematics—negotiating normative practices around their bodies in learning mathematics.

Both of stories emerged from the ethnographic studies that I (Takeuchi) conducted in urban cities in Canada and Japan. These studies were framed to capture a thick description of cultural practices and to reveal a "stratified hierarchy of meaningful structures" (Geertz, 1973, p. 7), which is produced, perceived, and interpreted in a particular social practice. Video and audio recording was employed, along with ethnographic fieldnotes, to conduct analyses of particular segments of interaction as well as interviews with participants. By retelling the stories of two learners, using the data obtained from these two studies, we will reconceptualize learner bodies in STEM education.

Through my previous ethnographic studies in urban cities in Canada and in Japan, I (Takeuchi) encountered May and Karim (pseudonyms). I met May when I conducted a study in an urban city of Japan that mainly focused on the continuity and discontinuity between schooling and out-of-school experiences for linguistically diverse students and families (for additional details and a fuller picture of this study, please see Takeuchi, 2018). The study focused on Filipina women who immigrated or migrated to Japan to work, and their school-aged children. I was involved in a local after-school program that was offered to academically support school learning, especially for those who used a first language or a home language other than the school instructional language. The study consisted of three phases: ethnographic observation, ethnographic interviews, and design of mathematics learning workshops with the participants.

I met Karim when I conducted a longitudinal ethnographic study of mathematics classrooms in a multilingual school within an urban city of Canada (for additional details and a fuller picture of this study, please see Takeuchi, 2015, 2016). The school had approximately 450 students, with representation from more than 30 different language groups; 23% of the students were born outside Canada, and for approximately 53% of the students, English was not the language spoken in their homes. The study focused on four newly arrived students who were labelled as "English language learners (ELLs)" and their trajectory of participation in classroom mathematics discourse and their development of identities. The study was conducted in two Grade 4 mathematics classes taught by Ms. Sally Wilson. The data collected through this study involved video data of classroom interactions focusing on the four students and interviews with Ms. Wilson as well as an English as a Second Language (ESL) class teacher.

11.3 Findings

11.3.1 Story of May

Let us introduce May. May was a student I (Takeuchi) met during my ethnographic study conducted in an urban city of Japan, where I worked with the communities of recent immigrants from the Philippines. May was born in Manila, the Philippines. She came to Japan when she was a Grade 4 student in the Philippines. When she started schooling in Japan, she repeated Grade 3 because of academic and linguistic gaps identified by her teacher, and thus she was older than other peers in her grade. When I met her, she was a Grade 6 student in a public elementary school. As for the most fluent language for her, she said, "I can speak Filipino (Tagalog), Japanese and English… but none is perfect." She emphasized that Japanese vocabulary associated with history and geography was particularly challenging. May's language practices could be positively seen as *translanguaging*—the fluid language practices unique to multilinguals (García & Wei, 2014); however, for her it was instead perceived as a deficit.

The following brief description of Japan's immigration policy would help contextualize May's story. In Japan, which is often perceived as "linguistically and racially homogeneous," some industrial areas are becoming much more ethnically and linguistically diverse, as represented in the percentages of registered foreign nationals in the following cities: Oizumicho, Gunnma (14.5%); Minokamo, Gifu (7.7%); and Kikukawa, Shizuoka (5.4%) (Committee of Localities with a Concentrated Foreigner Population, 2012). In addition to these cities, large cities such as Tokyo, Nagoya, and Osaka have relatively high percentages of "registered foreign nationals." Since the late 1970s, a significant number of Filipino women have come to work in Japan. Filipino migration in Japan is gender-biased: 77.6% of the Filipino population living and immigrated to Japan are women (Ministry of Justice, 2013). From my ethnographic interviews, I came to learn that some of these women came for an arranged marriage and worked in farm village areas, some of them worked as entertainers in urban cities, and, more recently, some of them worked as nurses, caregivers, and English language teachers and tutors. The immigration policy of Japan grants citizenship by parentage: children who are born into a family of a Japanese father or a Japanese mother are granted citizenship. However, a child like May or a child who was born in Japan from parents who are not Japanese citizens, cannot be granted permanent residency or citizenship in Japan. This political context placed many children in stressful and uncertain circumstances. May, for instance, could not picture where she would be living in the next year.

High drop-out rates from schools among children of migrant workers in Japan have started to be documented in recent years, in municipal and national government-issued reports. In one report (Shinjyuku-ku, 2012), the reasons why these children stopped going to school were listed as "I don't understand Japanese," "I don't understand the concepts discussed in class," or "I can't make friends." In the case of May, despite her uncertainty towards her future, she was viewed as putting much of her efforts to succeed towards her local community. Outside the school, she consistently attended the community after-school learning support program. There, she spent at least two hours in the evening, twice a week. She was resourceful—when she needed support for her studies, she had several friends and adult tutors to reach out to. She was also active in her school band wherein she played the role of band leader.

11.3.1.1 Mathematics Knowledge May's Body Embodied

One day, when I was at the community after-school program, I noticed that May was doing something with her fingers, under her desk, to solve a mathematics problem. I was curious but did not want to disturb her. I then heard from other tutors in the after-school program, some of whom were retired school teachers, that they were concerned about the use of fingers observed among students including May. They claimed that the use of fingers was often a sign of "immaturity" in knowing mathematics and that students should not be allowed to use fingers at school. After this conversation, I became more curious about what May was doing with her fingers and so I asked her about it.

It turned out that May was using her fingers as a tool for multiplication. Her explanation revealed an algorithm for multiplication using fingers, employed for the multiplication of numbers between six and nine (the algorithm can be slightly modified and extended to the multiplication of numbers between 11 and 15). For example, May used this algorithm when calculating 8×7. Each hand represented one factor. Five was represented by the closed hand, and any number above five was represented by the number of open fingers. In the case of 8×7, one hand showed three open fingers (and two closed fingers) and another hand showed two fingers (and three closed fingers). May added the number of open fingers and multiplied this number by ten [Product A; e.g., $(3 + 2) \times 10$ for the calculation of 8×7]. Then, she counted the number of closed fingers in each hand and multiplied these two numbers (Product B; e.g., 2×3 for the calculation of 8×7). Adding the Product A and Product B (e.g., $50 + 6$) provided the multiplication product (which in this example was 56).

Historically, this finger multiplication method was invented and employed in Florence, Italy, to deduct the multiplication table up to 10×10 to the multiplication table of 5×5, from a statement of the algebraic identity $(5 + a)(5 + b) = (5 - a)(5 - b) + 10(a + b)$ (Ball, 1888). By using the distributive property of multiplication, the algorithm can be also understood as: $(\chi - 5) \times 10 + (y - 5) \times 10 + (10 - \chi) \times (10 - y) = xy$ (where x and y are respective factors of multiplication) (Fig. 11.1).

May explained that she learned this finger multiplication method from her parents. When I interviewed her parents, May's mother explained to me that she would not encourage her children to use this method at school because using fingers would be considered illegitimate at school. Instead, May's mother told her children to memorize the multiplication table with the mainstream method taught in Japanese schools. May, herself, also reported that she would not use the finger multiplication method during mathematics quizzes or openly at school, because she thought that only computation strategies taught by the teacher were legitimate. In fact, when I first observed May's use of fingers for computation, she was hiding it under her desk. In this process, May's body was abjected (Butler, 1993).

Fig. 11.1 May's finger multiplication method (showing 8×9)

In the school context, multiplication operation skills were treated as one of the significant milestones in the early mathematics curriculum that impacted students' relationships with mathematics at school (such as a "slow mathematics learner") (Takeuchi, 2018). For a student like May who moved from another country and did not engage in a single "mainstream" technique, narrowly defining what is legitimate at school can constrain her capability. This is evidenced by May's narrative when she explained how she would hide her informal finger multiplication method during quizzes at school. May's story tells us how the informal mathematics knowledge she embodied came to be unintelligible and subjugated through the formal school curriculum and pedagogy that she experienced.

11.3.2 Story of Karim

Now let us introduce Karim. Karim was a student I (Takeuchi) met during my ethnographic study in an urban city of Canada. Karim was originally from Afghanistan. Before coming to Canada, he lived in refugee camps in Pakistan, with his family. He spoke Farsi at home to communicate with his parents and spoke English to communicate with his older brother and sister. Karim was receiving ESL support and attended ESL classes during language arts and social studies.

The Programme for International Student Assessment (PISA) by the Organization for Economic Co-operation and Development (OECD) examined the academic achievement gap between immigrant students and non-immigrant peers in mathematics, language arts, and science. A recent report revealed that immigrant students' performance in mathematics was lower than their native peers in many countries (OECD, 2013), although the gap between immigrant students and native students was smaller in Canada, Australia, New Zealand, and Macao-China. In Canada, second-generation immigrant students outperformed their native peers. However, a local assessment provides a different picture. For example, a provincial assessment of Ontario, the province with the largest immigrant population in Canada, indicated academic streaming in mathematics (Education Quality and Accountability Office, 2013). The majority of ELLs who pursued the university preparatory mathematics courses (i.e., "academic mathematics") met the provincial standard for mathematics. Yet, among ELLs who stayed in the mathematics courses that emphasized practical application (i.e., "applied mathematics"), only 35% of them met the grade-level expectation. Local school boards in Canada such as the Toronto District School Board (TDSB) conducted a study to grasp the picture of schooling and academic achievement gaps focusing on students from lower socioeconomic status and certain backgrounds. For example, the scores obtained by the students from Afghanistan indicated their challenges in reading, writing, and mathematics in school contexts (Brown, Newton, & Tam, 2015). Also in the report, the tendency for students from Afghanistan to pursue "applied mathematics," compared to "academic mathematics" was noted. In the case of Karim, Ms. Wilson and the ESL teacher both mentioned his limited and discontinuous prior schooling and lack of progress he was making. The report focusing on the students from Afghanistan in Canadian school contexts indicates systemic and socioeconomic challenges experienced by Karim.

During my school visits, in some days, Karim looked quite engaged; he was seated front and center and passionately raised his hands when the teacher posed a question to the whole class. On other days, he looked less energetic and disengaged. He was sometimes located to sit alone at the back of the classroom, staring down at his desk or looking out the window (Fig. 11.2). During group work, he was occasionally on the receiving end of authoritative interactions—his ideas were often denied without valid mathematical justification or he was excluded from the conversation in the formal space of school (Takeuchi, 2016). The teacher was constantly looking for ways to engage Karim by changing around the location he is positioned in the classroom. For example, if the teacher observed Karim being excluded from the group, she separated him from the group and provided him with individual support.

11.3.2.1 The Mathematics Knowledge Surfaced Through Karim's Gesture

As mentioned above, in the interactions with teacher-assigned peers, Karim's ideas were often denied without valid mathematical justification or he was excluded from the conversation. He was sometimes physically positioned at the corner of the classroom by himself as seen in Fig. 11.2. This used to be Karim's positional identity observed occasionally in mathematics classrooms. During my participation in Karim's mathematics class over an academic year, there was a moment in which I noticed a change in the way he participated as well as a shift in his "positional identity"—which refers to "the day-to-day and on-the-ground relations of power, deference and entitlement, social affiliation and distance with the social-interactional,

Fig. 11.2 Karim's position in the classroom. Note. Karim (farthest left) is working alone, facing the window

social-relational structures of the lived world" (Holland, Skinner, Lachicotte, & Cain, 1998, p. 127). This happened in mid-April. The following episode depicts how such shift came to light. In this episode, Karim's positional identity shifted in a subtle but significant manner.

In this introductory lesson to fractions, the teacher drew multiple shapes divided into equal pieces, where some pieces were colored in. Students were asked to name a fraction that corresponded to the colored pieces relative to the whole. After a series of similar interactions, the teacher drew a rectangle divided equally into four pieces, with three pieces colored. The teacher expected the answer to be $\frac{3}{4}$.

During this exchange, the teacher was sitting in a rocking chair and was drawing shapes on a flip chart. Students were sitting on the floor as a group and their bodies were therefore not bounded to desks (Fig. 11.3). Karim was sitting just in front of the teacher and was able to make his body visible to others. This position or positional identity was in stark contrast to a more marginalized position that he occupied in the classroom (Fig. 11.2). When the teacher asked the students to name a fraction showing the rectangle, Karim said "two out of eight." He then immediately corrected himself and said, "six out of eight." The teacher surprisingly repeated what Karim said, "six out of eight?" And she encouraged Karim to show where he saw six eighths. Karim pointed at how he saw a way to divide a rectangle into two equal parts; this gesture pointed out the equivalence between $\frac{3}{4}$ and $\frac{6}{8}$ (Fig. 11.4). The teacher acknowledged and took up Karim's contribution while saying "you're very clever," and took it up as the lesson of equivalent fractions by adjusting the original lesson plan. The teacher went beyond the curriculum expectation for the particular grade (the local curriculum did not require a lesson on equivalent fractions for Grade 4) and opened up the space for meaningful discussions.

In this interaction, Karim's bodily interactions with the object (that was expected to be interpreted in a normative way) *queered* the object. Going back to Ahmed's

Fig. 11.3 Classroom configuration during introductory fraction lesson

Fig. 11.4 Karim gesturing the invisible line (indicated with dots) and showing how he sees the equivalence between $\frac{3}{4}$ and $\frac{6}{8}$. *Note.* The invisible line is *halving* the rectangle

(2006) account on queering, this moment illustrates "how the strangeness that seems to reside somewhere between the body and its objects is also what brings these objects to life and makes them dance" (p. 163). What was overlooked in the rectangle representation of fractions got noticed and started to mobilize with a simple yet meaningful gesture of drawing a line halving the rectangle. In fact, Karim's answer was not what the teacher expected—but nonetheless the teacher praised Karim's idea as a vital contribution. Karim did not yet have the words to describe what he discovered, that is "equivalent fraction." However, he was able to "show" the concept by gesturing. Instead of discounting what Karim showed, by affirming his contribution, the teacher provided a new language, a new concept, and new positioning in this interaction.

At the beginning of this interaction, when Karim said "six out of eight," some of his peers whispered, "it's four" and tried to dismiss his contribution. However, when Karim showed how the equivalence between $\frac{3}{4}$ and $\frac{6}{8}$ simply drawing the invisible line to divide the rectangle, his peers said, "ohhhh," sounding surprised. The teacher took up Karim's gesture of the "invisible line" and let the class compare two rectangles showing equivalent fractions. By taking up Karim's contribution, the teacher not only moved beyond narrow mainstream mathematical expectations, but also leveraged Karim's positional identity into someone who can contribute to mathematical discussion. Karim tweaked the original rectangle model to a model that can possibly highlight other concepts such as the division or multiplication of fractions. In other words, Karim's gesture brought forth otherwise unnoticed affordance of the rectangle model. In North America, a frequently used representation of fractions used in the school context and textbooks is a circle model, which is often associated with pizza; however, the circle model does not afford reasonable representation of numerical operations, especially division or multiplication, with fractions (Watson & Mason, 2005). Just before this episode, the teacher also used the circle model of fractions. In his subtle gesture, Karim showed a way to advance the discussion mathematically with the alternative representation through this tweaks (e.g., equivalence of fractions visually, and potentially other more advanced numerical operations with fractions). In this sense, Karim's gesture could be read as the disruption of the dominant representation of fractions. The teacher acknowledged the central position that Karim took in this interaction by making his body more visible to others and thus opened up his capabilities.

11.4 Discussion

Rather than treating learners as human capital or disembodied entities, from the perspective of human capabilities, we shed light on learner bodies. Drawing from the integral theoretical perspective of sociocultural theory with queer theory and critical race theory, we conceptualized learner bodies as cultural and historical—and also the locus of negotiation of norms and power. By rethinking May and Karim's learner bodies in this light, we are able to better appreciate their vulnerable yet valuable mathematical contributions. Both May and Karim used their bodies in non-normative ways to engage with mathematics. May's finger multiplication represents embodied actions that contradict normative standards for engaging in mathematics, yet it could serve as tools that enable these learners to participate in mathematics discourse. By challenging the norms associated with the stabilized marginality he occupied in the classroom and mobilizing his body, Karim tweaked the given representation of fractions and brought forth its otherwise unnoticed mathematical affordance, through his use of gesture. Their stories highlight the need to reconceptualize learner bodies within the context of STEM education so that we honor the contributions and practices of all students, especially those historically left to the margins. These stories call for designing learning environments that value the whole learner and prompt to rethink STEM education as a tool for capitalist enterprise and reimagine it as a place for building human capability.

Both stories also prompt us to examine the design of equitable learning environments in STEM education. Design of the environment is powerful—in a sense that can either challenge, perpetuate, or create inequity in education settings. Panopticon, a prison architecture, designed by Jeremy Bentham that Foucault (1980) depicts, is a vivid example of the power of design. With carefully designed use of light and locations of prison cells and a guard, this design produced self-policing, self-monitoring, and self-control, just with (imagined) gaze. Similarly, the ways in which school curriculum, school architecture, classroom settings, and locations of a teacher and students all influence strongly the ways in which power produces its effects at the level of knowledge but also at the level of desire.

In the case of Karim, the material re-organization of the classroom allowed him to engage his body in the communication of mathematics, allowing the class to see his mathematics competence. The teacher rearranged the classroom space so that students were not bounded by desks and with this rearrangement, Karim was able to better mobilize his body and through which, normative mathematical representation was *queered*. In the case of May, the mathematical reasoning she embodied was unrecognized or even disciplined in the classroom space and by the rigid curriculum—with consequences for possible marginalization of her reasoning, her body, and possibly herself. May carefully monitored the ways she used her body to engage with mathematics. That is, she hid her finger multiplication method to fit into the mainstream mathematical practice of performing number operations. She tactically used the method, monitoring legitimate and illegitimate bodily acts in the classroom.

In school contexts, there is still a deeply rooted assumption that mathematical thinking has to be "a pure mental activity—something immaterial, independent of the body, occurring in the head" (Radford, 2009, p. 111). STEM education tends to be treated as politically neutral though it is inherently political and ethical, and discourse in STEM education can value certain bodies more than others (as demonstrated in Philip, Gupta, Elby, & Turpen, 2018). The assumptions about political neutrality and disembodied nature of learning can marginalize certain bodies of learners and hence their associated sense of identity with STEM disciplines. By incorporating queer theory and critical race theory, in this chapter, our attempt was to advance the conceptualization of learner bodies as the locus of negotiation of power, desires, and emotions. Such a reconceptualization of learner bodies can lead us to reframe what is considered as mathematics in STEM education, which is often perceived as "existing independently of the people who do it, and independent of their bodies, senses, desires, emotions, and aesthetics—everything that makes a person flesh and blood" (Greer, Mukhopadhyay, & Roth, 2013, p. 6). Rethinking learner bodies and challenging the traditional framework of teaching and learning is essential for STEM education; without which, the ways in which mathematics has been taught and conceptualized as a discipline will not be mobilized or renewed.

In discussing the materialization of bodies, Butler (1993) described the paradox of the subjectivation of bodies, where the formation of the subject is enabled or produced by the very norm that the subject is resisting. In this sense, agency is perceived as "reiterative or rearticulatory practice, immanent to power, and not a relation of external opposition to power" (p. xxiii). Agency conceptualized by queer theory is located with the chain of historicity; it is a power to "avow a set of constraints on the past and the future that mark at once the limits of agency and its most enabling conditions" (p. 174). This view of agency is insightful when we think of our agency to reimagine STEM education. Shanahan et al. (2016) maintained that "STEM education" can be best conceptualized as a boundary object. By liberating disciplined bodies, we could exercise our agency to include abjected bodies in traditional school mathematics and rearticulate STEM education. Such efforts have just begun. For example, Sengupta and Shanahan (2017) assembled otherwise-dispersed bodies through the learning environment of public computation. They demonstrated the possibility of extending the boundary of STEM education to public experience. By rearticulating the disciplinary boundaries afforded by STEM education, we can envisage mathematics learning which is more integral to learners' bodies, and which allows learners "to lead lives they have reason to value and to enhance the substantive choices they have" (Sen, 1997, p. 1959)."

Acknowledgements We would like to appreciate all the participants in the studies described in this chapter. A special thank-you goes out to the participants we called Karim and May. We also thank Dr. Lesley Dookie, Dr. Ayush Gupta, Dr. Pratim Sengupta, and Dr. Marie-Claire Shanahan for their comments and feedback on an earlier version of this manuscript. One of the studies described here was funded by the Grant-in-Aid for Scientific Research [12J02927]. Any opinions, findings, and conclusions expressed herein do not necessarily reflect the views of the funding agency.

References

Abrahamson, D., & Sánchez-García, R. (2016). Learning is moving in new ways: The ecological dynamics of mathematics education. *Journal of the Learning Sciences, 25*(2), 203–239.

Ahmed, S. (2006). *Queer phenomenology: Orientations, objects, others*. Durham: Duke University Press.

Ball, W. (1888). *A short account of the history of mathematics*. London: Macmillan.

Brown, R., Newton, L., & Tam, G. (2015). *The Toronto District School Board's student group overviews: Aboriginal heritage, Afghan, Portuguese-speaking, Somali-speaking, and Spanish-speaking students*. Retrieved from http://www.tdsb.on.ca/Portals/research/docs/reports/TDSB.StudentGroupOverviews.pdf.

Butler, J. (1993). *Bodies that matter: On the discursive limits of "sex"*. New York, NY: Routledge.

Bybee, R. W. (2010). What is STEM education? *Science, 329*(5995), 996.

Committee of Localities with a Concentrated Foreigner Population [Gaikokujin Shujyu Toshi Kaigi]. (2012). *Report on the 2012 Tokyo conference for the committee of localities with a concentrated foreigner population [Gaikokujin Shujyu Toshi Kaigi Tokyo 2012 Houkokusho.]* Retrieved from http://www.shujutoshi.jp/.

Committee on Science, Engineering and Public Policy. (2007). *Rising above the gathering storm: Energizing and employing America for a brighter economic future*. Washington, DC: National Academies Press.

Council of Canadian Academies. (2015). *Some assembly required: STEM skills and Canada's economic productivity*. Ottawa: Council of Canadian Academies.

de Freitas, E., & Sinclair, N. (2013). New materialist ontologies in mathematics education: The body in/of mathematics. *Educational Studies in Mathematics, 83*(3), 453–470.

Dixson, A. D., & Rousseau, C. K. (2006). And we are still not saved: Critical race theory in education ten years later. In A. D. Dixson & C. K. Rousseau (Eds.), *Critical race theory in education: All God's children got a song* (pp. 31–54). New York, NY: Routledge.

Education Quality and Accountability Office. (2013). *EQAO's provincial elementary school report: Results of the 2012–2013 assessments of reading, writing and mathematics, primary division*. Retrieved from http://www.eqao.com/Pages/launch.aspx

Esmonde, I., & Booker, A. N. (2016). *Power and privilege in the learning sciences: Critical and sociocultural theories of learning*. New York, NY: Taylor & Francis.

Esmonde, I., Brodie, K., Dookie, L., & Takeuchi, M. (2009). Social identities and opportunities to learn: Student perspectives on group work in an urban mathematics classroom. *Journal of Urban Mathematics Education, 2*(2), 18–45. Retrieved from http://ed-osprey.gsu.edu/ojs/index.php/JUME/article/view/46

Fernández, L. (2002). Telling stories about school: Using critical race and Latino critical theories to document Latina/Latino education and resistance. *Qualitative Inquiry, 8*(1), 45–65.

Foucault, M. (1980). *Power/knowledge: Selected interviews and other writings 1972–1977* (L. M. Colin Gordon, J. Mepham, K. Soper, Trans. G. Colin, Ed.). Vintage Books: New York.

García, O., & Wei, L. (2014). *Translanguaging: Language, bilingualism and education*. London: Palgrave Macmillan.

Geertz, C. (1973). *The interpretation of cultures*. New York, NY: Basic Books.
Greer, B., Mukhopadhyay, S., & Roth, W. M. (2013). Celebrating diversity, realizing alternatives). In W. M. Roth, L. Verschaffel, & S. Mukhopadhyay (Eds.), *Alternative forms of knowing (in) mathematics: Celebration of diversity of mathematical practices* (pp. 1–8). Rotterdam: Sense.
Gutiérrez, K. D., & Jurow, A. S. (2016). Social design experiments: Toward equity by design. *Journal of the Learning Sciences, 25*(4), 565–598.
Hoeg, D. G., & Bencze, J. L. (2017). Values underpinning STEM education in the USA: An analysis of the next generation science standards. *Science Education, 101*(2), 278–301.
Holland, D., Skinner, D., Lachicotte, W., Jr., & Cain, C. (1998). *Identity and agency in cultural worlds*. Cambridge, MA: Harvard University Press.
Hwang, S., & Roth, W. M. (2011). *Scientific and mathematical bodies: The interface of culture and mind*. Rotterdam: Sense.
Il'enkov, E. (1977). *Dialecticallogic: Essaysin its history and theory* (H. C. Creighton, Trans.). Moscow: Progress.
Japanese Cabinet. (2016). *Declaration for establishing a global IT nation*. [Sekai Saisentan IT Kokka Sozo Sengen]. Retrieved from http://www.kantei.go.jp/jp/singi/it2/decision.html.
Ladson-Billings, G., & Tate, W. (1995). Towards a critical race theory of education. *Teachers College Record, 97*(1), 47–68.
Langer, P. C. (2016). The research vignette: Reflexive writing as interpretative representation of qualitative inquiry—A methodological proposition. *Qualitative Inquiry, 22*(9), 735–744.
Lee, V. R. (2015). *Learning technologies and the body: Integration and implementation in formal and informal learning environments*. New York, NY: Routledge.
Leyva, L. (2016). An intersectional analysis of Latin@ college women's counter-stories in mathematics. *Journal of Urban Mathematics Education, 9*(2), 81–121. Retrieved from http://ed-osprey.gsu.edu/ojs/index.php/JUME/article/view/295
Ma, J. Y. (2017). Multi-party, whole-body interactions in mathematical activity. *Cognition and Instruction, 35*(2), 141–164.
Ministry of Justice [Homusho]. (2013). *Numbers of the registered foreign nationals: Nationalities, areas and purpose for stay*. [Kokuseki chiiki betsu zairyu shikaku mokuteki betsu zairyu gaikokujin]. Retrieved from http://www.moj.go.jp/housei/toukei/toukei_ichiran_touroku.html.
Moraga, C., & Anzaldúa, G. (1981). *This bridge called my back: Writings by radical women of color*. New York, NY: Kitchen Table.
Nasir, N. S., & Bang, M. (2012). Conceptualizing cultural and racialized process in learning. *Human Development, 55*(5–6), 247–249.
National Commission on Excellence in Education. (1983). *A nation at risk: The imperative for educational reform*. Washington, DC: United States Department of Education.
Nemirovsky, R., Tierney, C., & Wright, T. (1998). Body motion and graphing. *Cognition and Instruction, 16*(2), 119–172.
OECD. (2013). *PISA 2012 results: Excellence through equity: Giving every student a chance to succeed*. Retrieved from http://www.oecd.org/pisa/keyfindings/pisa-2012-results-volume-II.pdf.
Philip, T. M., Gupta, A., Elby, A., & Turpen, C. (2018). Why ideology matters for learning: A case of ideological convergence in an engineering ethics classroom discussion on drone warfare. *Journal of the Learning Sciences, 27*(2), 183–223.
Radford, L. (2009). Why do gestures matter? Sensuous cognition and the palpability of mathematical meanings. *Educational Studies in Mathematics, 70*(2), 111–126.
Rogoff, B. (2003). *The cultural nature of human development*. New York, NY: Oxford University Press.
Saxe, G. B., & Esmonde, I. (2005). Studying cognition in flux: A historical treatment of Fu in the shifting structure of Oksapmin mathematics. *Mind, Culture, and Activity, 12*(3), 171–225.
Science Technology and Innovation Council. (2009). *State of the nation 2008: Canada's science, technology and innovation system*. Ottawa: Science Technology and Innovation Council.

Science Technology and Innovation Council. (2015). *State of the nation 2014: Canada's science, technology and innovation system*. Ottawa: Science Technology and Innovation Council.
Sen, A. (1997). Human capital and human capability. *World Development, 25*(12), 1959–1961.
Sen, A. (1999). *Development as freedom*. New York, NY: Anchor Books.
Sengupta, P., & Shanahan, M.-C. (2017). Boundary play and pivots in public computation: New directions in STEM education. *International Journal of Engineering Education, 33*(3), 1124–1134. Retrieved from https://www.ijee.ie/contents/c330317.html
Shanahan, M.-C., Burke, L. E., & Francis, K. (2016). Using a boundary object perspective to reconsider the meaning of STEM in a Canadian context. *Canadian Journal of Science, Mathematics, and Technology Education, 16*(2), 129–139.
Shinjyuku-ku. (2012). *The study on the schooling situation of children with foreign origins* [Gaikoku ni roots wo motsu kodomo no jittai chosa]. Retrieved from http://www.city.shinjuku.lg.jp/content/000109321.pdf.
Solórzano, D. G., & Yosso, T. J. (2002). Critical race methodology: Counter-storytelling as an analytical framework for education research. *Qualitative Inquiry, 8*(1), 23–44.
Strong, L., Adams, J. D., Bellino, M. E., Pieroni, P., Stoops, J., & Das, A. (2016). Against neoliberal enclosure: Using a critical transdisciplinary approach in science teaching and learning. *Mind, Culture, and Activity, 23*(3), 225–236.
Sumara, D., & Davis, B. (1999). Interrupting heteronormativity: Toward a queer curriculum theory. *Curriculum Inquiry, 29*(2), 191–208.
Takeuchi, M. (2015). The situated multiliteracies approach to classroom participation: English language learners' participation in classroom mathematics practices. *Journal of Language, Identity & Education, 14*(3), 159–178.
Takeuchi, M. A. (2016). Friendships and group work in linguistically diverse mathematics classrooms: Opportunities to learn for English language learners. *Journal of the Learning Sciences, 25*(3), 411–437.
Takeuchi, M. A. (2018). Power and identity in immigrant parents' involvement in early years mathematics learning. *Educational Studies in Mathematics, 97*(1), 39–53.
Vygotsky, L. S. (1978). *Mind in society: The development of higher psychological processes*. Cambridge, MA: Harvard University Press.
Watson, A., & Mason, J. (2005). *Mathematics as a constructive activity: Learners generating examples*. New York, NY: Routledge.
Wertsch, J. (1998). *Mind as action*. New York, NY: Oxford University Press.

Chapter 12
Supporting Complex Multimodal Expression Around Representations of Data: Experience Matters

Victor R. Lee

Abstract Reporting of STEM education research tends to privilege the verbal statements that students and teachers make. While some of this is likely an artifact of how research is typically reported in print-based publications media that are optimized to handle text, it is important to note that the actions that take place when talking about and learning STEM content are actually multimodal. That is, they involve deployment of several modes of expression that may include verbal utterances that lend themselves to transcription but also include gesture, actions, and use of physical space. This chapter proposes that for producing multimodal expressions about data, direct and embodied experience with the creation of data is a valuable resource. Two examples are presented. One involves the description and re-enactment of what actions produced the distributional shape of a histogram. The other involves students discussing what bodily actions would produce different box-and-whisker plots. Each shows the competence students have for thinking through data representations when they are able to physically enact what motions are associated with those representations.

Keywords Multimodality · Gesture · Elementary statistics education · Embodied cognition · Inscription

12.1 Background

Reporting of STEM education research tends to privilege the verbal statements that students and teachers make. While some of this is likely an artifact of how research is typically reported in print-based publications media that are optimized to handle

V. R. Lee (✉)
Stanford University, Stanford, CA, USA
e-mail: vrlee@stanford.edu

P. Sengupta et al. (eds.), *Critical, Transdisciplinary and Embodied Approaches in STEM Education*, Advances in STEM Education,
https://doi.org/10.1007/978-3-030-29489-2_12

text, it is important to note that the actions that take place when talking about and learning STEM content are actually multimodal. That is, they involve deployment of several modes of expression that may include verbal utterances that lend themselves to transcription but also include gesture, actions, and use of physical space. Detailed multimodal analyses of interactions around STEM content and practices reveal how complex this coordination work can be, and remind us to be considerate of how multiple communicative modalities are at work when learning and communicating within a disciplinary domain (Kress, Jewitt, Ogborn, & Tsatsarelis, 2001). For example, physicists have been known to describe state transition such that they refer to themselves as people being blended into a state change graph (Ochs, Gonzales, & Jacoby, 1996) and in entomology research, gesture, inscription, and discourse may be deployed to introduce and appropriate new statistical concepts (Hall, Wright, & Wieckert, 2007).

The acknowledgment of multimodal interactions is important in STEM education because they can reveal what ideas are being communicated and appropriated by a student in ways that their words alone may fall short. For instance, a known phenomenon in gesture research is the speech-gesture mismatch in which students begin to show gestures and make deictic reference to appropriate quantities when they are about to develop in their mathematics understanding even though their speech may report incorrect information (Church & Goldin-Meadow, 1986). Additionally, non-literal uses of language such as when students describe graphs and diagrams in novel ways can reflect a level of sophistication in how meaning is constructed, even when they do not follow normative canon (Lee & Sherin, 2006). Work by Nemirovsky, Tierney, and Wright (1998) has illustrated this through detailed case analysis of a student manipulating a motion sensor and an immediately produced distance graph where the examined youth shifted perspectives and engaged in acts of "fusion" where they would situate their own activity into segments of the graph. These are arguably sophisticated ways of thinking through canonical scientific representations, and represent a contrast to literature that has identified difficulties in students' abilities to work with such representations (e.g., Cromley et al., 2013; Leinhardt, Zaslavsky, & Stein, 1990). While we do not dispute that students can make normative errors in interpreting representations, we believe that education research also needs to recognize what students can do well with representations and identify what conditions support greater competence.

This chapter proposes that when it comes to working with representations of data, direct and embodied experience with the creation of data is a valuable resource. In the science education literature, we have seen that familiarity with the data creation context matters for how prepared a class is to make sense of that data. When students work with data that was collected by outside entities, they spend much more time trying to understand how the data were created and what the data correspond to in the world (Hug & McNeill, 2008). The creation of data can be laborious. When orchestrated intentionally and thoughtfully, it can involve deliberate routines of measurement in which students create and critique ways of collecting data that lead to demonstrably improved normative understandings of measurement, error, and variability (e.g., Lehrer, Kim, & Schauble, 2007). What I propose here is that students can also leverage existing automated wearable data collection technologies to quickly obtain data that still provides them with direct experience with the phenomena being measured and allows for comparable explorations (Lee, Drake, & Williamson, 2015). To illustrate this, I present two brief episodes analyzed multimodally that involve students communicating around canonical data representations (histograms and box-and-whisker plots) of students' own physical activity data. Key to these episodes is how students connect the bodily experiences represented in the data to the data representations themselves and express and coordinate those through speech, gesture, and action.

12.2 Theoretical Framework: Multimodality in Interaction

Before proceeding further, I do wish to clarify my position on multimodality for this chapter. One of the most fundamental ways of recognizing multimodality is recognizing it as the deployment of more than one representational system. For instance, consider the example of multimodal communications that may come from a comic book or graphic novel. In comparison to a trader paperback bestseller, where written text is doing the bulk of the work to communicate intended meaning, images are also deeply involved in the comic book. Much of the same meaning and effect of a story in trade paperback can be retained whether or not the same pagination, font size, and spacing is used in a reproduction (such as conversion of a hardcover edition to paperback or ebook). The same does not hold true for graphic novels, where the specific placement of panels and text combine to create an effect and directionality to how one reads that printed material. Indeed, the subtlety of how a multimodal medium like graphic novels accomplishes communicative work has been subject to careful analysis (McCloud, 1993). A central point to be observed is that multimodal communications involve deliberate deployment of multiple systems of representation. In the case of comics, that involves a minimum of text and image (although arguably several more, ibid). Each system may have standard conventions when used alone that allow for generally accepted standards for interpretation and inference of intended meaning. Yet when deployed together, even in the case of text and image, the potentials for a multiplicity of possible meanings to be evoked grow rapidly (Lemke, 1998). On its surface, this multiplicity could seem problematic. However, we have techniques and pragmatic norms for constraining possible meanings, many of which have yet to be fully unpacked (Lee & Sherin, 2004).

For current purposes, the recognition that multiple systems are deployed is the most critical point to be understood. In the analyzed excerpts that follow, there are three primary modalities that are being considered: speech, gesture, and inscription. These are, knowingly, simplifications of all the features of interpersonal interaction that can do communicative work, but are privileged currently for both brevity of the written chapter format and because of some generally accepted delineations. Speech refers to verbal articulations and utterances. They are ephemeral in that unless special recording means are deployed, they do not maintain a presence in time and space after they are produced. They are primarily detected aurally, although visual and sometimes even kinesthetic sensory modalities may be involved. Paralinguistic features, such as prosody, stress, and rate of speech, will be associated with speech and are aspects that have been recognized as contributing more than what the transcribed words alone can do. Gesture refers to intentional bodily movement, which can serve a number of functions in communication. McNeill (1992) offered one of the most commonly invoked categorization schemes by identifying deictic gestures (those that involve resolving reference, as would happen with pointing), beat gestures (those that are used for emphasis, such as the thrust of a hand that coincides with a word to stress intensity), and metaphorical (those that enact or symbolize actions and relations, such as using one's swaying hand to mimic a car swerving on a fictive road). Inscription (Roth & McGinn, 1998), for current purposes, refers to representations and markings placed on a surface. Most commonly, those would involve graphic representations such as pictures, diagrams, graphs, tables, and charts, although arguably many other markings legitimately count as inscriptions.

As precedent for focusing on these modalities and these delineations, the reader is referred to two other book chapters published elsewhere that provide productive multimodal analyses of communications across these representational systems. Hutchins and Palen (1997) provided an analysis of meaning coordination using an airline cockpit display to explain how common reference is established and non-literal language can be made sensible and appropriate between multiple communicating parties. DeLiema, Lee, Danish, Enyedy, and Brown (2016) also provided an analysis of the deployment of speech, gesture, and inscription in a fine-grained analysis of how an undergraduate student made sense of and explained physical state change in a set of chemistry interviews. Both these cited chapters draw upon the work of Goodwin, and particularly relevant here is his discussion of semiotic lamination where multiple semiotic fields (modalities) with different properties are co-deployed to compactly convey complex relationships and meanings (Goodwin, 2013). It is through lamination that we can expect the multiplicity of meanings that come from multiple modalities (Lemke, 1998) to be constrained toward speaker intent, but also broadened to allow for evocation of ideas and reference that transcend what any single modality can do alone.

The contribution of the analyses in this chapter as a contrast to the aforementioned extant multimodal analyses resides in the reference back to students' known physical activity. Those are the actions that led to the production of particular mathematical inscriptions that they are discussing and learning in the excerpts that appear below. The implication is that direct, personal experience with those activities as part of a history of that inscription shapes how multiple modalities are deployed. This takes place immediately and fluidly for the speakers. Where gesture is invoked, there is an "echo" or re-enactment of what was the experienced activity. The inscriptions involved are understood and referenced in particular ways by the student speakers in ways that we can easily recognize as legitimate with respect to the task at hand by others in the classroom and as legitimate for demonstrating proficiency of mathematical and statistical understanding. The speech that is used aids in structuring and conveying intended meaning, but does not do so in isolation. The prior embodied experience that produced the inscriptions gets invoked through the use of multimodal communications. Where this chapter also contributes is in the proposal that this way of seeing mathematical and statistical interpretations and communications need not rely exclusively on experiences that have already happened. They can also, as the second excerpt demonstrates, involve hypothetical embodied experiences that are sensible still because they are plausible and relatable experiences to those that have already been invoked.

12.3 Research Context and Data Sources

The data for these accounts of multimodal expressions around data representations comes from the fourth iteration of a design-based research project that was focused on supporting elementary statistical understanding through examination of physical activity data obtained from wearable activity tracking devices. The underlying motivation for this project was that students would be able to bootstrap their knowledge of statistics and exhibit sophisticated ways of thinking about data if the data were opportunistically obtained from their routine school day activities and experiences (Lee, Drake, Cain, & Thayne, 2015). Thus far, this approach appears to be effective, in that students who participated in elementary statistics units that integrated students' physical activity data showed greater learning gains than those following more traditional classroom units (Lee & Thomas, 2011; Lee, Drake, & Thayne, 2016). Furthermore, the use of activity data invited students to generate and pursue novel questions about routine experiences, such as how different recess activities compared with one another in terms of activity level and whether a step-tracking device would register a jump as a walking step (Drake, Cain, & Lee, 2017).

The particular iteration from which the two excerpts below originate involved a class of sixth-grade students in a Title I rural school in the western United States who used step data obtained from commercial, wrist-based wearable activity trackers (Fitbit Flex devices) as part of a multi-week unit we designed to address a set of state standards related to statistics. Each student was provided with a wearable device that they used each day of the unit during school, and the data were automatically recorded and then downloaded. The primary quantity of interest was the number of steps taken, typically within single minute increments but also overall number of steps across a span of time. These step data were transferred into *TinkerPlots* data visualization software (Konold & Miller, 2005) for group inspection and discussion.

For each day of the unit, we had two video cameras recording classroom activity, with one focused on whoever was the primary speaker at the time and the other recording a full classroom view. The two cases come from the video footage of the first camera, and they had been selected as occasions where students were talking with one another about canonical statistical inscriptions in front of their entire class. We were interested in what observations students made, how they critiqued one another, and what they described about those inscriptions. In the course of reviewing those various episodes as they appeared throughout the unit, we noted these two episodes involved students making deliberate efforts to talk about not just the data inscription but also the underlying physical activity that would produce those inscriptions. Transcribing these episodes in such a way that they could be made sensible to someone outside of our research group required us to capture still images from the video and highlight some of the actions and gestures that students were making, which ultimately have been rendered as supportive illustrations in this chapter. While we believe that these can do more to communicate what happened in the classroom, our recording and transcription decisions inevitably include some biases about what were noteworthy phenomena (Hall, 2000; Ochs, 1999). Regardless, we believe that our efforts to capture and depict gesture and action along with student speech still serve to demonstrate how understanding of the data creation context and understanding of the data inscriptions were mutually supportive.

12.4 Findings

12.4.1 Episode 1: Explaining the Histogram

As part of the multi-week unit we had designed, students produced pooled step data from a planned walk back and forth across their playground. Each student in the class walked across the playground 12 times while tracking the number of steps required to make each trip. The point of the activity was to produce normally distributed data from counts of footsteps and also for each student to generate enough data about their individual walks to examine variability within individuals and across the entire class. Step data had become a familiar measure of activity for the class based on daily interaction with and inspection of time-ordered activity tracker data, which was how each lesson began. What was unanticipated in the implementation of this particular activity was that students were deliberately modifying their walking so as to produce deviant results. For instance, after a few passes across the playground, some students began to take small shuffle steps to see how large of a number they could produce. Other students began taking exaggerated leaping steps or hopping with their feet together.

These numbers ultimately introduced more variability in the data than had been expected in the design of the lesson and in other classrooms that had completed this unit. The resulting data were aggregated and viewed in *TinkerPlots* the following day, with a "long tail" of high step values being a distinguishing characteristic especially when compared to data that were obtained when other classes of students had done this activity in earlier iterations of the unit (Fig. 12.1). This was the first time that students had seen the entire class's step data organized as a histogram, and we had expected there to be some confusion on how to interpret it. Yet we observed students easily describing what was being shown and why. To illustrate the ease with which the students interpreted the data, consider the following short exchange

Fig. 12.1 The histogram showing the number of steps required to get across the playground that was produced from data obtained by the class while walking back and forth across their playground 12 times

that took place between the student, Melinda,[1] and the teacher, Ms. Hayley, with a response from one other student, Hudson.

> Ms. Hayley: So you saw what was happening out there. What do you think was affecting that shape?
> Melinda: I think right here people were taking like really giant steps to get less, and right here people were walking normal back and forth, and I think like Hudson who was over here was like taking really tiny steps.
> Hudson: I got the tiny steps on 159.

Taken alone, the speech transcription is limited in understanding what was being communicated. Melinda's frequent use of the deictic "here" is unresolved from speech alone. However, she did make some gestures that corresponded with the use of the word "here" to clarify what she meant. Just as physicists use language in a way that places themselves in a phase state representation (Ochs et al., 1996) and students may refer to their past movements in a motion graph (Nemirovsky et al., 1998), we see a non-literal use of language that links "here" with specific activities that Melinda re-enacts. These actions and their co-deployment with speech are depicted in Table 12.1.

[1] All proper names of participants are pseudonyms.

Table 12.1 Melinda's speech and actions while explaining the shape of the histogram

Speech	Body actions
"I think right here [A] people were taking like really giant steps [B] to get less…"	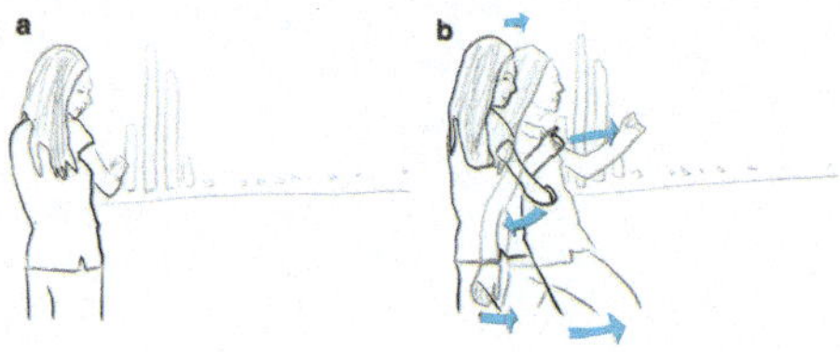 (In B, Melinda takes an exaggerated step forward)
"…and right here [C] people were walking normal back and forth [D]…"	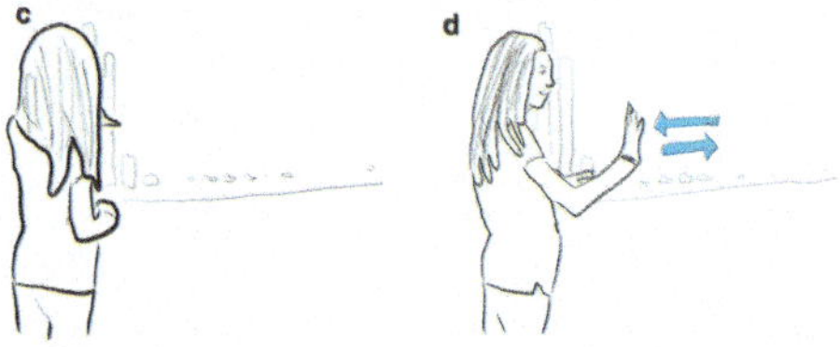 (In D, Melinda moves her open hand forward and back)
"…and I think like Hudson who was over here [E] was like taking really tiny steps [F]."	 (In F, Melinda moved her legs in shuffle steps and moved her arm back)

In her speaking turn, Melinda correctly identified segments that corresponded to different types of steps, with those who took large steps, and thus required fewer to get across the playground, being located to the left of the distribution and those who took small steps, and thus required more steps to get across the playground, being located to the right. Not only did she point to these on the projected histogram, but when talking about the "giant steps" and "tiny steps," she re-enacted them using her entire body. In some ways, her movements were a bit exaggerated, but it seemed appropriate to her as she was speaking to not only state that those kinds of steps were taken but also reproduce them herself. Interestingly, for those who were walking "normal," she did not re-enact steps. Instead, she used a single hand to simulate the back and forth walking. It seemed she was aware of what was already typical, and only felt the unusual steps needed to be re-enacted. Furthermore, she was aware of how many data points were in the various bins that she referenced. For the large steps and the normal steps, she referenced "people," but for the tiny steps, she singled out Hudson as the only person. Of additional note was the utterance from Hudson, who was seated and interjected "I got the tiny steps on 159." In stating that, he was correcting Melinda who was attributing the rightmost data point to him, while Hudson's data point was actually a different one. The attribution of the data, being obtained from the students themselves and reporting on their actions the previous days, seemed to matter to the students, both in this episode and others.

From this excerpt, we can see that Melinda was aware of what actions produced differences in steps from students, how those were positioned in the histogram, and how many people were involved in producing some of those results. Without any prompting, she physically re-enacted those movements. That was the seemingly natural way for her to express herself. This, I suggest, was enabled by her direct experience of being involved in the creation of the data and showed an ability to interpret the histogram immediately and without difficulty.

12.4.2 Episode 2: Manipulating the Box and Whisker Plot

The second episode also came from Ms. Hayley's class, after the first episode summarized above. Beyond being another multimodal account of how students talked about data representations, it also serves as a demonstration of how talk about physical activity was not exclusively retrospective. There were occasions when students were able to demonstrate enough familiarity with the kinds of activities that would generate the data that they could propose new activities that would generate data that could yield distributions with particular characteristics.

On the day of this episode, students presented their previous day's data of morning recess activity with each point being the number of steps taken in a single minute. Recess data had previously been identified as an especially useful space for inviting student commentary and questioning about their activity data (Lee & Drake, 2013). In Mrs. Hayley's class, the data points were stacked as density plots accompanied by box and whisker plots, automatically rendered in *TinkerPlots*. Most looked similar to Fig. 12.2a, with an identifiable box and identifiable whiskers. However, there was one students' plot in which the median and 25th percentile value were the same (Fig. 12.2b). This made the box in the plot appear to have only two vertical lines rather than the expected three lines. This played prominently in a portion of the discussion about prospective activities.

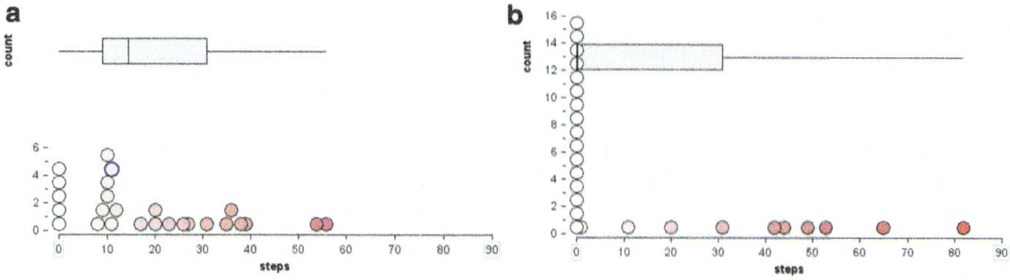

Fig. 12.2 (**a**) Lisa's recess data with a standard box and whiskers plot. (**b**) Emma's recess data with the median and 25th percentile overlapping

Having seen the different ways that box plots could appear, Ms. Hayley asked students to propose some possible challenges that students could pursue during recess to produce specific types of box plots. This episode began with Henry raising his hand to propose a challenge that referenced Emma's box plot (Fig. 12.2b). The entire episode, with student gestures, is shown in Table 12.2.

This set of transactions, all mediated by Ms. Hayley calling on students to take speaking turns, was noteworthy because students were modifying, elaborating, and critiquing each other's ideas in response to Henry's challenge. The activities to produce new recess data were speculative, but each student in this sequence, seated at their own desks, actively gestured to articulate their thinking. First, Henry suggested that the class try to produce a box plot that was a single line representing the interquartile range rather than a box. He never used words to explain his suggestion, and instead relied on the gesture of bringing his hands together to express what he was proposing. In that gesture, he was manipulating the box plot itself. Ms. Hayley offered specific language to articulate what he meant, to which he agreed.

Table 12.2 The transactions of four students about how to manipulate a box plot

Speech	Bodily actions
Ms. Hayley: What's your idea for a challenge? Henry: You know how those two lines are connected? [A] Try to get all of them, try to get all of them [B] Ms. Hayley: To get all of them on the same spot? Henry: Yeah	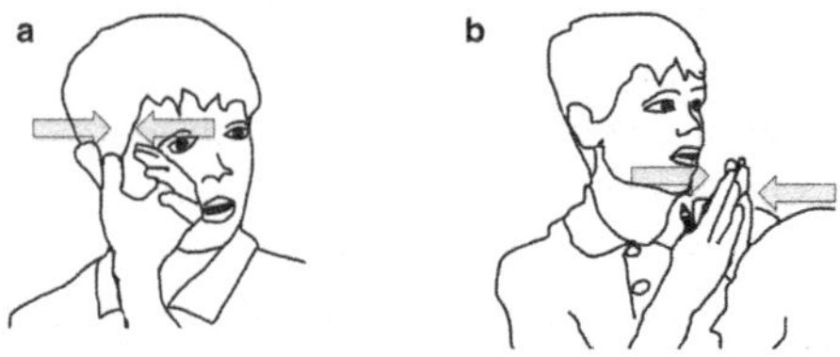
Ms. Hayley: How would you do that? Samantha? Samantha: You'd have to do one thing. Like you'd have to do the same thing the whole time pretty much. Yeah you have to sit still and just talk the whole time, or you'd have to keep running the same amount of steps every single minute. [C]… all of them in the same column [D]. So you need each minute you need to walk the same amount of steps. So you can sit there, and then like take 5 steps, sit down again, take 5 more steps, each minute, so that you have the same amount of steps in each minute [repeats C].	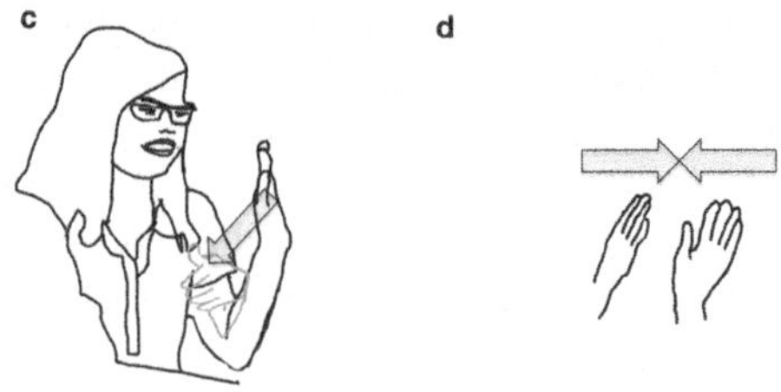 (In C, Samantha pulls her hand downward and repeats that motion)
Ms. Hayley: Okay, Carrie? Carrie: If you started running around at the first part of break, you should keep running around for the lines to get what Henry wants to get. But if you walk for like a minute [E], then sat down for a little bit [F], and then walked the exact same [E then F] for a minute more, and then sit down the same amount, and then stand up, it would be like Samantha says	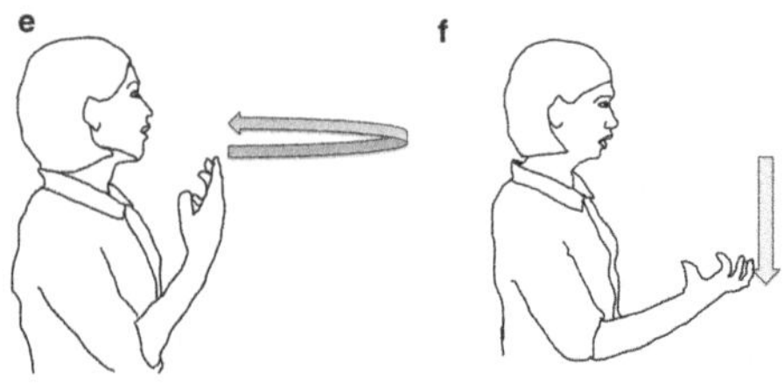

(continued)

Table 12.2 (continued)

Speech	Bodily actions
Ms. Hayley: Okay. Yes, Emma? Emma: Kind of like what Carrie's saying, and you think about it again, you have to have everything exactly the same, so like if you were running half of the break and sitting for half the break, there'd be two different chunks [G] so there's not two different box and whisker plots [H], there's one, they'd have to be the exact same [I] for the whole entire break	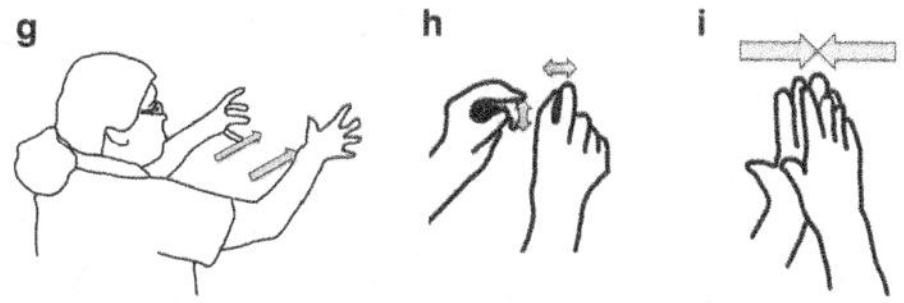

Samantha then offered a proposal for how to produce data at recess that would yield such a box plot. That involved doing the same activity the entire time ("you'd have to do the same thing the whole time pretty much"). She then elaborated that the most important factor would be making sure that each data point, representing a single minute of recess, had the same amount of steps. Gesturally, she expressed this with beat gestures to express the repetition. Then in gesture D, she explained how that would translate to the inscription for the data, with each point being in the same column in the density plot. Following that, she offered another version of a repeating activity, using a hypothetical five steps per minute with rest for the remainder of the minute and then repeating that cycle for all of recess. This was accompanied by a return to the beat gesture. With Samantha's speaking turn, she had moved fluidly between individual data points and the density plot in response to how to manipulate the box plot.

Carrie, who was sitting next to Samantha then took her speaking turn and offered two possibilities for realizing the scenario that Henry had proposed. She stated that if running was done at the beginning of the break, it should be continued to produce all the data points in one spot. That was largely consistent with what Samantha had proposed initially. Then, Carrie offered a second alternative which was to cycle walking the same distance repeatedly (as indicated by the circular motion of her finger, as if someone was running around the entire playground or the grassy field outside) with sitting down. Carrie then likened that second proposal to what Samantha had described just before. However, she had stated that sitting and walking could be alternated "the same amount." While it bore similarity to what Samantha had proposed with 5 steps and then rest, it actually would produce a different distribution.

Emma caught this statement and disagreed that Carrie's second proposal—walking for a minute or more and then sitting for the same amount of time—would produce the data display that Henry had originally proposed. Emma responded both verbally and gesturally that Carrie's second proposal would produce two different "chunks" of data in the data frequency plot. She extended both hands in front of her as if she were holding two different sets of data points or two different box plots. While each of these two "chunks" had a vanishingly small box when taken individually, Emma stated that there needed to be only one box and whisker plot. That is, the set of recess data needed to be taken as a whole. Making the same gesture of bringing hands together that Henry and Samantha had made before, Emma was reiterating that each minute would have to be the same value. In essence she was also taking up Henry's challenge, agreeing with Samantha, and challenging Carrie while talking in hypothetical and prospective activities and talking about how the data would be represented. Her manipulation involved talking about the aggregate set of data points in reference to what minute-by-minute activity behavior would have produced and then referencing the box plot. Thus, she was traversing three different ideational spaces in her turn. One was in referencing what would take place with bodies enacting research activity. Another space was related to how points would appear on the density plot. The last was related to the box plot representation and how lines would align with one another. This traversal of ideational spaces (recess activity, densities of points, and box plots) also applied collectively to this set of students and demonstrated a complex but substantive exchange about how box and whisker plots are produced and how recess activity can be translated into that inscription.

12.5 Discussion

The two episodes in this chapter, presented as cases of multimodal communications, show how students were demonstrating competence and sophistication with two canonical statistical inscriptions. These were both inscriptional forms that the students were only then learning, yet they were able to interpret them and articulate how a particular form of activity tracking data could yield those forms. The students were talking about a form of firsthand data that was about the movements of them and their classmates, and consistent with the literature on such data, were able to engage in interpreting and discussing those data rather than re-establish the prior context (Hug & McNeill, 2008). In that respect, the experience of being the source of the data mattered. The active gesturing and moving suggests that this was understood, not just as declarative information, but also in a way that was amenable to bodily expression.

In STEM education, there has been a recognition that embodiment can play a critical role in shaping how we understand different disciplinary ideas (Abrahamson & Lindgren, 2014; Hall & Nemirovsky, 2012; Lee, 2015). A core tenet of that emerging perspective is that how we physically engage with the world shapes the conceptual understandings that we eventually develop. To that position, this chapter establishes the importance of the communicative features that are expressed with the body in learning situations. The full meaning of what students are thinking and how much they understand can be better documented through acknowledgment of the multimodal nature of how we communicate, especially in learning situations. In the particular project from which these two episodes originate, students were deliberately supported in doing this by using bodily experience as the object of inquiry and as a means to think about canonical representational forms. The use of students' bodily movements as data privileged and supported the expression of their ideas in the ways that were presented above.

Considering some additional current discourses the body and STEM education, this elicits broader questions about where do meaning and understanding reside. In disciplinary educational research that has spotlighted analyses of gesture and bodies in action, some of the same authors have suggested that the manipulation of bodies in particular ways itself is actually the understanding that we seek to cultivate (Abrahamson & Sánchez-García, 2016; Nemirovsky, Rasmussen, Sweeney, & Wawro, 2012). In this view, involving the body is not a matter of teaching mathematics more efficiently or in a unique and clever way. Those ways of moving, speaking, and acting are fundamentally the mathematics that one comes to understand. That which we have historically privileged as mathematics and its understanding, such as what appears and is done on to figures and numbers on a sheet of paper, is a product of longstanding infrastructures common to classroom depictions of the discipline. When that infrastructure is disrupted, such as when geometry moves off a sheet of paper and is instead re-scaled to a large playfield, we can more readily see how mathematical understanding may appear to be different from what we traditionally treat as how one understands mathematics but still involves complex, coordinated understandings that still are fundamentally mathematical in nature (i.e., by addressing relations and properties of entities in relation to one another, in the case of Cartesian geometry) (Hall, Ma, & Nemirovsky, 2015).

The examples provided in the analyzed classroom excerpts above can be seen as illustrating that, albeit in a less dramatic fashion. Ultimately, the examples using canonical inscriptions and classroom conversations still serve to show some features of coordinated movement (expressed along with speech and inscription) as rich understandings. As we work toward new models of STEM education in the future, it would seem that providing students with direct bodily experiences and opportunities to reflect upon those could be one productive means for us to better equip students to participate in complex forms of expression in ways that are natural and accessible. By doing so, educators may be incrementally conveying that daily experience and seemingly straightforward ways of communicating are valued in the disciplines and are, in actuality, the basis from which any STEM discipline is ultimately accessed and understood.

Acknowledgement This work was supported in part by funding from the National Science Foundation under Grant No. DRL-1054280. The opinions expressed herein are those of the author and do not necessarily reflect those of the National Science Foundation.

References

Abrahamson, D., & Lindgren, R. (2014). Embodiment and embodied design. In R. K. Sawyer (Ed.), *Cambridge handbook of the learning sciences* (2nd ed., pp. 358–376). Cambridge: Cambridge University Press.

Abrahamson, D., & Sánchez-García, R. (2016). Learning is moving in new ways: The ecological dynamics of mathematics education. *Journal of the Learning Sciences, 25*(2), 203–239. https://doi.org/10.1080/10508406.2016.1143370

Church, R. B., & Goldin-Meadow, S. (1986). The mismatch between gesture speech as an index of transitional knowledge. *Cognition, 23*(1), 43–71.

Cromley, J. G., Bergey, B. W., Fitzhugh, S., Newcombe, N., Wills, T. W., Shipley, T. F., & Tanaka, J. C. (2013). Effects of three diagram instruction methods on transfer of diagram comprehension skills: The critical role of inference while learning. *Learning and Instruction, 26*, 45–58. https://doi.org/10.1016/j.learninstruc.2013.01.003

DeLiema, D., Lee, V. R., Danish, J., Enyedy, N., & Brown, N. J. S. (2016). A microlatitudinal/microlongitudinal analysis of speech, gesture, and representation use in a student's scientific explanation of phase change. In A. A. diSessa, M. Levin, & N. J. S. Brown (Eds.), *Knowledge and interaction: A synthetic research agenda for the learning sciences* (pp. 133–159). New York, NY: Routledge.

Drake, J., Cain, R., & Lee, V. R. (2017). From wearing to wondering: Treating wearable activity trackers as tools for inquiry. In I. Levin & D. Tsybulsky (Eds.), *Optimizing STEM education with advanced ICTs and simulations* (pp. 1–29). Hershey, PA: IGI Global.

Goodwin, C. (2013). The co-operative, transformative organization of human action and knowledge. *Journal of Pragmatics, 46*, 8–23.

Hall, R. (2000). Videorecording as theory. In R. Lesh & R. Lehrer (Eds.), *Handbook of reserach design in mathematics and science education* (pp. 647–664). Hillsdale, NJ: Lawrence Erlbaum.

Hall, R., & Nemirovsky, R. (2012). Introduction to the special issue: Modalities of body engagement in mathematical activity and learning. *Journal of the Learning Sciences, 21*(2), 207–215. https://doi.org/10.1080/10508406.2011.611447

Hall, R., Ma, J. Y., & Nemirovsky, R. (2015). Re-scaling bodies in/as representational instruments in GPS drawing. In V. R. Lee (Ed.), *Learning Technologies and the Body: Integration and Implementation in Formal and Informal Learning Environments* (pp. 112–131). New York, NY: Routledge.

Hall, R., Wright, K., & Wieckert, K. (2007). Interactive and historical processes of distributing statistical concepts through work organization. *Mind, Culture, and Activity, 14*(1–2), 103–127.

Hug, B., & McNeill, K. L. (2008). Use of first-hand and second-hand data in science: Does data type influence classroom conversations? *International Journal of Science Education, 30*(13), 1725–1751.

Hutchins, E., & Palen, L. (1997). Constructing meaning from space, gesture, and speech. In L. B. Resnick, R. Saljo, C. Pontecorvo, & B. Burge (Eds.), *Discourse, tools, and reasoning: Essays on situated cognition* (pp. 23–40). Berlin: Springer-Verlag.

Konold, C., & Miller, C. (2005). *TinkerPlots. Dynamic data exploration. Statistics software for middle school curricula*. Emeryville, CA: Key Curriculum Press.

Kress, G. R., Jewitt, C., Ogborn, J., & Tsatsarelis, C. (2001). *Multimodal teaching and learning: The rhetorics of the science classroom*. London: Continuum.

Lee, V. R. (Ed.). (2015). *Learning technologies and the body: Integration and implementation in formal and informal learning environments*. New York, NY: Routledge.

Lee, V. R., & Drake, J. (2013). Quantified recess: Design of an activity for elementary students involving analyses of their own movement data. In J. P. Hourcade, E. A. Miller, & A. Egeland (Eds.), *Proceedings of the 12th International conference on interaction design and children 2013* (pp. 273–276). New York, NY: ACM.

Lee, V. R., Drake, J., Cain, R., & Thayne, J. (2015). Opportunistic uses of the traditional school day through student examination of Fitbit activity tracker data. In M. U. Bers & G. Revelle

(Eds.), *Proceedings of the 14th International conference on interaction design and children* (pp. 209–218). Boston, MA: ACM.

Lee, V. R., Drake, J., & Williamson, K. (2015). Let's get physical: K-12 students using wearable devices to obtain and learn about data from physical activities. *TechTrends, 59*(4), 46–53. https://doi.org/10.1007/s11528-015-0870-x

Lee, V. R., Drake, J. R., & Thayne, J. L. (2016). Appropriating quantified self technologies to improve elementary statistical teaching and learning. *IEEE Transactions on Learning Technologies, 9*(4), 354–365. https://doi.org/10.1109/TLT.2016.2597142

Lee, V. R., & Sherin, B. (2006). Beyond transparency: How students make representations meaningful. In S. A. Barab, K. E. Hay, & D. T. Hickey (Eds.), *Proceedings of the seventh international conference of the learning sciences* (Vol. 1, pp. 397–403). Mahwah, NJ: Lawrence Erlbaum Associates.

Lee, V. R., & Sherin, B. L. (2004). What makes teaching special? In Y. Kafai, W. Sandoval, N. Enyedy, A. Nixon, & F. Herrera (Eds.), *Sixth international conference of the learning sciences* (pp. 302–309). Santa Monica, CA: Lawrence Erlbaum Associates.

Lee, V. R., & Thomas, J. M. (2011). Integrating physical activity data technologies into elementary school classrooms. *Educational Technology Research and Development, 59*(6), 865–884. https://doi.org/10.1007/s11423-011-9210-9

Lehrer, R., Kim, M.-j., & Schauble, L. (2007). Supporting the development of conceptions of statistics by engaging students in measuring and modeling variability. *International Journal of Computers for Mathematical Learning, 12*, 195–216.

Leinhardt, G., Zaslavsky, O., & Stein, M. M. (1990). Functions, graphs, and graphing: Tasks, learning, and teaching. *Review of Educational Research, 60*, 1–64.

Lemke, J. L. (1998). Multiplying meaning: Visual and verbal semiotics in scientific text. In J. Martin & R. Veel (Eds.), *Reading science*. London: Routledge.

McCloud, S. (1993). *Understanding comics*. Northampton, MA: Kitchen Sink Press.

McNeill, D. (1992). *Hand and mind: what gestures reveal about thought*. Chicago: University of Chicago Press.

Nemirovsky, R., Rasmussen, C., Sweeney, G., & Wawro, M. (2012). When the classroom floor becomes the complex plane: Addition and multiplication as ways of bodily navigation. *Journal of the Learning Sciences, 21*(2), 287–323. https://doi.org/10.1080/10508406.2011.611445

Nemirovsky, R., Tierney, C., & Wright, T. (1998). Body motion and graphing. *Cognition and Instruction, 16*(2), 119–172.

Ochs, E. (1999). Transcription as theory. In A. Jaworski & N. Coupland (Eds.), *The discourse reader*. New York, NY: Routledge.

Ochs, E., Gonzales, P., & Jacoby, S. (1996). "When I come down I'm in the domain state": Grammar and graphic representation in the interpretive activity of physicists. In E. Ochs, E. Schegloff, & S. Thompson (Eds.), *Interaction and grammar* (pp. 328–369). New York: Cambridge University Press.

Roth, W. M., & McGinn, M. K. (1998). Inscriptions: Toward a theory of representing as social practice. *Review of Educational Research, 68*(1), 35–59.

Chapter 13
Facilitating Enactment in STEM Teacher Education Within and Across Learning Spaces

Janette Bobis

Abstract Prompted by the desire to improve initial teacher education to better prepare novice teachers to support their future students to live and work in STEM rich societies in which critical thinking and reason are paramount, mathematics education has emphasized approaches focused on principles and practices of inquiry, student-centered and problem-based learning. However, the visions and practices of such approaches have consistently proven difficult for beginning teachers to translate into practice. In this chapter, I present findings of a study aimed at exploring prospective primary teachers' enactment of targeted practices for teaching mathematics in innovative ways when opportunities to approximate such practices were provided across university and school settings. Data were gathered from 54 prospective primary teachers as they undertook their first mathematics methods course and at the conclusion of their first professional experience. Analysis of written reflections from all 54 novice teachers, teaching observations and semi-structured interviews with four novice teachers, reveals that particular teacher education pedagogies, namely, rehearsals, video, and teacher educator modeling of practices and coaching during co-teaching opportunities in a range of designed settings, were particularly effective in supporting the enactment of targeted practices. I interpret the affordances of these pedagogies through an enactivist lens and, in so doing, highlight the benefits of applying embodied perspectives to explore and help develop a better understanding of learning to teach in STEM disciplines.

Keywords STEM teacher education · Practice-focused · Enactivism · Approximations of practice

---

J. Bobis (✉)
The University of Sydney, Sydney, NSW, Australia
e-mail: janette.bobis@sydney.edu.au

P. Sengupta et al. (eds.), *Critical, Transdisciplinary and Embodied Approaches in STEM Education*, Advances in STEM Education,
https://doi.org/10.1007/978-3-030-29489-2_13

Strengthening the science, technology, engineering and mathematics (STEM) disciplines is a vital and urgent goal of education (English, 2016). The increased interest in STEM has resulted in recent reconceptualizations of what students as future STEM knowledge users need to know and how they should learn. Preparing students for this type of knowledge work requires particular approaches to teaching and to the initial preparation of STEM teachers. Learning new ways of teaching, whether at preservice or inservice levels, is almost universally plagued by "problems of enactment" (Kennedy, 1999, 2016). Initial teacher education programs traditionally conduct lessons outside of classrooms, yet novice teachers are expected to enact new behaviors inside classrooms. Such expectations have proven overwhelmingly difficult for beginning teachers (Wideen, Mayer-Smith, & Moon, 1998) with the majority teaching the way they were taught (Bobis, Mulligan, & Lowrie, 2013). Echoing the views of Grossman and McDonald (2008), I argue that initial teacher education [ITE] programs need to incorporate *pedagogies of enactment* to facilitate the learning to teach process and help teachers "develop the conceptual sophistication needed" (Davis & Simmt, 2006, p. 293) to effectively enact their knowledge for teaching. One such pedagogy involves designing learning spaces for ITE to utilize both within university settings and across school settings.

In this chapter, teaching and learning in initial primary teacher education programs and the implications for how the problem of enactment can be addressed in STEM disciplines are explored through the lens of enactivism. Drawing upon international research that reflects changes in pedagogies and learning contexts in ITE programs, a case is made for designing a range of learning spaces as part of a comprehensive repertoire of pedagogies for facilitating enactment in the preparation of primary teachers of mathematics with a focus on *approximations of practice* and *rehearsals*. I discuss findings of a study aimed at exploring the enactment of targeted practices when opportunities of approximation were provided to prospective primary teachers of mathematics within the university and across school settings. My aim is to provoke discussion around theories and pedagogies of practice that facilitate enactment in STEM teacher education.

13.1 Background: New Pedagogies for Initial Teacher Education

Informed by constructivist theories of learning, recent approaches in mathematics education have been focused on principles and practices of inquiry, student-centered and problem-based learning (Anthony & Hunter, 2012). However, the visions and practices of such approaches adopted by ITE programs in the past have consistently proven difficult for novice teachers to translate into classrooms (Bobis, 2007; Kennedy, 1999; Lampert, 2010). In a study designed to explore how to better support preservice primary teachers learn to teach mathematics using an inquiry-based approach, Nicol (2006) provided extended opportunities for her preservice teachers to interact with grade 6/7 students in an embedded two-week field experience. Despite the support and encouragement provided throughout the course, student teachers resisted teaching in new ways. To the surprise of the researcher, the majority of preservice teachers found the reform-oriented visions and practices frustrating, problematic, and stressful to enact. Similarly, Aldridge and Bobis (2003) found that preservice primary teachers who received extensive experiences to interact with children that were embedded into their ITE program espoused beliefs and visions reflecting constructivist-informed teaching practices. However, once they transitioned to the classroom, the "realities of the classroom" caused beginning teachers "to make compromises" as to *if* and *how* they enacted these practices (p. 7). It appears that constructivist-based approaches to learning, including increased exposure to students and practice-based experiences during initial teacher education are, by themselves, insufficient to ensure that novice teachers can comfortably and effectively enact new practices when they transition to the classroom. The gap between what beginning teachers *know* and what they are able to *do* in a classroom context is a persistent problem of enactment that challenges all teacher education providers (Kennedy, 1999, 2016).

The past decade has seen an emerging body of research literature and a growing number of international mathematics and science teacher educators whose goals are to prepare teachers for "teaching that is more socially and intellectually ambitious than the current norm" (Lampert et al., 2013, p. 241). According to Forzani (2014), the goals of ambitious teaching are founded on the anticipation that all children will develop higher-order thinking, reasoning, and problem-solving skills—essential knowledge and skills for a STEM-focused society. In the pursuit of these goals, researchers are systematically exploring the effectiveness of various pedagogies that support beginning teachers translate the theory and knowledge they learn in their ITE methods courses into classroom practice. What each of these teacher education programs have in common is a focus on a set of core teaching practices—the teaching routines and strategies associated with high-quality instruction and considered essential for novice teachers to learn before assuming independent responsibility for their own students (Forzani, 2014; Grossman, Hammerness, & McDonald, 2009; McDonald, Kazemi, & Kavanagh, 2013).

Closely linked to the study of core practices in teacher education programs is the development of what Grossman et al. (2009) call *pedagogies of practice.* These practices are the strategies and approaches teacher educators apply to develop beginning teachers' understandings and effective use of core practices. *Approximations of practice* is one of these key pedagogies for teacher education. Approximations of practice are the learning experiences developed by teacher educators that provide novice teachers the opportunities to practice teaching in safe and approximated settings of reduced complexity.

13.1.1 Approximations of Practice

Approximations of practice are based on the belief that ITE programs need to provide multiple opportunities for novice teachers to "practice" teaching. Such opportunities occur in iterative cycles throughout practice-based methods courses and are not relegated solely to separate field placements. Approximations can occur in a variety of designed settings, "from more controlled settings in the university through the more authentic settings of classrooms" (Grossman et al., 2009, p. 284). For example, Ghousseini and Herbst (2016) used a transcript-writing task involving a constructed dialogue between a teacher and her students around a mathematical problem to approximate the practice of eliciting students' reasoning during problem-solving tasks. Parts of the teacher's dialogue were erased, with prospective teachers asked to complete the transcript by inserting moves a teacher might make to elicit students' mathematical thinking. Others such as Tyminski, Zambak, Drake, and Land (2014) provided prospective primary teachers opportunities to work in small groups of peers to plan and enact a brief mathematical activity with their classmates role-playing students. Approximations of practice used by Bobis (2007) provided opportunities for small groups of prospective teachers to co-teach a sequence of mathematics lessons with a small group of children at a local primary school while both the teacher educator and classroom teacher were present to provide feedback. Importantly, teacher educators deliberately design their settings in which the practices of teaching can be approximated with varying degrees of authenticity, generally becoming increasingly more authentic as the methods courses progress and novice teachers grow in skill and confidence to enact ambitious teaching practices.

It is through approximations of practice that *pedagogies of enactment* (Grossman et al., 2009) come to the fore in practice-based teacher education programs. Pedagogies of enactment are the strategies teacher educators use to assist novice teachers put their knowledge *about* practices *into* practice. Foremost among these pedagogies is *rehearsal* (Kazemi, Franke, & Lampert, 2009). Similar to its meaning in the performing arts, rehearsals allow novice teachers to practice particular instructional activities so as to improve their enactment in preparation for a public performance with children in the classroom. Critical to rehearsals in teacher education is the nature and processes used to provide feedback to teachers. Lampert (2010) argued that the complexity associated with teaching requires feedback to be given from a "knowledgeable other who could comment on aspects of the performance that would need to be improved" (p. 27). This means that rehearsals need to occur in settings that allow opportunities for knowledgeable others (namely, teacher educators or practicing teachers) to view and provide feedback. Furthermore, for the feedback to be effective, there must be mutual agreement as to the components of the practices or criteria by which the success of their enactment can be measured. This means that in the process of learning about core teaching practices, the components need to be clearly identifiable and articulated to novice teachers. Decomposition of a practice's components occurs when novice teachers critically reflect upon the practices they witness in each approximation of practice setting. Hence, the need for multiple opportunities to rehearse the same practice in a range of designed settings with increasing levels of authenticity.

In the remainder of this chapter, I report and discuss findings from a study aimed at exploring the impact of a mathematics methods course that provided opportunities for novice primary teachers to enact their teaching practices when opportunities of approximation were provided within university and across school settings. Prior to this, the conceptual framework underpinning the methods course at the center of the study is presented.

13.2 Conceptual Framework

Similar to Grossman et al. (2009), Ghousseini (2017) and others who are working to refine their own pedagogies of enactment, my work surrounding the development of ITE programs for primary mathematics was originally conceived from constructivist and situated perspectives of learning. In particular, a situated perspective acknowledges that all knowledge is situated, meaning that some types of knowledge are best constructed in one context rather than another and, the more authentic the context, the more effective the interaction between theory and practice (Aldridge & Bobis, 2001; Putnam & Borko, 2000). A situated perspective in teacher education does not imply that the learning of teaching practices can only occur in "real" classroom settings. As argued earlier, settings can be designed to approximate teaching contexts with various levels of authenticity in which specific practices can be rehearsed. In this way, knowledge for teaching is situated in practice—learnt while teaching rather than being taught as abstract or "inert" knowledge to be applied in a future field experience (Kennedy, 1999). However, as I progressed through analytical phases of the current study, novices' enactment of practices required interpretations beyond either a constructivist or situated perspective. While adequate to interpret novice teachers' responses to the social and intellectual challenges they faced in each new context, both perspectives failed to account for how non-cognitive knowledge (Begg, 1999) such as intuitions and emotions influenced their actions, particularly when children became involved. Enactivism provided the lens in which the intellectual, social, and emotional aspects of findings from this study could be more thoroughly interpreted.

Enactivism is a theory of cognition (learning) that views human knowledge and meaning-making as processes that involve the whole body, thus combining elements of constructivism and embodied cognition. Constructivism views learning and "coming to know" as predominantly an internal human construction that emerges as learners try to make sense of their experiences and environment; thought and actions are considered as separate. With enactivism, thought and actions are combined. Learners are not viewed as being situated within particular contexts, as in a situated cognition perspective, but inseparable. Learning is considered to be more than an intellectual activity; it is viewed as an embodied process, involving a complex web of interactions and incorporating intuitions, emotions, physical sensations, the environment, and the broader community (Sumara & Davis, 1997).

An enactivist perspective emphasizes the complexity of learning and of learning teaching (Davis, Sumara, & Kieren, 1996). As an interactional activity, the success of each instructional sequence is contingent upon the constant and complex interplay of ideas and interactions between teacher and student; they co-adapt to the context as one responds to the actions of the other. In the case of ITE, the different contexts designed to shape novice teacher knowledge and practices are therefore never static. The contexts themselves are constantly being reshaped as the interactions between teacher–learner and teacher–educator change each other's content knowledge and knowledge of teaching practices while practices are being enacted. This means that no matter how well a teacher may have planned and rehearsed an instructional activity, or how many times they teach the same activity, there will always be some level of uncertainty as each teaching situation is unique. An enactivist perspective helps to explain why the outcomes of teaching the same activity the same way to different students can never be entirely predicted—what might work in one context might not work in another. The point of rehearsals though, is that repetition of practice with opportunities for reflection will deepen novices' understandings of knowledge for teaching and better prepare them to respond to all the complexities of teaching. In terms of enactivism, routinizing many of the common core practices of teaching, teacher capacity is freed to interpret and respond to the non-routine student responses and intuitively restructure learning experiences "in-the-moment."

13.3 The Study

The aim of this study was to explore the enactment of targeted practices for teaching mathematics when opportunities of approximation with increasing complexity and rehearsals were provided to prospective primary teachers within the university and across school settings. In particular, I wanted to know if certain pedagogies employed in the course were effective in supporting novice teachers enact these practices. In the remainder of this chapter, I provide an interpretation of the study's findings through an enactivist lens. My intention is to explore how an embodied perspective can help develop a better understanding of effective pedagogies for supporting novice teachers enact ambitious teaching practices in STEM disciplines.

13.3.1 Context and Course Design

The context of this study was a twelve-week semester-long primary mathematics methods course in a University situated in a major capital city of Australia that was attended by 54 prospective primary teachers. The course was the first of two such methods courses in a postgraduate initial teacher education program that includes three professional field experiences, the first of which took place approximately six weeks after the methods course ended. The content of the course focused on the development of children's early number knowledge. In the fifth week of the course, novice teachers worked in pairs to conduct individual diagnostic interviews with two to three children aged 5–8 years from a local primary school. After analyzing student diagnostic interview data to determine number knowledge and mental strategy use in the four operations, three to four novice teachers collaborated to plan, rehearse, and teach a sequence of three lessons in weeks 10 to 12 of the course. Hence, the course was designed to not only assist novice teachers learn the knowledge and practices needed to progress children's arithmetical strategies, but to learn *how* to enact the knowledge and practices. The practices focused upon in the course were selected for their relevance to achieving this goal, including (but not limited to), eliciting and responding to students' mathematical reasoning, orchestrating productive group discussions, using mathematical representations, and teaching towards a clear mathematics instructional goal. The practices and their components targeted for scrutiny in this study are presented in Table 13.1.

Novice teachers were introduced to and learned to enact the practices by embedding them into instructional activities suitable for young children in their first three years of school. Without an actionable context for enacting the practices, they "remain an abstraction of the work of teaching" (McDonald et al., 2013, p. 382). Providing well-designed instructional activities allows novice teachers opportuni-

Table 13.1 Practices and their components targeted for exploration in the study

Practice	Practice components
Elicit, interpret, and respond to students' mathematical thinking	This practice requires teachers: (a) Know how to elicit students' reasoning/strategies via an individual diagnostic interview and whole class discussion (b) Know how to interpret students' reasoning according to research-based frameworks of children's cognitive development in number (c) Plan and teach appropriate instructional activities that will help progress students' reasoning to the next developmental level/stage
Orchestrate group discussions	This practice requires teachers: (a) Know how to implement talk moves (b) Establish and facilitate student–student and student–teacher interactions (c) Ensure all students are cognitively and emotionally engaged
Use mathematical representations	This practice requires teachers: (a) Represent mathematical ideas using different mediums (b) Know how to elicit and respond to student-generated representations (c) Assist students make connections between different kinds of representations

ties to trial new practices within fairly controlled settings, thus allowing them the mental space to focus on the more complex interactional work of teaching. For example, the practices of eliciting and responding to student thinking and orchestrating a class discussion were introduced to novices in the methods course through a range of instructional activities, including a targeted strategy discussion (Kazemi & Hintz, 2014), that were generally modeled by the teacher educator or explicated via a video of a classroom teacher enacting particular practices.

The methods course utilized three types of approximations of practice with each setting gradually increasing in its level of authenticity and complexity. The first setting occurred in the third week of semester (but was repeated on three other occasions using different instructional activities), when novice teachers role-played the part of students as their teacher educator modeled an instructional activity. Partway through the activity, modeling was paused and novices were provided with a partially completed lesson plan. The teachers were required to work in pairs and suggest the next series of moves the teacher educator could make to elicit children's thinking strategies and the likely responses from children as a result of each move. These moves and likely student responses were discussed and reflected upon with the whole class. Hence, from an enactivist perspective, the interactional relationship of teacher and learner was emphasized early in the course, and the potential for both teaching and learning to take several paths as a result became recognized as integral and inevitable to the learning environment.

The second type of approximation occurred in weeks 6, 8, and 9. At this stage, the teachers had already met and conducted individual diagnostic interviews with two or three children. Using one of the instructional activities modeled in university tutorials and adjusted to suit the needs of their targeted children, novice teachers planned and then rehearsed the implementation of a brief activity with peers role-playing the students. The teacher educator paused rehearsals at particular points to give feedback, and at times asked novices to repeat short segments to enact the suggested changes.

In the third approximation, small groups of novice teachers repeated the instructional activities they rehearsed in university tutorials, with the same group of children they had interviewed in week 5. These lessons took place in a primary school located near the university in weeks 10 to 12 of the course and were observed by the teacher educator and classroom teacher, both of whom provided feedback to novices and occasionally stepped into the role of co-teacher if needed. As part of their assignment, novice teachers collaboratively prepared lesson plans and collected records of the teaching practice including digital video and still images of children's responses. These data were used to reflect on their teaching in preparation for their impending field experience.

The whole cyclic process of learning to teach—from introducing and learning about the practices to preparing for rehearsals; from enacting the activities within a range of designed settings that gradually increased in their level of authenticity, to reflection on their enactments in preparation for teaching in real classrooms—is represented in Fig. 13.1.

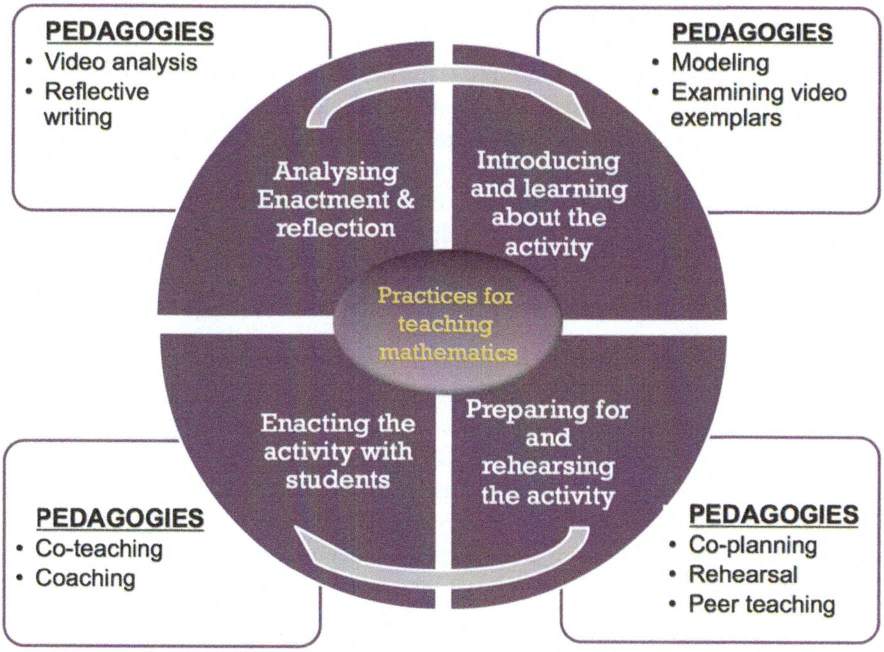

Fig. 13.1 Cycle of learning to enact targeted practices in the methods course (adapted from McDonald et al., 2013)

13.3.2 Data Sources and Analysis

Three types of data sources were used in the analysis: written reflections for all 54 novice teachers, focused observations during the co-teaching school setting, and semi-structured interviews with four case study teachers. The reflections were posted to an online discussion board in weeks 2, 6, and 10. The reflections provided insights into the novices' developing understandings of the practices and instructional activities learned and the value they attached to each one at three key points of the course—at the start of the course, immediately after conducting diagnostic interviews with children, and immediately after their first co-teaching experience with a small group of children. The four case study teachers were randomly selected for observation during the three co-teaching sessions—two co-teachers from each of the two tutorial groups. During co-teaching sessions, field notes were taken of observations that focused on the teachers' intentional planning for (evident in lesson plans) and enactment of the core practices. Semi-structured interviews with the four case study teachers took place immediately after their field placement. Each interview was approximately 1 h in duration, was conducted on the University campus and audio-recorded for later transcription. The interviews focused on novice teachers' perceptions of their capacities to enact specific practices during their recent field experience and aspects of the methods course they considered best prepared them to do so.

Data analysis occurred in three phases. The first phase focused on the online reflections to reveal novice teachers' developing understanding, valuing, and confidence to enact several practices that were explicitly introduced during the course. The second analysis phase focused on observation and interview data to capture evidence of the four case study teachers enacting the practices. In the third phase, all data sources were reviewed to search for evidence indicating the pedagogies considered most effective in supporting novice teachers to enact these practices, paying particular attention to the impact shifts in settings had during approximations of practice.

Each phase of analysis involved several waves of coding according to apriori and emerging codes (Strauss & Corbin, 1998). Apriori codes for phase 1 analysis were created based on the practices and their components listed in Table 13.1. Practices were identified when novices mentioned aspects of a particular practice, the instructional activities in which each practice was nested, or resources used in the activities (e.g., ten frames). For example, instances when novices acknowledged the significance of particular talk moves to encourage student engagement in class discussions were coded as "discussion: valuing [DV]" and when teaching observations revealed enactment of talk moves, data were coded "discussion: enactment [DE]." Similar coding processes were used for phases 2 and 3 whereby apriori codes reflected the foci for each phase of data analysis as previously outlined. An open coding process was then applied to all data sources to capture instances that revealed relevant aspects of the study but were not represented in any of the existing codes.

It quickly became obvious during the analysis process that novice teachers' growth in understanding, valuing, and confidence to enact the practices were inextricably linked to the pedagogies employed in the methods course as part of the cycle of learning to teach (Fig. 13.1). That is, understanding and valuing of the practices seemed to be enhanced through the actual opportunities provided to enact them—the settings especially designed by the teacher educator to support approximations of practice. In the following section, data sources are treated holistically as per phase 3 of the analysis process to reveal how and why these settings impacted novices' capacities to enact particular practices. Hence, reporting and discussion of findings is presented in relation to each of the approximations of practice: modeling and role-play, rehearsing while teaching peers, and co-teaching a small group of children.

13.4 Findings

13.4.1 Modeling and Role-Play

Modeling of practices in university tutorials involved the teacher educator modeling an activity as novice teachers role-played the students or via observation of a video set in a real classroom with an experienced teacher enacting the practices. In each case, the components of the practices were unpacked during follow-up discussions and novices were expected to anticipate possible student and teacher responses. This unpacking process assisted the novice teachers develop a deeper understanding of the practices and recognize their importance for use in the classroom. The importance of using and making connections between a range of mathematical representations was one of the first practices novices began to develop a deeper appreciation of. Online reflections in week 2 referred to how the instructional activities in tutorials helped novices "recognize the importance of making learning visible through concrete materials...they make learning enjoyable and engaging" for students. By the sixth week of the course, online reflective comments were completely dominated by novices exclaiming their "mind explosion" and "light bulb moment" after role-playing students in an instructional activity demonstrating fractions as division that was modeled by their teacher educator.

> I'm still reeling from the simple but incredibly clever fraction as division activity modeled in class. Who knew that by simply holding the plate of chocolates above our heads that the plates would become the "vinculum" of the fraction and we could clearly see how many parts the chocolate bars would need to be partitioned into to give each person underneath a fair share?! Amazing!! I've chosen my language carefully here as I've realised how important the language of fractions is to the process of learning them. (Andrew, Reflection Week 6)

> I had a light bulb moment during our tutorial last week... when I was told to raise the plate of chocolate bars above my head, I could instantly 'see' how fractions were just a way to work out how to divide. This highlights the importance of visual representations in mathematics when it comes to helping students understand key concepts. (Amanda, Reflection Week 6)

As demonstrated by the previous quotes, the physical act of lifting a plate of chocolates above their heads to model a fraction was a powerful "lightbulb" moment when mind and body intertwined and their knowing of fractions was deepened. Learning the mathematics while role-playing students reinforced the value of using materials to visibly represent mathematics concepts and of making connections between different kinds of representations.

Online reflective comments and interviews with the four case study teachers conducted after their field placement revealed how valuable "the videos in class were in providing a model to show me how to conduct a maths discussion..." (Cahlia, Reflection week 10). Novices considered the videos to be particularly "helpful to watch how those teachers taught those ideas, especially in a whole class setting..." (Joh, Interview). While talk moves and whole class discussions were regular parts of the methods course tutorials, it was the "real" class setting shown in the videos that novices found particularly powerful to assist their own enactment of conducting class discussions and asking questions to elicit student thinking.

13.4.2 Rehearsal and Peer Teaching

Rehearsals of lessons occurred with peers role-playing students in university tutorials one and two weeks prior to the co-teaching sessions in schools. The problem of enactment was evident even during rehearsals and peer teaching sessions. One novice reflected in her online posting that "despite learning and practicing these skills all semester, I still found that they didn't all flow… it was harder to implement than I thought." The complexity of teaching was a source of frustration to many novice teachers, with some turning to mental rehearsals prior to rehearsing with their peers. Jacklyn reflected:

> Teaching is problematic! Is my lesson too easy or too hard? Will I remember to use talk moves? I have been working on retraining my brain and my mouth! I visualize teaching students and using talk moves so they will come easier…

Jacklyn's comment highlights that building connections between *knowing* teaching practices and *doing* them takes effort and how mental and/or physical rehearsals provided opportunities for enacting pedagogical knowledge; in this sense, she realized that her mind and body needed to work in sync for her to effortlessly enact these practices. Teaching was acknowledged as "hard work" and the benefits of rehearsals were not immediately realized as indicated by Annicka in her week 10 reflection:

> I also found it helpful to watch other people rehearse their lesson plans, as there were some great ideas and valuable constructive feedback from [teacher educator], which we all benefited from. I walked away from that tutorial thinking about how I will effectively wrap up each lesson in a way that will help consolidate learning.

As Annicka's comment indicates, the benefits were often only recognized after multiple rehearsals or after viewing other groups rehearse and they had time to reflect on the feedback.

13.4.3 Co-teaching a Small Group of Children

> Throughout the semester we have been given the skills and techniques to be able to start to teach students mathematics. At times I've been overwhelmed with the responsibility I have…. However, today's school visit and teaching our four year 1 students has put it all into perspective, and not made my responsibility seem too overwhelming, because it's all been put into context. (Michael, Reflection week10)

In his comment, Michael captures the emotional and cognitive build-up and eventual release of tension many of the novice teachers experienced going into the co-teaching setting. Co-teaching a small group of children in a local primary school was without doubt considered to be the most powerful setting in which novices could approximate their teaching practices. Reflective comments from novices repeatedly referred to the co-teaching experience as: "Putting it into context," "working with *real* children," "cementing ideas," and "actual implementation." As indicated in Anthony's online reflection, a number of teachers commented on how the setting provided new insights regarding the valuing of certain practices:

> What struck me in particular was how important our 'talk moves' were when teaching the children. We covered talk moves in one of the first classes and revised them when rehearsing teaching, however it wasn't until we were with the children that I realised the importance of these strategies.

Another reflective comment from Rebecca in week 10 illustrates how she felt the new practices seemed quite easy to enact, possibly due to the repeated exposure in university-based tutorials:

> We were definitely using a lot of the 'talk moves'. I think I have found that they are used so frequently in our own learning in tutorials that they have become ingrained in us as teachers ourselves. They certainly don't feel like something that I have to force or fabricate and seem to work well with the students.

Rebecca's comment illustrates how exposure to and repeated opportunities to rehearse certain practices helped her enact them intuitively. Such instinctual responses, or knowing-in-action, provided Rebecca with the empowered feeling that "...I can do this!".

Observations of the four case study teachers during the co-teaching sessions revealed that they explicitly planned to elicit student reasoning through questioning. For instance, Olive and Bec asked questions during their interactions with children such as "Can you explain how you got your answer?" "Is there another way to find the answer?" and provided the children with mini-whiteboards to draw representations of and explain their strategies to each other. Bec later commented in her interview that "when we were taught to use questioning to elicit student thinking...I thought their main benefit would be during tests...Yet when teaching, I found them far more useful as classroom tactics to get the students to really think and explain their strategies."

One of the reasons Joh, a case study teacher, considered the co-teaching sessions with children to be so invaluable was the presence of the teacher educator being able to "come in and out of the lesson to show us how to re-engage a girl we were have trouble with. It helped me deal with a similar issue on prac...I knew what to do." Similarly, case study teacher Nikky reflected in her interview that the co-teaching with children was successful in supporting her learn many of the practices because it was "quite a controlled environment...you're not kind of managing the class and managing everything–you can focus on your teaching. On prac, I was doing the same practices but in a kind of scaled up version." In each of these cases, the previous exposure to similar circumstances, albeit of less complexity, provided novice teachers not only with the knowledge of what to do in the classroom, but the actions.

Two other aspects of the co-teaching setting that novices considered helpful in improving their teaching were the opportunities to observe and receive feedback from their co-teachers and the opportunity to film "myself teaching." Olive commented in her interview that watching "others teach, being filmed and receiving feedback from my co-teachers really helped to fine-tune some of my questioning skills and my pacing" of lessons. She even sought to "copy many elements of their teaching styles" in her lessons on field placement.

Overall, teachers' responses regarding the co-teaching setting illustrate how they felt that the co-teaching approximation of practice allowed them to bring all the practices that they had been learning together and enact them in a relatively safe setting. They felt that the complexities of teaching were relatively controlled, but still provided them with sufficient experiences of everyday teaching problems, such as that faced by Joh, to enable them to manage similar problems when on field placement.

13.5 Drawing It All Together: From Decomposition to Complex Teaching

The initial focus of tutorials was on introducing and decomposing the practices of teaching. While the decomposition of practices has met with some resistance and even skepticism in the field of teacher education, it is most likely resulting from an incomplete understanding of the pedagogy or confusing its intent with previous approaches in initial teacher education programs that focused on isolated micro-skills with no accompanying enactment pedagogies such as rehearsals to assist the translation of such skills to the complexity of real classrooms (see Forzani, 2014). In the current methods course, decomposition of practice ensured novices not only developed a deeper understanding of the components but were able to approximate these practices in small "chunks" during carefully designed settings so as not to overwhelm them immediately with the complexity of a real classroom. To achieve this, the design and selection of settings in which novices were provided to enact these practices were critical; the gradual increase in authenticity and complexity, however, must be balanced with the underpinning conceptual knowledge—why is this practice important? How can this practice enhance student learning? Nikky shared her appreciation of being provided the opportunity to deconstruct, discuss, and reflect upon each practice in her interview:

> Everything that I did in that classroom while I was on prac was influenced by what I did in that course. Everything. As I learn things at university, I have to understand them at a conceptual level…doing something in practice doesn't make sense to me unless I've made sense of it in my head first. So everything that we learnt in that course came together and helped me develop a pedagogical approach that I could take into the classroom.

In terms of enactivism, Nikky's comment reminds us of how teaching actions/practices must make intellectual sense for them to "come together" and be adapted for each new classroom environment. Other case study teachers' comments confirmed the same sentiments of feeling "well prepared" to teach mathematics using the practices focused upon in the course. Joh excitedly recounted in her interview, "When I went into a real classroom, I was doing it—and it worked!" Meanwhile, Olive was pleasantly surprised because she entered field experience "expecting a little bit more of a confrontation between what I learnt in university and what it was like in the real classroom. But for maths, it seemed to work really perfectly, especially when I could get the kids to talk about strategies." Enactivist and embodied perspectives refer to this capacity of teachers to intuitively enact practices as knowing-in-action, implying a deep understanding and coherence between the mind and body. Such knowledge-in-action of common practices frees up cognitive space, allowing teachers to deal with the more cognitively sophisticated aspects of teaching (Davis & Simmt, 2006).

> Ultimately, it was not just the settings—within the university and across school contexts—that empowered these novice teachers to begin enacting new practices in real classrooms. The settings, by themselves, have all been "done" before in ITE. It is the effect of combining the settings with some complimentary pedagogies such as modeling, rehearsal accompanied by coaching from a knowledgeable other and co-teaching.

13.6 Facilitating Enactment: An Enactivist Perspective of Teacher Education

The methods course and associated pedagogy at the center of this study were originally conceived from constructivist and situated perspectives of learning with the aim of addressing the perennial problem of enactment. The pedagogy of ITE has, until recently, been mostly hidden partly due to the fact that teacher education has not only lacked a common language around which to describe, share, and debate effective pedagogy, but as noted by Grossman et al. (2009) and Lampert (2010), has also lacked a comprehensive theory to better understand initial teacher learning. Attempts to analyze novices' enactment of practices in the current study revealed the need for a theory beyond constructivism or a situated perspective—one that could account for the intuitive or knowing-in-action aspects of novice teachers' knowledge that became prevalent as opportunities to approximate practices became more authentic in context. Enactivism provided the lens by which findings from this study could be more coherently interpreted and has ignited a re-thinking of how my novice teachers learn to teach in all aspects of my courses.

Reframing and re-examining my pedagogies involving approximation of practice and rehearsals from an enactivist perspective helped highlight the potential of embodied perspectives for better understanding effective pedagogies that support novice teachers' enactment of ambitious teaching practices in STEM disciplines. I am not yet able to declare a definitive relationship between the enactment of certain practices by novices and their pedagogical effectiveness for teaching mathematics to children. However, it is clear that there is still much to learn about effective pedagogies for ITE, particularly those that facilitate enactment and position novice teachers in roles of greater agency when learning teaching.

> Hey I'm a teacher! I think it's the first time I've ever said that, and it feels fairly appropriate, ha! (Nikky, Interview)

References

Aldridge, S., & Bobis, J. (2001). Multiple learning contexts: A vehicle for changing preservice teachers' mathematical beliefs, knowledge and practices. In J. Bobis, B. Perry, & M. Mitchelmore (Eds.), *Proceedings of the 24th Annual Conference of the Mathematics Education Research Group of Australasia Inc* (pp. 43–49). Sydney: MERGA.

Aldridge, S., & Bobis, J. (2003). Implementing beliefs, knowledge and practices: A beginning teacher's story. In L. Bragg, C. Campbell, G. Herbert, & J. Mousley (Eds.), *Proceedings of the 26th Annual Conference of the Mathematics Education Research Group of Australasia Inc* (pp. 65–71). Geelong: MERGA.

Anthony, G., & Hunter, R. (2012). (Re)Thinking and (Re)Forming initial mathematics teacher education. *New Zealand Journal of Educational Studies, 47*(1), 145–151.

Begg, A. (1999). Enactivism and mathematics education. In J. M. Truran & K. M. Truran (Eds.), *Proceedings of the 22nd Annual Conference of the Mathematics Education Research Group of Australasia, Adelaide* (pp. 68–75). Sydney: MERGA.

Bobis, J. (2007). Empowered to teach: A practice-based model of teacher education. In J. Watson & K. Beswick (Eds.), *Proceedings of the 30th Annual Conference of the Mathematics Education Group of Australasia* (pp. 61–70). Adelaide: MERGA.

Bobis, J., Mulligan, J., & Lowrie, T. (2013). *Mathematics for children: Challenging children to think mathematically*. Sydney: Pearson Education.

Davis, B., Sumara, D., & Kieren, T. (1996). Cognition, co-emergence, curriculum. *Journal of Curriculum Studies, 28*(2), 151–169.

Davis, B., & Simmt, E. (2006). Mathematics-for-teaching: An ongoing investigation of the mathematics that teachers (need to) know. *Educational Studies in Mathematics, 61*(3), 293–319.

English, L. (2016). STEM education K–12: Perspectives on integration. *International Journal of STEM Education, 3*, 3. https://doi.org/10.1186/s40594-016-0036-1

Forzani, F. M. (2014). Understanding "core practices" and "practice-based" teacher education: Learning from the past. *Journal of Teacher Education, 65*(4), 357–368.

Ghousseini, H. (2017). Rehearsals of and opportunities to learn mathematical knowledge for teaching. *Cognition and Instruction, 35*(3), 188–211.

Ghousseini, H., & Herbst, P. (2016). Pedagogies of practice and opportunities to learn about classroom mathematics discussions. *Journal of Mathematics Teacher Education, 19*(1), 79–103.

Grossman, P., Hammerness, K., & McDonald, M. (2009). Redefining teaching, re-imagining teacher education. *Teachers and Teaching: Theory and Practice, 15*(2), 273–289.

Grossman, P., & McDonald, M. (2008). Back to the future: Directions for research in teaching and teacher education. *American Educational Research Journal, 45*(1), 184–205.

Kazemi, E., Franke, M., & Lampert, M. (2009). Developing pedagogies in teacher education to support novice teachers' ability to enact ambitious instruction. In R. Hunter, B. Bicknell, & T. Burgess (Eds.), *Proceedings of the 32nd Annual Conference of the Mathematics Education Research Group of Australasia* (Vol. 1, pp. 11–29). Palmerston North: MERGA.

Kazemi, E., & Hintz, A. (2014). *Intentional talk: How to lead productive mathematical discussions*. Portland, ME: Stenhouse Publishers.

Kennedy, M. (1999). The role of preservice teacher education. In L. Darling-Hammond & G. Sykes (Eds.), *Teaching as the learning profession: Handbook of policy and practice* (pp. 54–85). San Francisco, CA: Jossey Bass.

Kennedy, M. (2016). How does professional development improve teaching? *Review of Educational Research, 86*(4), 945–980.

Lampert, M. (2010). Learning teaching in, from, and for practice: What do we mean? *Journal of Teacher Education, 61*(1–2), 21–34.

Lampert, M., Franke, M., Kazemi, E., Ghousseini, H., Turrou, A., Beasley, H., & Crowe, K. (2013). Keeping it complex using rehearsals to support novice teacher raining of ambitious teaching. *Journal of Teacher Education, 64*(3), 226–243.

McDonald, M., Kazemi, E., & Kavanagh, S. (2013). Core practices and teacher education pedagogies: A call for a common language and collective activity. *Journal of Teacher Education, 64*, 378–386.

Nicol, C. (2006). Designing a pedagogy of inquiry in teacher education: Moving from resistance to listening. *Studying Teacher Education, 2*(1), 25–41.

Putnam, R., & Borko, H. (2000). What do new views of knowledge and thinking have to say about research on teacher learning? *Educational Researcher, 29*(1), 4–15.

Strauss, A. L., & Corbin, J. M. (1998). *Basics of qualitative research: Techniques and procedures for developing grounded theory*. Thousand Oaks: Sage.

Sumara, D., & Davis, B. (1997). Enactivist theory and community learning: Towards a complexified understanding of action research. *Educational Action Research, 5*(3), 403–422. https://doi.org/10.1080/09650799700200037

Tyminski, A. M., Zambak, V. S., Drake, C., & Land, T. J. (2014). Using representations, decomposition, and approximations of practices to support prospective elementary mathematics teachers' practice of organizing discussions. *Journal of Mathematics Teacher Education, 17*(5), 463–487.

Wideen, M., Mayer-Smith, J., & Moon, B. (1998). A critical analysis of the research on learning to teach: Making the case for an ecological perspective on inquiry. *Review of Educational Research, 68*(2), 130–178.

[illegible] Handbook of [illegible] (pp. [illegible]). [illegible]

[illegible], [illegible], Kaczmarek, [illegible] (2016). [illegible] differences [illegible] during [illegible] use of [illegible] in [illegible] common [illegible]. [illegible] (pp. [illegible]).

[illegible] (2019). [illegible] In [illegible] (Eds.), [illegible] (pp. [illegible]).

Pintrich, P. R. [illegible] (2002). The role of [illegible] knowledge in learning, teaching, and assessing. [illegible], 41(4), [illegible].

[illegible] (2017). [illegible]. [illegible]: Sage.

[illegible] (2017). [illegible]. [illegible] https://doi.org/10.[illegible]

[illegible]

[illegible] (pp. [illegible]).

Chapter 14
Narrative Co-construction of Stances Towards Engineers' Work in Socio-Technical Contexts

Ayush Gupta, Chandra Turpen, Thomas Philip, and Andrew Elby

Abstract As part of their jobs, professional engineers engage in ethical, environmental, social, and economic negotiations with other engineers, managers, and with the public. Therefore, they need to understand the social impact of new technologies in a global context. However, research on students' developing sense of engineering ethics often emphasizes the "microethics" of research, mentoring, and publications. In comparison, only limited research has explored how future engineers understand "macroethics" pertaining to the social, ethical, environmental, economic, and political impact of their scientific and technological contributions. In this chapter, we present a thick description of an interview between Ayush, an engineering faculty who is also an engineering education researcher, and "Tom" (pseudonym), an engineering undergraduate student. The interview conversation focused on the macroethics of designing weaponized drones. Drawing on tools from narrative analysis and interaction analysis, we model how Ayush and Tom co-construct stances pertaining to the ethics of technology use and design. Specifically, we show how the co-construction of Ayush's and Tom's roles in the conversation as question-asker and responder, respectively, constrained sense-making and entangled with the reproduction of the social-technical divide in the conversation. The fine-grained modeling of how these roles crystallized in conjunction with the emergence of the socio-technical divide provides insight into how activity systems should be designed to foster heterogeneous meaning-making in conversations on socio-technical issues.

Keywords STEM education · Engineering education · Macroethics · Justice · Engineering ethics

A. Gupta (✉) · C. Turpen · A. Elby
University of Maryland, College Park, MD, USA
e-mail: ayush@umd.edu

T. Philip
University of California, Berkeley, CA, USA

P. Sengupta et al. (eds.), *Critical, Transdisciplinary and Embodied Approaches in STEM Education*, Advances in STEM Education,
https://doi.org/10.1007/978-3-030-29489-2_14

14.1 Introduction

In their jobs, professional engineers engage in ethical, environmental, social, and economic negotiations with other engineers, managers, and the public. Therefore, they need to understand the social impact of new technologies in a global context (Bucciarelli, 1994; Herkert, 2005). Responding to this need, engineering programs increasingly offer engineering ethics education (Haws, 2001). Though a few courses take a different approach (e.g., Colby & Sullivan, 2008; Doorn & Kroesen, 2013; Philip, Gupta, Elby, & Turpen, 2018), most engineering ethics courses focus on micro-ethics—codes of professional conduct, treatment of co-workers, responsibility to clients, etc.—to the exclusion of broader ways of thinking about the beneficial and detrimental role of engineering products and solutions (Herkert, 2005). This distinction between a focus on individual engineers and their relations with clients and co-workers versus broader socio-political entanglements of technology has been captured in various ways: professional versus social ethics (McLean, 1993), and ethics "in" versus "of" engineering (De George as discussed in Roddis, 1993), and, perhaps most popularly, "micro" versus "macro" ethics (Ladd, 1980; Herkert, 2001). Real-life scenarios demand professional engineers to negotiate between personal and professional values and to balance responsibilities towards self, team, employer, and society (Bucciarelli, 1994; Hauser-Kastenberg et al., 2003; Herkert, 2005; Riley, 2008). Engineers design technologies—social media apps, cybersecurity technologies, big-data analytics, and others—that reinforce or alter the socio-political-economic landscape, for better and/or worse.

To reason deeply and responsibly about these issues during the design process, engineers need to understand a cluster of concepts drawn from sociology, anthropology, and science and technology studies (STS) called ***social constructivism*** (Bijker, 2001; Bijker, Hughes, Pinch, & Douglas, 2012). *Social constructivism* is the idea that society and technology form a complex system in which societal and technological developments influence each other in multiple direct and indirect ways, and hence issues of power, privilege, policy, and politics in a globalized world shape and are shaped by technology (Bijker, 2001). A social constructivist framework pushes back against common narratives about engineering and technology—that technologies and other engineered solutions should be judged only by their intended and obvious uses and effects, that certain kinds of technological "progress" are inevitable and a net good, and that engineers are ethically walled off from the complex long-term effects of their work. Although not all long-term beneficial and detrimental effects can be anticipated, good design thinking includes integrating technical and moral reasoning to consider the short *and* long term, intended *and* unintended effects, direct stakeholders *and* those affected indirectly (Banks & Lachney, 2017; Leydens & Lucena, 2017).

Our educational systems, however, have made limited progress towards empowering engineering students to engage in socially responsible design practices, especially with respect to social constructivism. Courses on engineering ethics and/or single lectures in engineering courses have not produced enough of an impact on learners' ethical formation as engineers (Bielefeldt & Canney, 2016; Cech, 2014). Cech attributes this in part to an engineering culture at universities that upholds "three ideological pillars: the ideology of depoliticization, which frames any 'non-technical' concerns such as public welfare as irrelevant to 'real' engineering work; the technical/social dualism, which devalues 'social' competencies such as those related to public welfare; and the meritocratic ideology, which frames existing social structures as fair and just" (Cech, 2014, p. 45). Slaton (2015) and Riley (2008) similarly pose technocracy, the valuing of technical as separate and superior to the social aspects of any domain, as a cultural aspect of engineering that limits the integration of ethics in engineering education.

These critiques resonate with insights from science and technology studies documenting that engineering culture is dominated by technological determinism (Smith & Marx, 1994), a loose cluster of cultural narratives stating that

- Technological development inevitably leads to progress.
- Technical experts know best how to govern new technologies (technocracy).
- Technology homogenizes cultures.
- Society adapts to technology rather than shaping it.

An engineering education that takes seriously the "ethical" formation of engineers will need to be designed to challenge the narratives of technological determinism and strengthen learners' resources for understanding and practicing social constructivism. However, as research in discipline-based education has shown reliably, responsible and effective design of curriculum on a topic requires a deep understanding of students' thinking on that topic (McDermott & Shaffer, 1992; Redish & Hammer, 2009). While some progress has been made in quantitative studies of engineering students' ethical stances (Bielefeldt & Canney, 2016; Cech, 2014), qualitative studies on modeling students' reasoning about socio-technical issues has been limited. Our paper seeks to build towards filling that gap.

In this chapter, we present a thick description of an interview conversation on the ethics of designing weaponized drones between Ayush, an engineering faculty who is also an engineering education researcher, and "Tom" (pseudonym), an engineering undergraduate student. The conversation weaves in Tom's trajectory into engineering, the genesis of his course project on the ethics of weaponized drones, and explorations into the ethics of engineers designing these drones. The stances that emerge during the interview conversation are not a "read-out" of Tom's ideas; rather, these stances are shaped by both speakers, Ayush and Tom, and we model them as such (Gupta, Elby, & Sawtelle, 2016; Wortham, 2000, 2001). Through this modeling we hope to shed light on how engineers might co-construct stances towards engineering ethics in the context of socio-technical issues. These co-construction processes inform the design of research and classroom environments, as we discuss towards the end.

14.2 Data Collection and Methodology

Our data corpus consists of nine videotaped interviews with junior/senior engineering majors and engineering graduate students recruited through targeted outreach to the Science, Technology, and Society Scholars Living Learning Community (STS-LLC), Engineering without Borders (EwB), and Women in Engineering (WiE) at the University of Maryland, College Park (UMD). Participants were compensated $20 for their time.

We designed the interviews to be open-ended and ill-structured, following themes salient to the participant. So, the interview "prompts" were intended mostly to get students talking about some of these themes. Example prompts included asking students about their journey into their chosen major, their experience with STS/EwB/WiE programs (as relevant), their interest in socio-scientific issues, and how they saw the role of ethical reasoning in their future profession as an engineer.

Brief field notes taken after each interview guided preliminary data selection. We hoped to help expand a research landscape where qualitative analysis of learners' discussions on socio-technical issues is very limited. Many interviews, we felt, lent themselves to this theme. The topics of conversation ranged from life journeys to religious beliefs, and the technologies that came up also covered a wide range—prosthetics, digital technologies, agricultural technologies, new materials, weaponized drones, and more.

Two of the students, Tom (a junior-year engineering major at the time of the interview) and Matt (a junior-year computer science major at the time of the interview) talked about weaponized drones as part of their interview. As sophomores, they had written about this topic as part of a capstone research project in the STS-LLC program. We were intrigued by this topic, partly because of our own interest in the ethics of engineers participating in the design of military technologies, and partly because we were simultaneously analyzing a classroom discussion on the ethics of weaponized drones (Philip et al., 2018). Also, Tom and Matt were very willing to talk and share their views. And at least at that time, we felt that the interview allowed for exploring their views and individual histories in good depth. (Below, however, we argue that the interview missed several opportunities to expand the sphere of possibilities that could have been explored.) Given the limitations of personnel time and resources, we could conduct fine timescale conversation analysis for a limited number of interviews. We chose Matt's and Tom's interview for initial analysis presented at the Annual Conference of the American Society for Engineering Education (Gupta, Elby, & Philip, 2015). For this chapter, we further restrict our analysis to Tom, given length considerations and space required for turn-by-turn conversation analysis.

We had the interview professionally transcribed. Then, Gupta generated a content log of the interview, i.e., a summary of every ~2–3 min of the interview conversation. After this, we viewed the video in our research meeting as a group (Derry et al., 2010) with the four authors as participants. We aimed to understand what constructs contributed to Tom's construal of engineers' responsibility. Later, David

Tomblin, the Director of UMD's STS-LLC program, also participated in some analysis meetings. We also presented the preliminary analysis to a wider local community of physics and science education researchers for feedback. We took notes of our conversations at the meetings and generated analytical memos to capture our preliminary model of the constructions of engineers' responsibility in the context of weaponized drones. Subsequently, we reviewed the data systematically looking for instances that confirmed or disconfirmed our emerging understanding (Engle, Conant, & Greeno, 2007). In our analysis, we loosely draw on tools from microgenetic analysis (Siegler & Crowley, 1991), interaction analysis (Jordan & Henderson, 1995), and narrative analysis (Wortham, 2000, 2001).

Following Wortham (2000, 2001), we take the orientation that the form and content of the storytelling that unfolds in an interview is a co-construction between the interviewer and interviewee (see also, Gupta et al., 2016); and, that particular utterances simultaneously place the speaker in a specific relationship to the listener as well as to the characters and action in the story itself. In doing so, the speaker is also locally constructing themselves in that particular moment and context. Understanding the dynamics of these simultaneous positionings is important to understand the construction of the narrative. In our case, the story that we follow foregrounds how aspects of social constructivism or technological determinism are constructed through the conversation between Ayush and Tom. Since these stances often emerge over a single or a few turns of talk, our orientation from micro-genetic analysis (Siegler & Crowley, 1991) is to engage in analysis at a time scale shorter than that emergence so we can "see" the processes of construction. As such, we engage in line-by-line analysis of the transcript. From conversation analysis (Schegloff, 2007), we draw on the orientation that conversations proceed via turn-taking in which a turn of talk responds to the prior context while also altering/shaping the context in which the next turn of talk occurs. Thus, we try to analyze short segments of talk to show how qualities of their conversation, such as when we see the emergence of social-technical dualism, are being co-constructed by Ayush and Tom. We also draw on some other tools of conversation analysis such as attending to pauses (Button, 1987), repairs (Schegloff, Jefferson, & Sacks, 1977), evaluative words (Wortham, 2001), and hedge words (Kärkkäinen, 2003) to analyze how Tom and Ayush co-construct the conversation, whether and how stances taken up are weighed, and to what extent they are trying to make sense of the socio-technical scenarios in the moment.

14.3 Profiles of These Conversational Partners

Before we present the conversation between Tom and Ayush, we want to briefly present profiles of Tom and of Ayush, so that we could better understand their positionalities in the conversations. While we do not tie specific features of their histories to the bits of conversation analyzed, we feel that the profiles will give the reader a sense of who these people are, which could help in following the conversation.

The information on Tom is gleaned from his conversation with Ayush during and before the interview. The paragraph on Ayush is autobiographical.

14.3.1 Introducing Tom and Ayush

Tom responded to an email recruiting students who had "graduated" from the 2-year Science, Technology, and Society Scholars Living Learning Community (STS-LLC) to participate in interviews on their thinking about socio-technical issues. At the negotiated interview time, Tom came to Ayush's office and they walked down to the interview room. As they walked, Ayush engaged Tom in conversation about how his semester was going, what classes he was taking, etc. Tom was finishing his junior year in college as a mechanical engineering major. In his first year, Tom had joined the engineering college without declaring a major. He wanted to do something related to math and science, having succeeded in those courses in high school. He was also drawn towards design, which he distinguishes at various points in the interview from engineering because his father, who was an engineer, worked in graphic design. His mother, also an engineer, worked in marketing. He noted in the interview that being in STS-LLC with other engineering students helped make up his mind to both stay in engineering and pursue a mechanical engineering major. Tom shared that, through internships and undergraduate research experiences, he is still trying to figure out where exactly his interests lie. He wanted a design component in whatever he did, attributing this interest to being inspired by his father. He expressed that he loved the technical side of engineering but was convinced, after an internship, that he did not want project management to constitute a big part of his work. Towards the end of the interview, he said that while he thinks engineers often focus only on the technical aspect of the product, he would personally like to engage with the entire product development cycle from conceptualization to prototyping to the design of the final product.

Through this brief introduction to Tom, we want you to get a glimpse into his relationship with engineering and how that is entangled with his relationships with peers and family and his experiences in various learning environments inside and outside the classroom. During this time and later in the interview, neither person talked explicitly about race, nationality, or gender identification. We can glean from information in the interview conversations that both Tom was brought up in the United States. To Ayush, Tom presented as a white male.[1]

Ayush, the interviewer, was a faculty in physics and in engineering at UMD. He was leading the data collection in this project. So, the emails for recruitment had

[1]About 8 months after the interview, Tom participated in several group conversation sessions on socio-technical issues, also facilitated by Ayush and another facilitator. At the conclusion of those sessions (~1 year after the interview), he (and other focus group participants) filled out a survey. On that survey, Tom self-identified as male, Caucasian, and a US citizen.

come from him, and students interested in interviewing met him at his office in the Physics building. Ayush was doing electronics engineering major in college, which he finished in India. After graduating, Ayush immigrated to the United States to pursue a Ph.D. in electrical engineering. His dissertation focused on laser-plasma physics. After his Ph.D., Ayush shifted focus to work on physics education research and engineering education research. At the time of the interview, Ayush had taught a first-year introductory course on engineering design for 3 years. Through some of the interactions during the interview, we glean that Tom positions Ayush as an engineer and Ayush reinforces that positioning. At the time of the interview, Ayush also self-identified as a gay, cis-gender, South Asian male, who had been in the United States for 14 years. Going into the interview, Ayush had been thinking about how engineers think about ethical aspects of socio-technical issues, and especially how students think about whether ethical issues are entangled with the technical work engineers do. However, some of the language we now use in the manuscript, such as social constructivism, socio-technical duality, or technological determinism, was not familiar to Ayush at that time. Ayush was also not aware that the conversation during this interview would turn to military drones. However, it is important to note that Ayush was (and is) critical of military interventions as of utility in the world, and he moved from laser-plasma physics to physics education partly because he wanted not to work in industry or research related to the defense establishment. This made the topic of engineers' responsibility in the context of weaponized drones especially interesting to Ayush. Given that our protocol allowed for improvizationally following where the conversation took us, within the broad fence of talking about socio-technical issues and engineers' responsibility, Ayush could pursue engaging Tom in considering engineers' responsibility with respect to designing and producing weaponized drones. As we will see later, Ayush's lack of prior attention to social constructivism of technology limited his ability to pursue certain lines of reasoning or word questions in a manner that might have expanded the scope of the conversation. This, in turn, limited the potential for the space to be a more expansive learning experience for Tom and Ayush.

14.4 Results/Data

We first present the coarse-grained content log of the 1-h videotaped conversation between Tom and Ayush (See Table 14.1). In this chapter, we focus on the segments of conversations where weaponized drones and the responsibilities of engineers become the central focus. These segments of the conversation are in bold font in Table 14.1.

Table 14.1 Coarse-grained content log of the videotaped conversation between Ayush and Tom

Intro/Getting Started
Journey of life. Major choice. Family influence. STS influence
Negotiating the meaning of "journey of life" question
Internship experiences and future career choices; does not want to be a project manager. Project management is not engineering but lots of CE do it. Wants "some design"—connection to dad
Design as part of career; distinctions between design and technical work; also between the full product development versus the math/science calculations part. Design gets tied to the aesthetics of product
Tom changed his major from arts to engineering and then got into STS program. STS program helped build community/make friends. Describes STS experiences
Describes the STS capstone project. **First mention of the thesis of autonomous warfare "dehumanizing us."** Excitement about the project experience
How Tom chose the project, family reference, TED talk reference, describes TED talk content as drones are "going to dehumanize people"
Feelings towards the topic of drones
Campus spaces to have (or not have, to be precise) socio-technical discussions. How it is difficult to have these kind of discussions though some opportunities with fellow STS students
Is technology morally neutral? social impact of technology; social responsibility of engineers
Thinking about solutions to the "dehumanizing" problem. Accuracy is an issue
Role of engineers in mediating/regulating technology use
Social responsibility of future self as engineer; notion of engineering mindset (hearkens to earlier social-technical divide)
Back to mom/dad references and career, elaboration on design-engineering divide to say that design and engineering "mesh together" (distinct but connected). Separation between product design and marketing
Going back to whether and how design connects to the use of the technology
Whether engineers should stop designing drones. National affiliations. Pride in country. Weapons as deterrent to other nations
Engineers' agency in regulating drone use—limited agency; but "should have some say"
Smart phone issue
Social responsibility of engineers in the context of smartphone use and app design (unsure how to fully parse this)
Nanotech in low SES areas, causing pollution, scenario

The portions of the conversation that are analyzed in more detail in this manuscript are in bold

14.4.1 Segment-by-Segment Analysis of the Conversation Between Tom and Ayush

While describing his experience in the STS-LLC, Tom mentions enjoying doing the Capstone project. Ayush asks him to say a little bit more about the Capstone project. Tom first describes how the Capstone course is situated within other curricular requirements of STS-LLC and then, says:

1	Tom:	Mine happened to be...what was it...it was the effect of...it was basically the effect of autonomous warfare on our society and I think my thesis was along the lines of like dehumanizing us and making us less sensitive to war because they have drones with cameras on them and the footage is being shown. People are seeing war and death all the time. It's just going to dehumanize society ultimately.

Initial hesitations in Tom's speech at the start of this utterance suggest he might be trying to recall the project details. Tom's recounting of the project foregrounds the dehumanizing effect of weaponized drones on American ("our") society. Tom then shares a bit more about the STS-LLC poster session where students present their project reports. Ayush asks him for more details about how he explored this topic:

2	Ayush:	So how did you go about your research?
3	Tom:	Um the first thing I wanted to do was just find something that interested me. I have a couple...not family members but relatives who are in the military. They always tell me all these stories about like the new developments in technology that's coming out with our army and our navy and whatnot. So I did some research on that. I still didn't know what topic I wanted to pick 'cause there's not many like issues going... well there aren't many issues but like one that would be relevant to the class. There's actually this really interesting Ted Talk which was directly related to my thesis that basically talked about you know how these autonomous drones with no people inside them have no problem doing suicide bombings with somebody behind a keyboard thousands of miles away. It's just going to dehumanize people. It's just going to make things less sensitive. You're not actually like performing the action and seeing this trauma like right before you. Uh so that was the whole idea behind the Ted Talk and I think that ultimately made my decision.
4	Ayush:	To be [inaudible]
5	Tom:	Yeah. It was a really interesting topic. I mean there's a lot of research going on with it. Just because it's emerging so fast these new technologies, I was able to do a lot of research on that which I actually enjoyed. I think they're...I don't want to say I think they're cool 'cause they're not exactly a good thing but um it was interesting to research just from an engineering standpoint.
6	Ayush:	And you're saying it's not a good thing because of this dehumanization.
7	Tom:	Yeah.

Two things emerge in Tom's narration (#3): (a) stories of military technologies helped mediate Tom's relationship with some family members, (b) potential for autonomous drones to dehumanize people, and (c) looking across this and the previous utterance, Tom's capstone project is closely aligned with the TED Talk that he saw. In #5, Tom expresses some hesitation in labeling the military drones as cool, even while acknowledging that he enjoyed doing research on the topic. In this utterance, we see a glimmer of the separation between technical and moral considerations: drones that dehumanize society (#3) are described as interesting to research "from an engineering standpoint" (#5).[2] Ayush's next utterance (#6) does not take this up, however, but revoices a stance towards drones as "not a good thing" or

[2] This separation of the moral and technical is connected to the history of development of military drones (Whittle, 2013).

contributing to "dehumanization." Next, Ayush segues the conversation into whether Tom finds opportunities to have such discussions in other spaces on campus or with friends. After that, Ayush asks Tom how engineers should think about their ethical responsibility.

8	Ayush:	Tell me like um here's a...sometimes I hear things from the engineering students when I talk to them. Here's a view I've heard a couple times is the technology like as an engineer I'm designing technology. The technology itself is free from ethics. You know like sure people make [inaudible] for good uses or bad uses but as an engineer I'm trying to push the boundary...the technological boundary.
9	Tom:	Right. That's the dilemma isn't it? I mean engineers...the whole idea behind engineering is to build technology that makes our life simpler. That's the whole idea behind it. That's what we're trying to do with the whole information age but you know the whole dilemma is that if you're an engineer and you're putting these technologies...when do you cross the line where you're responsible for its impact on society? Just because you're building it and you know whatever you're signing off on that you know you're not responsible for how people use your technology. That's kind of part of the debate. I'm not too well versed on like you know the different technologies besides autonomous warfare that are coming out that can have potential serious impacts on our society. Um I'm sure there's plenty of them. You probably know more than I would.
10	Ayush:	So how do you think about this issue. Like where are the boundaries of the engineer's responsibility? Can you talk a little bit more?

Before interpreting this segment, we want to note that so far, it is mostly Ayush who has asked questions and Tom has responded to those questions, a trend that mostly holds during the interview. This has consequences. It legitimizes their relative positioning as information seeker (*Questioner)* and information provider (*Answerer).* Tom's response (#9) suggests that he is trying to make sense about this landscape, framing it as a "dilemma," and possibly inviting Ayush to share his views ("You probably know more than I would"). Ayush does not take up that bid to share his own views, supporting the construction of the *Questioner-Answerer* roles for Ayush and Tom, respectively. However, through these roles, both Ayush and Tom still contribute to the substance of the conversation, co-constructing the stances that emerge. Here, for example, Ayush's question (#8) introduces engineers' technical responsibility in a more abstract way rather than concretely drawing on drones. It also introduces the notion of ethics-free technology, and the separation between designers and users into the conversation, which Tom had not explicitly brought into this conversation previously. Here, we can observe the socio-technical divide being co-constructed over multiple turns of talk by both speakers (Tom and Ayush). Tom also raises the notion that there might be some threshold at which technology cannot be considered ethics-free, which allows for Ayush (#10) to probe deeper on this issue (but also functioning to position Tom as the *answerer*):

10	Ayush:	So how do you think about this issue. Like where are the boundaries of the engineer's responsibility? Can you talk a little bit more?
11	Tom:	Um well there is an engineering code of conduct, right? Um I'm not sure exactly what that states. You know I've never seen it before. Um but I think it does hold engineers liable for the things that they build. I'm not...see I think what the debate is if an engineer builds a bridge or something and it collapses that's obviously they're fault. If an engineer builds like you know a weapon of mass destruction and they distribute that, whose responsibility is that if they use it in like a bad way?
12	Ayush:	So so you're saying that uh there are engineers who are designing these weaponized drones and in some sense you're talking about how this is having an effect on dehumanizing the population. What I see you're expressing some sort of uh like it isn't clear whether the engineers who are designing these bear some responsibility towards that. You're placing the question mark around that. Is that what I'm...
13	Tom:	Um hum. Right I mean these engineers are building these drones but you know they're not the ones using them. So it's kind of a question of who's responsible. Is it the one who's actually using them or is it because you know they'd never be using them if these engineers hadn't made them.
14	Ayush:	So of course this is a sensitive topic. People have strong opinions on both sides. There are people who talk about engineers really being responsible. There are others who would say no they're not responsible.
15	Tom:	Right.
16	Ayush:	And how to decide between which position reigns truer?

In response (#11), Tom notes explicitly that the conversation is venturing into spaces that he might not have thought out fully before, suggesting that he is trying to make sense about this situation in the interview context, rather than sharing pre-formed or stable conceptions of engineers' responsibility. In doing so, Tom contrasts the "weapon of mass destruction" with the case of a collapsing bridge. To Tom, the latter seems a simpler case in which the engineer should be liable for their work. But for the weaponized drones, Tom frames the question of responsibility by drawing, again, on the relationship between designers and users of technology. Tom's utterance (#13) suggests that he's struggling between the stances that the engineer, as uncoupled to the user, is not responsible for the unethical impact of weaponized drones, and the counter-stance that the very act of creating drones is what leads to its use and impact. Ayush's next move (#14 & #16) frames the "dilemma" as a debate between these two stances. This framing has consequences: it precludes, or at least makes more difficult to emerge in the interview context, the examination of more nuanced positions that explore the connections between the different stakeholders of weaponized drones (and the emergent impact) rather than the dichotomy of engineers as responsible or not. This dichotomy is linked in the discourse to the dichotomy between creators (engineers) and users of technology; if engineers aren't morally responsible for adverse effects of technology, it's *because* the user—"the one who's actually using them," in Tom's words—is responsible instead, with the word "actually" underscoring the distinction between creating and using. This twofold dichotomy gets further reified in the following conversational segment:

16	Ayush:	And how do decide between which position reigns truer?
17	Tom:	I'm a little biased because I am an engineer. Um I don't even know I mean... I didn't really touch on that in my research. I more just don't like the effect on our society. I'm not much on responsibility. I just touched on the actual debate behind it. Um like personally I think it's hard to blame the engineers because ultimately they're doing like I said before they're building technologies to help us improve as a society. They don't have any mal-intentions with the technology they are building. If someone takes their technology and uses it in a bad way I think that's more the person who used it. That's my personal I mean you may disagree.
18	Ayush:	No no no I just want to kind of get a sense of where you know how you see the issues.
19	Tom:	Right. Like I said I'm obliged because I'm an engineer.
20	Ayush:	[chuckles] Me too.
21	Tom:	Yeah exactly. I think you know I don't know how relevant it is to like the autonomous warfare thing but let's say an engineer builds like you know a gun and someone goes out and uses that gun. You know you can't really blame the engineer for them using that.

Responding to Ayush's dichotomous question about responsibility, Tom (#17) refers to the intent of engineers that engineers are trying to help society without bad intent. (This had come up earlier in line #9.) His response (#17, #21) also retraces the separation between the engineer as the designer of technology and those who use that technology. Tom notes that his stance might be informed by his own identification with engineers; but the verbal hedges in his utterance also suggest some distance from this stance or discomfort in stating this stance, as does his explicit acknowledge that "you may disagree" (#17). Ayush's utterances (#18, #20) might be aimed at alleviating some of that discomfort. The ending of Tom's utterance in #17 could also be understood as a bid by Tom to open up the space for Ayush to share his opinions; Ayush's reply in #18 functions to not take up that bid, reinforcing Ayush's control over the direction of the conversation.

Ayush's next question asks Tom to think about what would be a responsible way forward for engineers who are building drones. Tom's response highlights the inaccuracies of the drone technology and its political ramifications:

22	Ayush:	So what practice do you see forward for, um, this now opinion. We're using these weapons which are dehumanizing society. So what practice are there moving forward?
23	Tom:	Um what do you mean by that?
24	Ayush:	So like how do we sort of address this issue of dehumanizing society?
25	Tom:	I wish I had my research paper in front of me.
26	Ayush:	[chuckles]

27	Tom:	Um I think what it's going to take is something big. I think I touched on that too. I mean at this point like we're just kind of getting into these drones. You know they're not perfect at all. In fact I have a whole section in my research paper about this one type of drone you know it's autonomous obviously. I think you can control it remotely but it's very inaccurate. They've been using it...I can't remember the country. I want to say Pakistan but I'm not sure. Don't quote me on that. They've been using it. It's been killing civilians because of its inaccuracy and I think there's going to come a point where people at least our government is just going to have to take a stand. Obviously the government of Pakistan is not going to tolerate that. It's going to come a point where we might be in a potential war because of these drones in like you know their impact on not only our society but other people's societies.

Tom expresses concerns for the civilian deaths caused by the inaccuracy of drone attacks. He links that to the geopolitics between the United States and the countries where the US military is operating drones as well as to the impact on non-US societies.

Ayush next asks Tom that, given the high stakes, could engineers have a say on how the technology is used. The conversation there explores briefly whether engineers have any agency towards regulating technology use. Ayush refers to the earlier part of the conversation where Tom had expressed an interest in professionally participating in the whole design cycle of a product (not just the technical aspects) and asks how he might think about his own professional responsibility going forward within a future job context:

27	Ayush:	Projecting forward into the future imagine what kind of job you're doing and then thinking about whether in your job you would have to grapple with this issue of the impact of your actions on society. Do you think that could be more you know how you feel about that?
28	Tom:	Um I mean I think it's something that you know every engineer is kind of … It's kind of a slippery slope. I think every engineer knows that they are partially responsible but none of them kind of...I don't think any of them...I wouldn't say any...I don't think a lot of them kind of grasp the true impact of what they're making could have on society or a particular group of people in general. Um just because you know engineers can be very closed minded. Um you know we're very technically oriented. We don't think a lot about our impacts like that. We don't really think in that way. I read that somewhere when I was doing my research too is that you know as engineers we don't really think about social impacts. We kind of just stay focused on what we're doing and then we kind of leave that all to whoever we manufacture it to.

Prompted to think about his own personal responsibility, we see a slight shift in Tom's stance from before. Now Tom says that all/most engineers know that they bear some responsibility for the societal impact of their design; the user doesn't bear *all* responsibility. But this acceptance of responsibility is balanced by his next statement where Tom says that a lot of engineers do not grasp the "true impact" of their designs, the "but" connecting "[engineers] are partly responsible" to "...[don't] grasp the true impact of what they're making" positions the ignorance of societal effects as countering the idea that engineers are "partially responsible." And this countering coheres with Tom's earlier statement (#17) that it's "hard to blame the engineers" partly because they "don't have any mal-intentions with the technology they are building"; if engineers don't fully grasp the potential adverse effects of their designs, they couldn't have intended those adverse effects to occur. Returning to turn #28, we also note the hedging and multiple starts around this utterance. These again suggest that Tom, rather than presenting previously thought-out ideas, is trying to seek some coherence between ideas that appear to be in conflict. Tom's sense-making here also draws on perceptions about how engineers think, simultaneously constructing self and engineers (as in, "...you know as engineers we don't really think about social impacts...").

Subsequently, when asked about himself, he says that given the influence of his parents, he is more disposed towards thinking about the social impact of products, beyond the "engineering mindset." Ayush prompts him if he sees the work of engineering as compartmentalized into that of engineers who design the technical prototype and those who take that prototype towards a designed marketable product. Tom clarifies that at times, as in computer-aided design, design and engineering can merge together, but the packaging and marketing of the product usually remain separate. He notes that the software used by the engineers can help test the safety and reliability of the products, but notes again that the end users of a product will often use the product with their own intent, which might differ from the engineers' intent.

At this point, Ayush asks him if engineers stop working on weaponized drones, given that drones can be misused:

29	Ayush:	So you know you mentioned about improving the accuracy of the drones. So what about the view that you can't control how people use these things. You mentioned about how technology drones get into other hands. Even the stakes are so high. Poor countries don't do war kind of a thing. Um would you say engineers shouldn't even design drones in the first place?
30	Tom:	That goes into a whole different topic...That goes into like you know our nation's defense and us being prepared for...I mean that's a whole other debate. I mean just building up our arsenal of weapons. Many people have debates you know are we spending too much money on that? Are we spending too little money? I think it's important to have this kind of technology so that if we ever need to use it we can but I don't think we should be using it entirely for the purposes that we are, basically using it accidentally...misusing it I guess. Not intentionally but I don't think we should be using it to the extent that we are because it's causing a lot more problems than helping I think.
31	Ayush:	You're referring to the civilian deaths?
32	Tom:	Yeah. I do think it's important to have them and to keep developing these technologies because that's the only way we are going to stay you know a nation well respected. You know what I'm saying like you know one that's prepared for anything.
33	Ayush:	Why is that important?
34	Tom:	Why is what?
35	Ayush:	Being a nation that is prepared?

36	Tom:	I mean because I know we're not the most respected nation and we get into things that sometime we shouldn't. If another nation is continually developing their technology and their technology is more advanced than ours and we get into war with them, we're kind of screwed if we don't have that kind of technology to counteract theirs.
37	Ayush:	So it's sort of like keeping these technological developments on levels alive.
38	Tom:	Right. It's kind of like a cold war so to speak. There's actually not any tension. It's more just like at least all the first world nations are just developing this technology constantly in preparation for anything that could happen with another first world nation.
39	Ayush:	So you think that is okay like don't misuse it, don't kill civilians.
40	Tom:	Right. I think it's definitely a [inaudible]
41	Ayush:	But it's important.
42	Tom:	Um hum. Yeah I think it's important to market that as well. Let other nations know that you know hey we have these weapons so that they know not to mess with us.
43	Ayush:	Sort of like you're saying use it simply as a piece. Don't actually use it kind of thing. Okay we have these things. Don't mess with us.
44	Tom:	Yeah I mean I think more defense standpoint. I mean I don't see any reason to use it for purposes that aren't completely necessary. So if we're being threatened by someone else I think it's important to use these weapons or at least let them know we have these weapons so that they don't you know start a war or whatnot. I don't think it should be used for much more than that personally. That's another debate.

Ayush poses a stark choice for Tom to consider: should engineers participate in the design of drones at all? In sense-making about this, Tom draws on notions of national security but also signals that he sees this as a different topic than what was being discussed before. In his response, he characterizes the current use of drones by the military as "misuse" but also notes that it is important to develop this technology. He poses two interconnected reasons: defending the country in a war with another "first world nation," and deterring that scenario in the first place. He connects both of these ideas to being a nation that is "well-respected" and "prepared for anything."

Ayush asks if engineers should have a say in determining military policy towards using drones, and Tom responds that he thinks they should but is unsure about to what extent that is feasible. The conversation then moves on to other socio-scientific issues.

14.4.2 Analysis: Patterns Across Segments of the Conversation

In this section, we look across multiple segments of the conversation between Ayush and Tom to extract some of its salient features.

Co-construction of Ayush as question-asker/interrogator and Tom as responder/position-taker: Through most of the interview, Ayush asks questions and Tom responds to them. Ayush's questions and responses to Tom's answers set up possibilities within which Tom is sense-making about the scenarios. In multiple ways, the constraints on the structure of the conversation are jointly achieved. For example, Ayush's questions often frame a dichotomy: are engineers responsible or not for the negative consequences of weaponized drones? Should engineers build weaponized drones? Tom's responses tacitly take up these dichotomies, without challenging them. The interaction pattern constrains the possibilities for more nuanced explorations of how engineers are embedded in broader social-political-economic networks and how to think about (a) responsibilities as distributed within these networks and (b) the impact of technologies as emergent from the interactions within these networks.

The pattern of Ayush asking questions and Tom responding also limits joint sense-making around these issues. Ayush affirms or accepts some of Tom's responses and probes further using questions, but doesn't jump into offering original ideas for sense-making around these issues. At various points, Tom hesitates from answering the questions or acknowledges limitations of his knowledge about the topic or the plausibility of other viewpoints (e.g., #9, #17). As shown above, however, Ayush rejects these bids to join Tom in sense-making, in a way that reinscribes his position as questioner. The questioner-answerer positioning limits the range of ideas that can arise in this space, as compared to if Tom and Ayush had collaboratively contributed to making sense of the scenario.

Co-construction of a divide between the engineer and user: A recurring theme in the interview is the relationship between the engineer who designs a product and the user of the product. Over several segments of talk, a separation—even a dichotomy—emerges between the engineer/designer and the user. In most utterances, the engineer is constructed as being responsible for the technical design of the product. In the case of drones, the negative impact of their use was constructed as resulting from their misuse by agents/users other than the engineer. As constructed in the interview, this stance did not include a pathway by which the social impact of weaponized drone technology would affect the work of engineers who design these technologies. This separation between the engineer and the user constructs the work and responsibility of the engineers as limited to technical design and its evaluation based on technical considerations: sound structural analysis of bridges, improving the accuracy of drones, etc. Engineers are held accountable to failures on these fronts, but not for broader social impacts of their designs. This stance of separation also limits the "democratic control of technology" (Feenberg, 1991).

However, there are also moments when counter-stances emerge. In the interview, Ayush notes that Tom is putting a "question mark" on whether the engineers designing weaponized drones bear responsibility for their impact. To this, Tom poses the dilemma that the users of weaponized drones wouldn't be able to use them if engineers hadn't designed them. This dilemma not only acknowledges that engineers *could* reduce the adverse impacts of drones, but also hints at the possibility that users' intents and actions are likely influenced by technologies that exist, thereby connecting engineers' work to user intent. In the segment where Tom expresses the importance of designing drones to deter war, Tom and Ayush also tacitly construct the enterprise of technology design as entangled with the broader social-political machinery. Later, when Ayush asks Tom about his own trajectory, the question possibly allows for conceptualization of the issues in a more personal context. There, Tom expresses a desire to participate in the entire process from the technical design to determining social impact of that technology. These brief moments allow us to imagine possibilities of divergence from the stance of engineer-user divide. But these moments were limited within the specific context of this interview. An interview or a group-discussion setting structured towards generating and pursuing a diversity of stances could likely have led away from convergence on the engineer-user divide.

Under-explored themes for future exploration: Family, Gender, Militarism, National Discourse, and Camaraderie: There are a variety of aspects of the conversation that we haven't analyzed in depth. We explore here, with unsure footing, some of these themes in the hopes that we (or other scholars) can pursue them in the future.

At the start of the interview, we see Tom noting that his interest in the topic was partly spurred through conversations with family members who have military associations. Tom's interest in engineering design as a future profession is also guided by his perceptions of and experiences with his parents. What role might these associations and Tom's "history-in-person" (Holland et al., 2001) have played in shaping the ideas that emerge in this conversation?

In the context of the interview conversation, Tom seemed to identify with American interests and the idea that drone-based technologies are needed to protect American supremacy in the world. The idea of nationalism, patriotism, and militarism is, however, intimately connected with gender. Our military and political systems tend to draw power from patriarchy; are male-dominated spaces; and embody stereotypical notions of masculinity (Di Leonardo, 1985; Nagel, 1998). The identification of masculinity with the role of protection overlaps with how militarism is also seen as serving the function to protect. This segment of the conversation, then, can be interpreted as the playing out, in this instance, the broader social connection between gender and militarism. However, alternative interpretations are also possible (Toktas, 2002). We need to explore further the interview as a microcosm where the gender-nationalism connection that shapes the public discourse around militarism and patriotism is being played out.

The stances that emerge in the conversation are reminiscent of public discourse in the United States around drone warfare and the general ways in which the purpose of engineering is seen in society: drones have a dehumanizing effect on society (Docksey, 2013; Healy, 2013), the need for the United States to protect against terrorist threat (Byman, 2013; Petesch, 2018), engineers as serving society, government as the only entity responsible for regulating, etc. Thus, the interview conversation between Tom and Ayush can be seen as embedded within the national discourse in the United States around war and engineering and as an example of that national phenomenon playing out in a local, private setting. In the future, we would like to explore the possibly causal micro-macro connections between talk in meaning-making about socio-technical issues and broader societal narratives.

At various points in the conversation, we get a sense of camaraderie between Tom and Ayush. For example, they both acknowledge their identity and affiliation as engineers, and they share laughs and smiles. This camaraderie could have roots in disciplinary identity (as engineers) and/or some (presumed) gender identity. What role might this have played in structuring the conversation? Could this have served to limit some of the directions that could have been pursued, or was this a missed opportunity to leverage as a footboard for launching more divergent thinking around the ethics of weaponized drones? In the future, we would like to explore how the qualities of interaction patterns such as friendship or camaraderie can affect the dynamics of the substance of talk.

14.5 Discussion

In this section, we zoom out further from the details of the conversation between Tom and Ayush, to draw out implications for the design of research and/or learning environments focused on discussions of socio-technical issues.

Questions probing socio-technical issues may unintentionally reify socio-technical divide: As discussed in the Introduction, very limited research has addressed the conversational dynamics that emerge in discussions of socio-technical issues. In this chapter, we have carefully documented the co-construction of stances within an interview setting, arguing that the interview wasn't a space for simply "eliciting" the interviewee's thoughts; rather, the stances that emerged were actively shaped by all the participants in the conversation (Gupta et al., 2016). This suggests that in educational settings where instructors or discussion leaders might think that they are eliciting students' views on socio-technical issues, the instructor's prompt itself may shape the landscape for discussion. Our analysis of how dichotomies emerged in this interview context suggests that a fruitful direction for future work may be to examine how discussion prompts in educational settings, such as asking students to take a stance on whether society influences technology or technology influences society may unintentionally reify the socio-technical divide.

Conversational roles of Questioner-Answerer within interview setting may constrain sense-making: Across multiple segments of this conversation, we consistently saw Ayush taking on the role of question-asker, and Tom taking on the role of question answerer. Within these roles, Ayush did influence the form the narrative takes (in ways discussed above), but he does not engage in active co-reasoning with Tom. In this way, Ayush remains more peripheral from the interpretative work and position-taking with respect to specific socio-technical issues. These stable roles within this activity system constrain the possibilities for sense-making about ethics within this setting. Tom may have explored very different narratives and stances if there was another active participant sharing different perspectives. So, the interview context did not provide for expansive or liberatory possibilities—and this should be interpreted as an emergent characteristic of the activity system that was built in this interview context rather than as a shortcoming of Tom. For these reasons, this interview context does not tell us how to facilitate heterogeneous meaning-making (Rosebery, Ogonowski, DiSchino, & Warren, 2010).

Research should explore how activity systems could be designed to encourage heterogeneous meaning-making: The limitations of the interview space that become visible through our analyses point to the need to design novel activity systems that would allow for heterogeneous meaning-making (Rosebery et al., 2010) and divergent thinking in the context of socio-technical issues (Haws, 2001). Activity systems with multiple participating learners, such as a focus group setting or classroom discussion, could allow for more divergent perspectives to emerge. However, here too there is the possibility of early convergence to socio-technical duality if the facilitation moves aren't designed to monitor for and gently nudge the discussion away from early convergence (Philip et al., 2018). We need to explore the design of facilitation prompts that do not tacitly embed socio-technical duality, perhaps using a version of the "hypothetical disagreement" type of question common in physics education research-based curricular materials (e.g., McDermott & Shaffer, 1998). Such questions present a brief discussion among hypothetical students who partly agree and partly disagree about a topic, with each student laying out their stance. The real students using these materials are then asked to reason through the different positions. In the context of socio-technical issues, this "hypothetical disagreement" style of prompt could be used to seed the discussion space with a variety of contradictory perspectives, with learners asked to meaning-make about these positions, explore the implications of each, and examine whether and why some positions seem more sensible or plausible to them.

Some of the counter-stances we observed in the interview emerged when the conversation allowed for the exploration of personal histories and meaning-making in the context of Tom's past and future trajectory. We need to explore how group learning environments could be designed so as not to truncate this kind of deeper exploration of individuals' histories.

14.6 Conclusion

Through this case study, we want to highlight how participants in an interview reason about engineers' ethical responsibility in an affectively charged ethical context such as autonomous warfare. For many engineering and science students, these contexts will not be mere hypotheticals in their future professional lives. As such it is important that engineering learning spaces provide opportunities for students to grapple with these issues. For instance, we feel that engineering ethics courses should include more discussions of such emotionally charged and geopolitically complex ethical issues since many of our students are likely to face these issues in their professional lives. As engineering students become more diverse, engineering classrooms will likely include a greater diversity of backgrounds and perspectives. And it is likely that discussions around these topics will be emotionally charged for students and draw on various aspects of their identity. In order to design instructional strategies to facilitate such discussions in a safe yet productive way, we need more research into understanding the socio-cognitive dynamics of how engineering students talk and think about complex socio-scientific issues. And we need to design these learning environments with the goal of allowing for divergent thinking about ethics and social responsibility of engineers (Haws, 2001). We have ventured to speculate on how the gender-nationalism-militarism connection and public narratives around drone warfare and engineering are yet other lenses through which we can view the interview interactions, but these are exploratory interpretations that we hope to flesh out in our future work.

Acknowledgments We thank the participants who volunteered to be part of the research study and instructors in the Science, Technology, and Engineering program for helping us recruit participants. We thank Michelle Porcelli, an undergraduate researcher on the team who helped with data collection and initial analysis. We thank members of the Engineering Education Research and the Physics Education Research groups at UMD, David Tomblin, and Angela Little for valuable discussions. This material is based upon work supported by the National Science Foundation under Grant Nos. NSF SES #1338700 and NSF SES #1338753. Any opinions, findings, and conclusions or recommendations expressed in this material are those of the author(s) and do not necessarily reflect the views of the National Science Foundation.

References

Banks, D. A., & Lachney, M. (2017). Engineered violence: Confronting the neutrality problem and violence in engineering. *International Journal of Engineering, Social Justice, and Peace, 5*, 1–12.

Bielefeldt, A. R., & Canney, N. E. (2016). Changes in the social responsibility attitudes of engineering students over time. *Science and Engineering Ethics, 22*(5), 1535–1551.

Bijker, W. E. (2001). Understanding technological culture through a constructivist view of science, technology and society. In S. H. Cuttliffe & C. Mitcham (Eds.), *Visions of STS* (pp. 19–33). New York, NY: State University of New York Press.

Bijker, W. E., Hughes, T. P., Pinch, T., & Douglas, D. G. (2012). *The social construction of technological systems: New directions in the sociology and history of technology*. Cambridge, MA: MIT Press.

Bucciarelli, L. L. (1994). *Designing engineers*. Cambridge, MA: MIT Press.

Button, G. (1987). Answers as interactional products: Two sequential practices used in interviews. *Social Psychology Quarterly, 50*, 160–171.

Byman, D. L. (2013). *Why drones work: The case for Washington's weapon of choice*. Published online on Brookings. Retrieved June 17, 2013, from https://www.brookings.edu.

Cech, E. A. (2014). Culture of disengagement in engineering education? *Science, Technology, & Human Values, 39*(1), 42–72.

Colby, A., & Sullivan, W. M. (2008). Ethics teaching in undergraduate engineering education. *Journal of Engineering Education, 97*(3), 327–338.

Derry, S. J., Pea, R. D., Barron, B., Engle, R. A., Erickson, F., Goldman, R., … Sherin, B. L. (2010). Conducting video research in the learning sciences: Guidance on selection, analysis, technology, and ethics. *Journal of the Learning Sciences, 19*(1), 3–53.

Di Leonardo, M. (1985). Morals, mothers, and militarism: Antimilitarism and feminist theory. Feminist Studies, 11(3), 599–617.

Docksey, L. (2013). Old and new wars: "Dehumanizing" war. Armies facing armies no longer happens? *Global Research*, January 8, 2013.

Doorn, N., & Kroesen, J. O. (2013). Using and developing role plays in teaching aimed at preparing for social responsibility. *Science and Engineering Ethics, 19*(4), 1513–1527.

Engle, R. A., Conant, F. R., & Greeno, J. G. (2007). Progressive refinement of hypotheses in video-supported research. In R. Goldman, R. Pea, B. Barron & S. J. Derry (Eds.), Video research in the learning sciences, pp. 239-254. Mahwah, NJ: Lawrence Erlbaum Associates.

Feenberg, A. (1991). *Critical theory of technology* (Vol. 5). New York, NY: Oxford University Press.

Gupta, A., Elby, A., & Philip, T. M. (2015). How engineering students think about the roles and responsibilities of engineers with respect to broader social and global impact of engineering and technology. In *Proceedings of the 2015 ASEE Annual Conference & Exposition*, Seattle, WA. https://doi.org/10.18260/p.24192

Gupta, A., Elby, A., & Sawtelle, V. (2016). Bridging knowledge analysis and interaction analysis through understanding the dynamics of knowledge in use. In A. A. diSessa, M. Levin, & N. J. S. Brown (Eds.), *Knowledge and interaction: A synthetic agenda for the learning sciences*. New York, NY: Routledge.

Hauser-Kastenberg, G., Kastenberg, W. E., & Norris, D. (2003). Towards emergent ethical action and the culture of engineering. *Science and Engineering Ethics, 9*(3), 377–387.

Haws, D. R. (2001). Ethics instruction in engineering education: A (mini) meta-analysis. *Journal of Engineering Education, 90*(2), 223–229.

Healy, J. (2013). Dehumanizing with drones and torture. *Huffington Post*, May 11, 2013.

Herkert, J. R. (2001). Future directions in engineering ethics research: Microethics, macroethics and the role of professional societies. *Science and engineering ethics, 7*(3), 403–414.

Herkert, J. R. (2005). Ways of thinking about and teaching ethical problem solving: Microethics and macroethics in engineering. *Science and Engineering Ethics, 11*(3), 373–385.

Holland, D. C., Lachicotte Jr, W., Skinner, D., & Cain, C. (2001). *Identity and agency in cultural worlds*. Cambridge, MA: Harvard University Press.

Jordan, B., & Henderson, A. (1995). Interaction analysis: Foundations and practice. *The Journal of the Learning Sciences, 4*(1), 39–103.

Kärkkäinen, E. (2003). *Epistemic stance in English conversation: A description of its interactional functions, with a focus on I think* (Vol. 115). Amsterdam, The Netherlands: John Benjamins.

Ladd, J., 1980, The quest for a code of professional ethics: an intellectual and moral confusion. In R. Chalk, M. S. FrankelL & S. B.Chafer (Eds.), AAAS Professional Ethics Project: Professional Ethics Activities in the Scienti.c and Engineering Societies (Washington, DC: AAAS), pp. 154–159..

Leydens, J. A., & Lucena, J. C. (2017). *Engineering justice: Transforming engineering education and practice*. New York, NY: Wiley.

McDermott, L. C., & Shaffer, P. S. (1992). Research as a guide for curriculum development: An example from introductory electricity. Part I: Investigation of student understanding. *American Journal of Physics, 60*(11), 994–1003.

McDermott, L. C., & Shaffer, P. S. (1998). *Tutorials in introductory physics*. Upper Saddle River, NJ: Prentice Hall.

McLean, G. F. (1993). Integrating ethics and design. *IEEE Technology and Society Magazine, 12*(3), 19.

Nagel, J. (1998). Masculinity and nationalism: gender and sexuality in the making of nations. *Ethnic and Racial Studies, 21*(2), 242–269.

Petesch, C. (2018). Drones and armored vehicles are protecting US troops in the terror fight in Africa. Published online on *Business Insider*, July 31, 2018, 10:57 AM.

Philip, T. M., Gupta, A., Elby, A., & Turpen, C. (2018). Why ideology matters for learning: A case of ideological convergence in an engineering ethics classroom discussion on drone warfare. *Journal of the Learning Sciences, 27*(2), 183–223.

Redish, E. F., & Hammer, D. (2009). Reinventing college physics for biologists: Explicating an epistemological curriculum. *American Journal of Physics, 77*(7), 629–642.
Riley, D. (2008). Engineering and social justice. *Synthesis Lectures on Engineers, Technology, and Society, 3*(1), 1–152.
Roddis, W.M.K. (1993) Structural failures and engineering ethics. *Journal of Structural Engineering, 119*, 1539–1555.
Rosebery, A. S., Ogonowski, M., DiSchino, M., & Warren, B. (2010). "The coat traps all your body heat": Heterogeneity as fundamental to learning. *The Journal of the Learning Sciences, 19*(3), 322–357.
Schegloff, E. A. (2007). *Sequence organization in interaction* (Vol. 1: A primer in conversation analysis). Cambridge: Cambridge University Press.
Schegloff, E. A., Jefferson, G., & Sacks, H. (1977). The preference for self-correction in the organization of repair in conversation. *Language, 53*(2), 361–382.
Siegler, R. S., & Crowley, K. (1991). The microgenetic method: A direct means for studying cognitive development. *American Psychologist, 46*(6), 606.
Slaton, A. E. (2015). Meritocracy, technocracy, democracy: Understandings of racial and gender equity in American engineering education. In *International perspectives on engineering education* (pp. 171–189). Cham, Switzerland: Springer.
Smith, M., & Marx, L. (1994). Does technology drive history? : The dilemma of technological determinism. Cambridge, Mass.: MIT Press.
Toktas, S. (2002). Nationalism, militarism and gender politics: women in the military. *Minerva: Quarterly Report on Women and the Military, 20*(2), 29–45.
Whittle, R. (2013). The man who invented the predator. *Air & Space Magazine*, 2013.
Wortham, S. E. (2000). Interactional positioning and narrative self-construction. *Narrative Inquiry, 10*(1), 157–184.
Wortham, S. E. F. (2001). *Narratives in action: A strategy for research and analysis*. New York, NY: Teachers College Press.

Chapter 15
Moving Beyond the Singular: A Deconstruction of Educational Opportunity in Science Through the Lens of Multiples in an Era Marked by Globalization and Neoliberalism

Jrène Rahm

Abstract In this chapter, I challenge the functionalist view of informal science education and instead, through "a lens of multiples," attend to youths' diverse forms of meaning making of science and self in science; and how these processes are charged by and grounded in placemaking (entanglement of feelings with materials, bodies, and multiple ways of knowing, being, and becoming in STEM). I do so through two case studies, first, a video production project in *ArtScience*, a club that is part of a Saturday school that reaches out to elementary school level children and families with histories of recent immigration; and second, a joint video project about a girls-only afterschool program by now young women of color who no longer participate in that program. I show how the two projects took for granted the heterogeneity of forms of engagement with science and identities as insiders to science and thereby became critical sites of critique and transformation of informal science education and visions of who can do and be in science, mediated in part also by the researcher who as a collaborator contributed to that transformation. As such, the chapter challenges visions of colored youth as disposable through a discourse on multiples.

Keywords Youth · Learning · Identity · Mobility · Informal Science · Video Production · Heterogeneity · Critical Science Literacy

---

J. Rahm (✉)
Faculté des sciences de l'éducation, Département de psychopédagogie et d'andragogie, Université de Montréal, Montreal, QC, Canada
e-mail: jrene.rahm@umontreal.ca

P. Sengupta et al. (eds.), *Critical, Transdisciplinary and Embodied Approaches in STEM Education*, Advances in STEM Education,
https://doi.org/10.1007/978-3-030-29489-2_15

In 2009, the National Research Council in its report about informal science learning put forth the argument that "informal environments can have a significant impact on science learning outcomes for individuals from non-dominant groups who are historically underrepresented in science" (Bell, Lewenstein, Shouse, & Feder, 2009, p. 301). Supported by research at the time, it spurred the development of design studies in informal science across a vast array of settings, assumed by researchers and communities who still struggle today to offer clear evidence in support of that kind of a functionalist argument. While the direct impact of the informal science field on youths' interest in science and pursuit of science careers is difficult to measure, its implicit objective of documenting "non-dominant students' mastery of dominant forms" is what needs critique. That lens to learning and identity work in science further perpetuates a narrow vision of what science learning implies and who can do science. It leaves unexplored the potential of informal educational settings in transforming "historical inequities and political structures that substantively shape learning" (Bang & Vossoughi, 2016, p. 175). It neither addresses nor questions deep assumptions about who can become an insider to science and scientist, and what we recognize and value as science. It leaves unquestioned assimilative narratives of participation in science.

In this chapter, I challenge that functionalist vision and its focus on the singular. Having pursued research on science literacy development in an array of informal settings in underserved communities for many years, I show instead what "a lens of multiples" can reveal about emergent learning opportunities in informal science practices owned by youth in an era marked by globalization and neoliberalism. Grounded in sociocultural-historical theory, I document expansive and transformative learning and identity work, which in the end takes me often away from science to other places and discourses that constitute the social futures of the youth I have worked with. It naturally also leads to questions about "how race and power operate in learning settings, especially as they may relate to privilege and marginalization" (Nasir & de Roystone, 2013, p. 266). Essentially, I re-engage with the study of creative ways of becoming, interpreting, and learning that Vygotsky had so much to say about and which the commodification of education has so eloquently marginalized or erased altogether. Re-engagement with creative learning also makes possible an unpacking of socio-material practices "analyzing agency 'non-anthropocentrically, as a situated process in which material culture is entangled'" (Knappett & Malfouris, 2008, p. xii, cited in McKenzie & Bieler, 2016, p. 14). It makes possible the documenting of "situated or practical experiences" that function as "pedagogical pivot points in enabling critical learning and social change" for learners often forgotten about in the current market driven education system (McKenzie & Bieler, 2016, p. 16).

It also calls for a focus on the affective dimensions of informal learning and, in particular, placemaking and belonging, key dimensions for understanding learning and identity work in science in informal and formal settings (Ehret & Hollett, 2016). I assume that youths' affective life is tied up with learning and becoming in science in complex ways. I also wonder how youth mobilize "feelings" or "our emotional relations to others and our emotional reactions to events—for constructive purposes" (Ehret & Hollett, 2016, p. 252).

Finally, I build on the call by Kress (2012), to engage deeply, and possibly in new ways, with "signs of learning" in our times of complexity and mobility. It takes me beyond a focus on the multimodal and multisensory in learning and identity work toward the unpacking of "agency evident in semiotic work" and hence, a new reading of affordances and possibilities (Kress, 2012, p. 129). I attend to the affective life and underlife of informal educational settings, and in what ways the coming together through social relations over time leads to the emergence of learning opportunities that matter, driven by deep emotions and shared affect, and tied to placemaking and belonging in ways empowering and potentially supportive of transformation and change at multiple levels, resulting in voice and agency (Ehret & Hollett, 2016). Given my work with immigrant youth, I also ground the work in the vast literature on transnational youth and youth circulations, implying travel both real and imagined among cultures and languages that we take as a rich toolkit and constitutive of who they are and are becoming (Lamarre, 2013). As such, I position the youth I work with as agents of their learning and identity in science and distance myself from neoliberalism's vision of youth as disposable (Giroux, 2012).

I begin the chapter with a brief overview of the theoretical framework and a selective literature review of the use of video production in science education. I then present case 1, the making of a video project in Art-Science, a club we ran for 2 years within a Saturday school organized by a community organization reaching out to immigrant families. Case 2 offers an analysis of a joint-video production among a group of six young women about *Les Scientifines*, an afterschool science program for girls only that they all participated in. In the conclusion, I return to the challenge raised by Bang and Vossoughi (2016), namely how to design for learning and identity in science in ways supportive of "sustainable and transformative change" deeply committed to and open to multiples.

15.1 Theoretical Framework

In line with a theoretical grounding in sociocultural-historical theory, it is assumed that learning opportunities emerge from interactions among youth and materials in place. Hence, learning is understood as a process of making meaning through doing, talking, and becoming in action and place (Wells, 1999). The latter offers affordances for certain forms of learning and becoming that, once appropriated by participants, result in multiple learning outcomes and as such can be understood as expansive (i.e., as building on prior forms of knowing, doing, and becoming) and transformative (i.e., resulting in new agentive ways of knowing, doing, and being) (Vygotsky, 2004). Meaning making and becoming in place are also understood as intertwined with the affective in that through interaction in place, participants develop social and material attachments to such programs, activities, and each other. As noted by Ehret and Hollett (2016), "the affective intensities of bodies moving, feeling, and generating social connections to each other, to place, and to the common goals of change-making" (p. 250) are key for understanding informal learning environments' multiple contributions to youths' lived experiences and future selves.

Meaning making and identity work emerges from and is grounded in placemaking, entailing the "active engagement of human beings with the places they inhabit" (Fettes & Judson, 2010, p. 124). As well said by Duff (2010), "to experience place is to be *affected by place*" (p. 881).

Placemaking is also an anchor for youths' identity development (i.e., assumed identity) and identity work (i.e., the making of new identities). It is through youths' participation and contribution to a community of practice and its affective force that youth can assume their identity in place while simultaneously forge new identities from place. If youth experience places in positive and empowering ways, engagement results in agency and voice and new imagined possible selves. Yet, affect like identity do not reside "in individual places or individual bodies but rather in the dynamic and relational interaction of places and bodies" (Duff, 2010, p. 886). Building on the work of Holland, Lachicotte, Skinner, and Cain (1998), I take identity work to imply an interplay of figured worlds (i.e., realms of interpretations of STEM and self in STEM), positional identities (i.e., how the system and others position the youth in STEM or in the world), the authoring of selves (i.e., how youth think of themselves in light of the former), and the making of worlds (i.e., the creation of new meanings and selves through this dynamic and ongoing process). The two case studies offer insights into the dynamic between meaning making and identity work in STEM and beyond.

15.2 Joint Video Productions with Youth in and About Science

> Any human act that gives rise to something new is referred to as a creative act, regardless of whether what is created is a physical object or some mental or emotional construct that lives within the person who created it and is known only to him. (Vygotsky, 2004, p. 7)

To facilitate creative forms of engagement with science, we engaged youth in video productions. Building on the work of Furman and Calabrese Barton (2006), we were curious in what ways video can become a means for youth to tell stories about science and reconfigure their relationship with science and "to communicate on their own terms" (p. 670) their understandings of science. Since we worked with youth with histories of immigration, video also seemed a promising tool to express understandings of science and selves in science in multiple ways other than through language alone and essentially engage youth in the creative use of media (de Block & Buckingham, 2007). In prior work (Gonsalves, Rahm, & Carvalho, 2013), we used video to engage in joint-questioning about science in the lives of girls and their peers in an afterschool program. The girls we worked with opted to interview others about science and its role in their lives, and then produced a rich story about figured worlds of science of urban youth. The production process led to rich discussions about science and what counts as science in different settings, and how these multiple ways of understanding science constituted their identity as learners of science. It also led to discussions about the manner engagement with science is marked by social status, power, and gender. While the produced video made evident youths' funds of knowledge and everyday practices of science, the girls discounted it as not being about "real science."

In these studies, video production became a means to speak back to dominant visions of science and stereotypical images of doers of science (Luttrell, 2010). Video production supported "hybrid, unsanctioned literacy practices" and encouraged youth to pursue multimodal representations of science and their positioning in science with a critical gaze (Rogers, Winters, LaMonde, & Perry, 2010). Video production also implied deep questioning of their role and place in science and engagement in a critical reading of science, a reading that could then become transformative and agentive, leading to new ideas about how to promote more equitable engagement and futures in science. An interest in these kinds of processes and goals drove our projects, and the use of video led to a focus on the following: First, youths' meaning making of science (the construction of new science knowledge in light of their past ways of knowing and understanding) and self in science (how they perceived themselves in relation to STEM); and second, how these processes were charged by and grounded in placemaking (entanglement of feelings with materials and bodies and multiple ways of knowing, being, and becoming in STEM). Case 1 speaks more closely about challenges second language learners experience in education, whereas Case 2 attends more closely to the being and becoming in STEM over time of young women of color.

15.3 Case 1. Video Productions in *ArtScience*: Stories About Languaging, Meaning Making, and Becoming in Science

15.3.1 Context

ArtScience was embedded in a community program, *Aspiration*, reaching out to immigrant families. In the context of their Saturday tutoring program that they run in collaboration with six elementary schools within an ethnically diverse underserved community in Montreal, we co-designed *ArtScience,* a science club we ran for 2 years from 2009 to 2011. Its goal was to create a space for student interest-driven science activities, animated in part by a science major, a graduate student in education, and myself. The design of the club was inspired by a previous work that explored the effects of "doing science" on language minority students' learning and becoming (Rosebery, Warren, & Conant, 1992). Inquiry science was understood as a tool for language and STEM literacy development. We worked with two groups of 14 youth, aged between 8 and 12 years, primarily from the Philippines, Sri Lanka, Bangladesh, Morocco, China, and the Caribbean. All activities were recorded on video given our goal to document with them student-owned engagement with science and a science practice responsive to their needs (Vossoughi & Escudé, 2016).

15.3.2 Emergent Learning Opportunities, Affordances, and Transformation

French was the language of instruction in *ArtScience*, as mandated by the community organization and the language charter of Quebec which declared French as the official language of Quebec in 1977. However, most youth in *ArtScience* struggled academically due to that language charter and the kind of language discontinuities they experienced between their home and school. Most youth who participated in the club were at ease in English, had oral fluency in their native language, but struggled with French. They were still developing strategies to manage the "language obstacle course" they faced daily (Lamarre, 2013). In the club, we encouraged students to mobilize their entire language repertoire. Hence, they switched forth and back constantly between English, French, and native languages, when working in their teams, while whole group dialogue mediated by us was typically conducted in French. I focus on a video production of three youth, Vasu (11 years old), his brother Viskar (9 years old), and Sami (11 years old). They were born in Canada to parents from Sri Lanka and spoke primarily Tamil and English at home and with each other, and French at school. Viskar was sent to the program by his teacher to work on sentence structures and reading, Vasu to work on his attention and to develop effective working strategies, while Sami was described by his teacher as very hardworking yet in need of more opportunities to engage with others in French. Together, they pursued a video about volcanoes as shown in the timeline in Fig. 15.1. Their

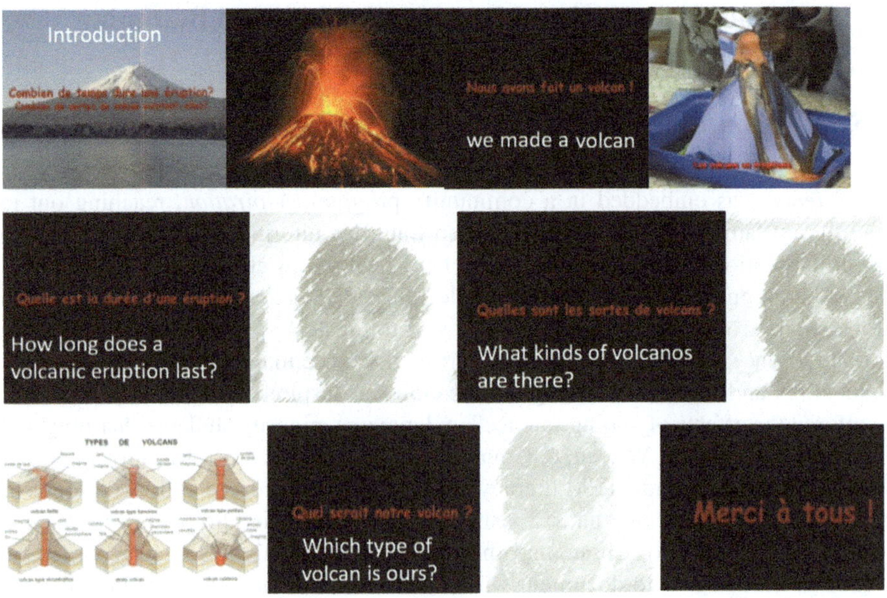

Fig. 15.1 Visual depiction of the storyline of the video

video entailed some video footage of a simulation of an eruption of a volcano with images, sound, and strolling text, and periods of talk by each one of them looking directly into the camera, conveying some scientific information and terminology about the kinds of volcanos that exist and forms of eruptions. Sami responded to the question about how long an eruption might last, while Vasu offered a list of the scientific terms of the different volcano types, and Viskar explained what type of volcano they had constructed for the simulation.

As shown, their video production was multimodal, weaving together images, movie clips, presentations assumed by them, text, and sound, attesting to much creative joint work. Taking a closer look at their work, I was struck by the energy and time the team put into the recording of their voices. Each one of them had a handwritten note of scientific information that they had copied from the web and reformulated somewhat with the help of the instructor who encouraged them to use their own words. The team struggled appropriating the scientific terms and pronunciation. The recording called for concentration and patience, but at the same time, was supported by respectful relationships among youth and the instructor as shown below:

Talk	Image	Activity
Sami: An eruption lasts one to six months in 10% of all cases of volcanic eruptions, six months to an year in 12% and 5-10 years in 2%, less than 10 minutes in 10% of all cases		Sami is reading his note that he placed on the right side of the screen. We can see his face being recorded in the middle, and his finger on the keyboard, controlling the beginning and the end of the recording.
Sami: Yeah, done, finished!		Calling out loud with a big smile
Ray: Wait, no, it cut the last ten percent, Sami, can you redo, sorry...		Instructor checks recording and notes that the beginning was cut off
Sami: What?		
Ray: Yes, see here...		
Sami: I do not hear a thing...		
Ray: We miss the five...		

Talk	Image	Activity
Sami: Oh no, not again…	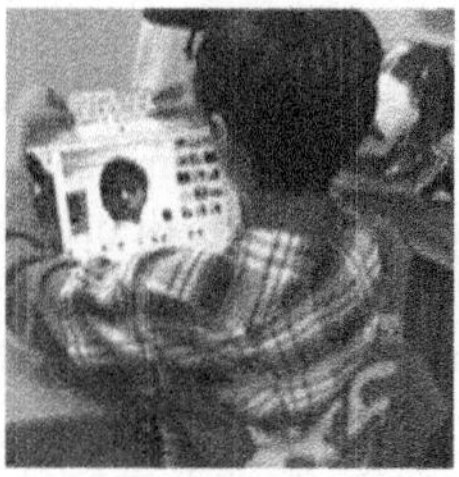	Third trial. Straight back signals level of concentration by Sami

Ray encouraged Sami to try one more time. The peers of Sami stepped back and practiced their portion of the talk, giving him the space needed to begin the recording anew. Video self-recordings are challenging, but even more so for second language speakers whose accents and struggles with pronunciations are readily evident in such recordings. Listening to oneself on video can also be emotionally charging. Yet, the team did not shy away from trying it. Other groups resorted to the inclusion of written text only. This team essentially experienced and lived "the gift of confidence" that Mahn and John-Steiner (2002) discussed, in that they knew that it was safe to try. The instructor's insistence on redoing it and getting it right made evident the high expectations he had of them yet also confidence that they could succeed. These kinds of affectively charged moments led to the development of a sense of belonging to *ArtScience*. Essentially, play with language in this manner constituted placemaking (Lamarre, 2013). It implied teamwork and solidarity among the youth, as the following excerpt makes evident:

Talk	Image	Activity
Sami: One, two, three, go!		Sami is standing behind Vasu.
Vasu reads: There are six types of volcanos, the fissure, the shield, the dome, the ash-cinder, the composite, and the caldera		Ray, standing on his right, holds the paper. Vasu in the middle reads the script and records himself.

Note above and also in Fig. 15.2 (left) how the team worked together with Vasu looking on while Ray and Sami held up the poster board with the questions they had developed, thereby also ensuring an artistically interesting background. They all supported Viskar's recording who was nervous about recording and doing it well. Following the recording, the team sat together with Ray to check the video for accuracy and potential glitches as shown in Fig. 15.2 to the right.

Fig. 15.2 Affectively charged moments recording (left) and reviewing their video clip (right)

What was at stake is evident from their facial expressions. They were anxiously watching the recorded clip together with the instructor on the left. Notable are the signs of relief on the right, once they judged their video a success. It makes evident their emergent and yet still developing identity as speakers of science but also masters of technology and French. The video project gave them the opportunity to author selves as "being able" to engage with science in French, their third language. In essence, they were on their way toward "finding a voice, or controlling a new discourse" (Rosebery et al., 1992, p. 92) to which they had little access to elsewhere.

The excerpts above also make evident what it takes to support students' transformation of feelings in ways supportive of constructive purposes (Ehret & Hollett, 2016). Initially, the level of stress was high, with some teams abandoning the idea of videotaping themselves. By persisting and taking up the challenge to film oneself in this manner, however, these teams developed an affinity that mediated placemaking and belonging in *ArtScience*. In many ways, the case makes evident "how identities are formed (improvised) in the flow of historically, socially, culturally and materially shaped lives" (Rogers et al., 2010, p. 300). Improvisation and risk-taking constituted in important ways the participants' engagement in the activity. The challenges video production implied in this context (e.g., mastery of technology, mastery of science content, language), and the kind of embodied learning it supported, also led to the "thickening" of relationships. As such, learning was more than a cognitive act. Instead, as Linds et al. (2015) describe it, "learning is felt" and "is a sensation" (p. 6). Those kinds of feelings and sensations then led to empowering images of selves and possible future selves—as youth who *can* achieve and be successful. And it is that kind of identity work which challenged the functional imperative typically associated with informal science clubs. The club was not about STEM per se. In fact, the club barely had them engage with and think scientifically through the video project. Yet, it certainly opened up a small part of the world of science to them. In that process, heterogeneity was valued in terms of the languages youth could use to talk science, in terms of the format the video production could take, and the kind of multimodal modes they were encouraged to leverage to convey meaning (Rosebery, Ogonowski, DiSchino, & Warren, 2010). Hence, the project empowered youth to come to see themselves in new ways, as youth who can succeed despite their complex histories of immigration and struggles with language, two dimensions still too often seen as barriers rather than strengths for meaning making and identity work in science and beyond.

Fig. 15.3 A timeline of the introduction

15.4 Case 2. What the Joint-Creation and Sharing of a Documentary About a Girls' Afterschool Science Program Reveals About Being and Becoming in Science

15.4.1 Context

Les Scientifines is an afterschool science program in Montreal, serving girls from two elementary and one high school in its community. They offer hands-on science activities, the opportunity to pursue science fair projects, and a science newsletter activity. The video production project I report here took place in the winter of 2013 and is an extension of a previous 2-year video ethnography of the afterschool program and its scientific newsletter activity that took place from 2009 to 2011. My research assistant and myself worked with six girls, some of whom were no longer with the program given their age and having moved on to college. We met six times to produce a video about the program for an international science conference. The six participating girls then had a chance to share their video production and exchange with youth from other informal science programs at an international science education meeting via video conference. The six girls had complex immigration histories. Their parents came from Bangladesh, Trinidad, St Lucie, and Sri Lanka. Five of the six girls were born in Canada. All of them had traveled back to their home country for family gatherings and were tightly connected with the families and culture of their country of origin, often dialoguing with family members through the Internet.

15.4.2 A Lens of Multiples, Science, and Being in Science

The joint video production about *Les Scientifines* is an interesting mix of multiple discourses. The video begins with the program logo and a read aloud by one girl of an introduction in English as follows:

Les Scientifines is a non-profit organization founded in 1987. Its goal is to promote science and develop life skills in young girls. They hold different free activities every day afterschool from the journalism workshop to the science fair.

After the logo, two girls enact "doing science" in stereotypical ways by mobilizing powerful markers of science. It begins with the mixing of colors in beakers, pursued by two of the girls who are wearing lab coats and eye protection. The two girls mix two colors (red and blue). One color is in an Erlenmeyer flask, whereas the other liquid is in a beaker. Then both the colors are mixed in a bigger-sized beaker in front of the camera. While mixing happens, the camera briefly zooms in on the mixing and the emergent color "brown" and then backs out and closes in on the two girls who are engaged in the doing of science, and who show much excitement about their accomplished experiment. Achyntia jumps up in the air, while Saliha then takes the lead in mixing vinegar with baking soda in a test tube, resulting in a bubbling substance that then spills out of the small test tube, followed by expressions of excitement and waving by Achyntia, marking the end of that staged performance. Figure 15.3 offers a timeline of that introduction:

The rest of the movie shows the girls sitting in front of the camera See Fig. 15.3, image on the lower right), first introducing themselves, and then presenting the program. They basically respond to staged questions by the research assistant and offer prepared answers, a script they co-created and practiced. As with the youth in *ArtScience*, the taping of their staged dialogue was a challenge and implied many trials to get it right. Unlike the other group, the girls were fluent in English and French, leading them to introduce the program in English given the audience they were targeting (science educators at an international conference), while the informal dialogue that followed was pursued in French, and later translated, and subtitles added. The girls exhibited flexible multilingualism in that they moved among languages constantly in their informal talk but knew when boundaries between languages mattered (Lamarre, 2013).

Yet, the documentary about the program was about "things thicker than words" (Rogers & Schofield, 2005). It was about a program that most of the girls considered as their second home and that helped them develop aspirations for their future by identifying and being encouraged by other women who were successful and by engaging in meaningful science activities with others who shared an interest in becoming educated:

One of the goals of the program is to get girls interested in science, since science, at least traditionally, was not for women. Pursuing a career in the sciences was also not something women thought about traditionally. So, giving girls the taste for science so they can pursue their futures in or beyond science is important. And the women working in the program, and all the invited guest speakers are all good role models for the girls in the program. The participating girls can be inspired by them and become like them. It gives all of us a taste to learn more about science, to be curious, to ask questions, to become better adults… for me, Les Scientifines is an inspiration. [Alana, Group interview, 2010]

Participation in the program supported the development of "a science affinity-identity" (Adams, Gupta, & Cotumaccio, 2014, p. 15)—in that the girls learned about science but also contributed to science by becoming the next generation of role models as women in science, and as women pursuing nontraditional careers in the sciences—goals the program was designed for. Yet, that vision was not as readily articulated and apparent in their video, where they presented themselves primarily as youth living in an underserved community and in need of a safe place to go after school, and as needing additional educational opportunities of high quality. Take for instance the first part of the dialogue that followed the introduction in the documentary:

Audrey:	Personally, what did you get out of Les Scientifines?
Sari:	Mmh, doing better academically, for instance, in French, the writing. That's it, it also improved our understanding, or for doing research, if they gave us a project at school, where we had to do research, we knew from Les Scientifines how to do it and also knew about some scientifically sound websites that we could then consult.
Achyntia:	Me, here, me, here, what I like at Les Scientifines is the fact that I feel safe here, 'seriously' (to emphasize she mixed English 'seriously' into her French) I feel really at ease and safe here (giggling)
Alana:	It's also, it's also for the parents, they know that you go some place serious after school, a safe space, and that you are not about to just hang out in the street.
Achyntia:	Yes, exactly!
Alana: *Later*	So it's really good for the parents also. You feel like being at home here

The dialogue positions the actors and the program in an underserved community which might not be safe for hanging out on the street. The program offers safety to girls who need it given how busy their parents are. In doing so, the program helps the girls' parents to "feel good" despite the fact that they cannot offer their children an education in a safer neighborhood. It positions the girls and their families as in need of programs that help them manage their lives and integration in an educational system that is new to them. Interestingly, the girls themselves contributed to the maintenance of such a discourse given the manner they presented the program, science, and themselves in the documentary. At the same time, the program supported the emergence of a collective identity and "a sense of group membership with like-minded peers" (Adams et al., 2014, p. 16), in that it was a safe space to show interest in science and in becoming somebody.

Later, when they were asked to find one word to describe Les Scientifines, they referred to it as "fun" while others added, "you do things you would not do otherwise outside of school" or another, "we have access to experiences that are really interesting." That discourse hints at the value of making such quality programs accessible to girls in underserved communities. Simultaneously, it makes evident some of the many contradictions that marked the girls' everyday lives, such as not having access to quality education as youth growing up in an underserved community or being shameful about their community and sense of self, as became apparent through informal talk during a work session:

> To say the truth, it is only recently that I realized that *Les Scientifines* is a program that aims to support youth in underserved communities. And I lived through a period when I was ashamed to be from that community. At my current high school, there are girls from many different neighborhoods from Montreal, and most are not poor, and so it was really embarrassing for me to admit that I came from this neighborhood. Yet, now, I can say that I am from an underserved community and it is what made me become who I am now, and I am very proud about who I have become… and thanks also to having participated in *Les Scientifines*, I realized that I do not have be ashamed about living in an underserved community, it is just a fact, and it actually helped me to become better, more open-minded, and better able to understand others who are ashamed about their place of origin. [Mohini, Informal conversation, 2013]

In the end, through the video production, opportunities emerged to work through some of those contradictions by naming contradictions and the positionings they implied of them, and build on each other's experiences. As such, the program was about much more than science, the initial depiction of it up-front in their video.

15.4.3 Storying of Selves, Selves in Science, and Science

Through five editing sessions, some parts of the script were cut out or readjusted, other episodes were filmed a second time or readjusted, a time-consuming and not always trivial process inherent to video editing as shown in Fig. 5. Hence, that tedious yet collaborative effort over a couple of weeks not only led to a video production, but also supported the girls' reflections about selves, selves in science, and science (Gonsalves et al., 2013). As shown in the brief exchange below, it led to the development of deep affinities given the affectively charged work the video production process implied, and placemaking in ways we described in case 1.

Talk	Image	Activity
Achyntia: Ah, c'est laid *(Agh, it's awful)*		Watching video clip together and deciding how to edit it.

The storying of selves through the making of the documentary and the video production was tied up in complex ways with the kinds of resources and tools youth had at their disposition. Yet, the documentary also became both, "a window to the world" the girls lived in (see quote before by Mohini) and a "window to identity." The latter is evident in the following dialogue where some of the girls position themselves as science savvy and doers of science (see also Fig. 15.3):

Achyntia:	And there is the newsletter too, where you write articles.
Alana:	That's it, you do your research.
Achyntia:	That's it!
Alana:	And as you say, you go to many websites and find information, it helps you become critical, to know when something is credible.
Achyntia:	That's when I decided to become a writer, because of the article I wrote.
Sari:	Yes, its like this, you read, and when you do the research on internet, you improve your capacity to make a summary in your own words, you do the research, you use synonyms and other words.
Alana:	You learn to popularize science.
Sari:	It's popularizing science and all, that makes your whole life simpler, when you do oral presentations, when you do research, you no longer just copy things.

Essentially, the program supported the development of an identity as a writer and communicator of science. It enlarged their figured worlds of science and sense in science, leading to the authoring of new selves and worlds not always aligned with the manner the system positioned them. Through the program activities, the girls were essentially offered opportunities to "try on" and play with other types of identities that then positioned them as insiders to science, at least momentarily. It is in this manner that the program offered opportunities for meaning making in science and the storying of selves, selves in science, and an introduction to the multiple discourses of science.

15.5 Discussion

15.5.1 Critical Science Literacy and the Case for Multiples

What do we mean by critical STEM literacy? And how do the two cases speak to critical science literacy? In what ways do the two cases transcend power differences between researchers and youth, between what counts as science in mainstream and what may count in an informal setting? In what ways do they contribute to the refiguring of who can become somebody in science?

As suggested in the report by the National Research Council (Bell et al., 2009), informal science programs may play a particularly important role in offering especially youth in underserved communities with opportunities to critically reflect upon their relationship with science and build the confidence needed to come to see themselves as insiders to science. The two cases certainly suggest that video production became a tool to engage with science into them meaningful and new ways, leading to rich reflections about science and who can be an insider to science. In doing so, the video production projects also challenged the singular view of what science is, what form participation in science takes, and what an identity as an insider to science looks like. The projects took for granted the heterogeneity of forms of engagement with science and identities as insiders to science. Those diverse forms and practices were not seen as a problem in need of fixing, but instead, as fundamental to learning and as a richness to mobilize (Rosebery et al., 1992; Rosebery et al., 2010).

It is in this way that the case studies are about multiples. They were told in ways to highlight different notions of doing science and being in science. The visual ethnography and participatory video projects were a means for "working with things, objects and artefacts" (Mitchell, 2011, p. 37). The process of co-production was also a powerful means to support youth's placemaking and develop a sense of belonging. These affective dimensions are essential features of a practice supportive of student-powered learning and transformative forms of participation and identity work in science. Both programs and emergent practices supported students' engagement in learning in ways they valued and could come to own. The two practices also encouraged critical reflections about that learning and identity work. The joint creation of videos led the teams to engage in deep reflections about themselves as learners, as learners of science, and also about their future selves in and beyond science. The projects helped youth refigure who they are and who they thought of becoming. The study of youths' editing choices and processes of co-creation also make transparent youths' bids for recognition of selves and programs in ways aligned with imagined images and discourses of STEM that ensure public recognition (Halverson, 2010). Essentially, they can be read as "trying on" identities in STEM and as such are about local agency and voice. They also orient the participating youth toward "new" or "different" social futures and aspirations for which they initially did not feel entitled to or knew about.

Counter-stories as those offered in this chapter are essentially about relationships between non-dominant youths' everyday experiences and discourse practices and "the everyday practices of professional disciplines" (Rosebery et al., 2010, p. 324). And as shown here, these everyday practices "can be mobilized productively in learning and teaching" (ibid). For instance, had we insisted on the use of French only in *ArtScience*, as mandated by the current language policies in Quebec, we would have compromised youths' participation. By valuing the tool kits and language repertoires they brought to the program, we literally multiplied opportunities for learning. It made the teams take up the challenge to tape their science talk, even if it took them multiple trials and was emotionally tasking. Clearly, the narrow vision of language development and learning and current language policy in place needs to be revised (Lamarre, 2013). As is, that policy undermines the complexity of current language practices of multilingual youth in Montreal, and their skills at navigating a vast array of educational practices that all have different stakes attached to them in terms of the language that needs to be spoken, and in terms of the kind of knowledge that counts. Attending to that multiplicity and youths' incredible navigation skills of such a complex and politically charged educational landscape can only help us move forward toward a more inclusive and empowering system of education. It would offer rich insights into what STEM as practices of critical literacies could imply and how we may design for it in a vast range of educational practices together.

In sum, the stories in this chapter make evident that we need new ways to study learning and identity work in STEM, that we need to reposition programs as those described here as sites of critique and transformations, and our own position as researchers from collaborators to co-constructors of such change-making over time. We need longitudinal studies with youth and programs to develop the kind of affinity and emotional safety that is needed to work together on voice, agency, transformations, and new social futures. As researchers, we also need to move *with* youth and be open to scale making with them while contributing to it—the latter makes our work naturally political and steeped in practice with them. It might also move us away from STEM as we know it toward an appreciation of multiples, and most important, beyond a discourse of youth as disposable.

References

Adams, J., Gupta, P., & Cotumaccio, A. (2014). Long-term participants. A museum program enhances girls' STEM interest, motivation and persistence. *Afterschool Matters*, pp. 13–20.

Bang, M., & Vossoughi, S. (2016). Participatory design research and educational justice: Studying learning and relations within social change making. *Cognition and Instruction, 34*(3), 173–193.

Bell, P., Lewenstein, B., Shouse, A. W., & Feder, M. A. (Eds.). (2009). *Learning science in informal environments: People, places, and pursuits*. Washington, DC: The National Academies Press.

de Block, L., & Buckingham, D. (2007). *Global children, global media: Migration, media and childhood*. London, England: Palgrave Macmillan.

Duff, C. (2010). On the role of affect and practice in the production of place. *Environment and Planning D: Society and Space, 28*(5), 881–895.

Ehret, C., & Hollett, T. (2016). Affective dimensions of participatory design research in informal learning environments: Placemaking, belonging, and correspondence. *Cognition and Instruction, 34*(3), 250–258.

Fettes, M., & Judson, G. (2010). Imagination and the cognitive tools of place-making. *The Journal of Environmental Education, 42*(2), 123–135.

Furman, M., & Calabrese Barton, A. (2006). Capturing urban student voices in the creation of a science mini-documentary. *Journal of Research in Science Teaching, 43*(7), 667–694.

Giroux, H. A. (2012). *Disposable youth: Racialized memories and the culture of cruelty*. New York, NY: Routledge.

Gonsalves, A., Rahm, J., & Carvalho, A. (2013). "We could think of things that could be science": Girls' refiguring of science and self in an out-of-school-time club. *Journal of Research in Science Teaching, 50*(9), 1068–1097.

Halverson, E. R. (2010). Film as identity exploration: A multimodal analysis of youth-produced films. *Teachers College Record, 112*(9), 2352–2378.

Holland, D., Lachicotte, W., Skinner, D., & Cain, C. (1998). *Identity and agency in cultural worlds*. Cambridge, MA: Harvard University Press.

Kress, G. (2012). Recognizing learning: A perspective from a social semiotic theory of multimodality. In I. de Saint-Georges & J. Weber (Eds.), *Multilingualism and multimodality* (pp. 119–140). Rotterdam, Netherlands: Sense Publisher.

Lamarre, P. (2013). Catching "Montréal on the move" and challenging the discourse of unilingualism in Québec. *Anthropologica, 55*(1), 41–56.

Linds, W., Goulet, L., Episkenew, J.-A., Schmidt, K., Ritenburg, H., & Whiteman, A. (2015). Sharing the talking stones: Theatre of the oppressed workshops as collaborative arts-based health research with indigenous youth. In D. Conrad & A. Sinner (Eds.), *Creating together: participatory, community-based, and collaborative arts practices and scholarship across Canada* (pp. 3–19). Laurier: Waterloo, ON.

Luttrell, W. (2010). 'A camera is a big responsibility': A lens for analysing children's visual voices. *Visual Studies, 25*(3), 224–237.

Mahn, H., & John-Steiner, V. (2002). The gift of confidence: A Vygotskian view of emotion. In G. Wells & Claxton, G. (Eds.), Learning for life in the 21st century: Sociocultural perspectives on the future of education (pp. 46-58). Malden, MA: Blackwell Publishing.

McKenzie, M., & Bieler, A. (2016). *Critical education and sociomaterial practice. Narration, place, and the social*. New York, NY: Lang.

Mitchell, C. (2011). *Doing visual research*. Los Angeles, CA: Sage.

Nasir, N. S., & de Roystone, K. (2013). Power, identity, and mathematical practices outside and inside school. *Journal for Research in Mathematics Education, 44*(1), 264–287.

Rogers, T., & Schofield, A. (2005). Things thicker than words: Portraits of youth multiple literacies in an alternative secondary program. In J. Anderson, M. Kendrick, T. Rogers, & S. Smythe (Eds.), *Portraits of literacy across families, communities and schools* (pp. 205–220). New York, NY: Routledge.

Rogers, T., Winters, K.-L., LaMonde, A.-M., & Perry, M. (2010). From image to ideology: Analysing shifting identity positions of marginalized youth across the cultural sites of video production. *Pedagogies: An International Journal, 5*(4), 298–312.

Rosebery, A. S., Ogonowski, M., DiSchino, M., & Warren, B. (2010). "The coat traps all your body heat": Heterogenity as fundamental to learning. *The Journal of the Learning Sciences, 19*(3), 322–357.

Rosebery, A. S., Warren, B., & Conant, F. R. (1992). Appropriating scientific discourse: Findings from language minority classrooms. *The Journal of the Learning Sciences, 2*(1), 61–94.

Vossoughi, S., & Escudé, M. (2016). What does the camera communicate? An inquiry into the politics and possibilities of video research on learning. *Anthropology & Education Quarterly, 47*, 42–58. https://doi.org/10.1111/aeq.12134

Vygotsky, L. S. (2004). Imagination and creativity in childhood. *Journal of Russian and East European Psychology, 42*(1), 7–97.

Wells, G. (1999). *Dialogic inquiry. Toward a sociocultural practice and theory of education*. New York, NY: Cambridge University Press.

Chapter 16
Critical Transdisciplinary STEM: A Critical Numeracy Approach to STEM Praxis by Urban Environments and Education Research Coven

Atasi Das and Jennifer D. Adams

Abstract Aligned with the call to view STEM as critical literacies, we will describe the Crit-Trans approach (Strong et al., *Mind, Culture, and Activity,* 2016) where we emphasize research, teaching, and learning centered on the lives and geographies of learners, their local knowledges, and place experiences and accentuate critical, decolonizing, and desettling frameworks to provide learners and educators tools necessary for critical civic participation in STEM. The Crit-Trans (Strong et al., *Mind, Culture, and Activity,* 2016) heuristic emphasizes a critical numeracy that critiques and connects the form and content of mathematics education to struggles and realities of learners, particularly those at the K-16 levels. Critical numeracy emerges from experiences, reflections, politicization, and research into public schooling as a site of tracking (Oakes, *Keeping track: How schools structure inequality* (2nd Ed). New Haven: Yale University Press, 2005), exploration (Dewey & Small, *My pedagogic creed.* Battle Creek, MI: E.L. Kellogg & Company, 1897), death (DeJesus, http://www.pennlive.com/news/2016/05/carlisle_indian_school_repatri.html, 2016), and possibilities (Giroux, *Teacher Education Quarterly, 31,* 1, 2004), indicating an urgency to organize, teach, and practice toward human emancipation particularly as State violence continues to intensify forms of occupation in a settler colonial and anti-Black society (Patel, *Decolonizing educational research: From ownership to answerability* (1st ed.). New York: Routledge, 2015; Hudson & McKittrick, *The CLR James Journal, 20*(1/2), 233–240, 2014).

Keywords Crit-Trans · Critical numeracy · Decolonizing · Politicized knowledge · Mathematics education

---

A. Das (✉)
The Graduate Center, City University of New York, New York, NY, USA
e-mail: adas@gradcenter.cuny.edu

J. D. Adams
Department of Chemistry and Werklund School of Education, University of Calgary, Calgary, AB, Canada

P. Sengupta et al. (eds.), *Critical, Transdisciplinary and Embodied Approaches in STEM Education*, Advances in STEM Education,
https://doi.org/10.1007/978-3-030-29489-2_16

16.1 Introduction

In order to move toward STEM as critical literacies, we must first address the Eurocentricity and presumed objectiveness that underlies STEM and subsequently STEM education. This rests in understanding the coloniality embedded within science, which Maldonado-Torres (2007) defines as the "long-standing patterns of power that emerged as a result of colonialism, that define culture, labor, intersubjective relations, and knowledge production" (p. 243). It becomes necessary to focus on the socio-historical-political context of knowledge production in STEM, enabling the underlying coloniality of science and mathematics education to be unveiled, thereby creating a more nuanced understanding of pedagogy and curriculum. It is about decolonizing colonized objects, in this case the European ownership of scientific and mathematical knowledge, in order to promote an inclusive and empowering vision of science education (Adams, Luitel, Afonso, & Taylor, 2008).

Furthermore, with numbers being a basis of the presumed objectivity of STEM and the foundation for tracking and creating hierarchies of who gains entry to the STEM fields, we need to problematize the meanings of numbers. The Crit-Trans (Strong et al., 2016) heuristic emphasizes a critical numeracy that critiques and connects the form and content of mathematics education to struggles and realities of learners, especially those at the K-16 levels. Strong et al. (2016) outline the parameters of Crit-Trans heuristic,

> which are neither fixed nor hierarchical: (a) Contextualize and historicize knowledge; (b) Challenge assumptions of neutrality and objectivity through critical inquiry; (c). Decenter hegemonic notions of knowledge production; (d) Situate place and space; (e) Privilege process over product; and finally (f) Promote participatory teaching, learning, and research. We view these parameters as useful challenges in guiding educators in their teaching and scholarship to help unsettle neoliberal reforms in science education. (p. 3)

A Crit-Trans stance forwards the idea that we can work collectively to collaboratively critique and reconstruct educational research and practice despite diverse lived experiences and historically intersectional subjectivities.

Transdisciplinarity, embedded in the aforementioned Crit-Trans heuristic, draws on a number of works contesting the created, reproduced, and hegemonic boundaries of disciplines. Osbourne (2015) historically contextualizes the use of transdisciplinarity, highlighting the basis of discipline, or institutionalized methods of transmitting intellectual practices, and academic disciplines, or the rules governing the production, reproduction, and socialization of knowledge. The Crit-Trans heuristic builds with Pratt-Clark's (2010) articulation of transdisciplinarity underscoring a purpose and politics of knowledge production in that multiple theories, approaches, and frameworks are employed to "understand, strategize, and implement transformative initiatives in society" (p. 19).

Critical numeracy emerges from experiences, reflections, politicization, and research into public schooling as a site of tracking (Oakes, 2005), exploration (Dewey & Small, 1897), death (DeJesus, 2016), and possibilities (Giroux, 2004), indicating an urgency to organize, teach, and practice toward human emancipation particularly as State violence continues to intensify forms of occupation in a settler colonial and anti-Black society (Hudson & McKittrick, 2014; Patel, 2015).

The Crit-Trans approach is the conceptual and practical space to center what Wynter calls a science of being human (*scientia*); one that "unsettles our familiar (Darwinian, objective, racist, sexist) governing cores of scientific thought…and seeks to locate [scientific] knowledge-making as connected to the human lived experience and recoding science through representational and biological feelings- it is, as noted earlier, a project *scientia*. This is an interdisciplinary and collaborative task, one that allows us to think about how the creative narrative can and does contribute to what is otherwise understood as "laws of nature,…" (McKittrick, 2015, p. 154).

In this piece, we contend critically with science and math, in that the disciplines themselves are problematic in many ways but also there continues to be many unrecognized or silenced aspects of knowing. A critical engagement with STEM would have to interrogate these marginalized or dismissed knowledges. A recent piece by the Politics of Learning Writing Collective (2017) has urged researchers to contend with the rise of US nationalism in relation to learning environments. Connected to this call, we mean to engage in a deep critique of number while also actively engaging and transforming in our communities through our teaching and research.

16.2 Background

Criticality: explicating meaning and practices. Jennifer: I have been a science educator in both formal and informal context, from secondary to postsecondary education. It was not until my work at the American Museum of Natural History and my interactions with the artefacts in the "people" halls that I began to rethink STEM and question the hegemony of the Western Modern Science presented in the science halls versus the lesser emphasized people halls. For me, the people halls provided notions of science that extended well-beyond Western ways of knowing and demonstrated deep and complex knowledges of the natural world. This prompted me to view these artefacts as technology and rethink how I approached science in formal education settings and everyday learning. I also brought this line of thinking into our research group, the Urban Environment and Education Research Coven as a way to rethink how we have been framing and positioning STEM, especially in relation to people's lived experiences and connections to places.

Atasi: Having been an educator in an elementary classroom for a number of years, I have been thinking about science and math for some time. However not just science and math, but these "subjects" as integrated in the larger spaces of teaching and learning. Critical literacy in STEM provides the possibility of radically rethinking and redoing how we engage and how we understand science and math and for the purpose of improving the conditions of all human life. Ongoing collaborations with fellow witches have helped shape and make clearer my emerging framework for critical numeracy.

We are both witches in the Urban Environment and Education Research Coven where we have converged our multiple experiences and analyses toward articulating and enacting a critical and transdisciplinary stance to STEM teaching, learning, and research. In the following sections, we will explicate some of the ways we have made meaning and practiced criticality. In this chapter, Atasi will explicate her work on numeracy by socio-politically situating its discourse and practice and assert a critical numeracy as potential to transforming collective forms of agency. The first person will be used throughout the piece to forefront Atasi's practice and voice in the process of making meaning of critical transdisciplinarity.

Further the meaning and practice of criticality has been centered in numerous dialogues and practices. In our Coven, we have had lengthy discussion about criticality and how it influences our research, practical work, and activism. Critical pedagogy, stemming from the Frankfurt School to the works Freire and Illich, has been one major avenue. We have made the concerted effort to draw on scholars of color and/or from the Global South in order to shape our notions of criticality. As such, we pull from scholars such as Sylvia Wynter, Samir Amin (1989, 2014), HLT Quan (2012), Cedric Robinson (2005), and W.E.B. DuBois, among many others. These scholars resituate the entry point into activities of knowing and being outside of confines of the overly reified concept of modernity. This is key in extending the conversation so that criticality can thus be considered from the historical to the present phenomena, one that deeply acknowledges the relationship between past and future in the very present. Criticality in this sense links both material conditions and knowledge production in direct opposition from Cartesian separations of subject-object.

The consequences of this form of criticality in relationship to localized knowledges is that all notions which are taken for granted are called into question. It is unsettling for some yet liberating for others. Some of the questions a praxis of criticality poses are: How do we know what we know in the deepest of ways? How can we develop our engagements of knowing and doing in ways that ground, critique, extend, and transform knowledge and material production? How is numeracy connected to or divorced from these activities?

Diverse forms of critical literacy and critical pedagogy. Within recent scholarship of learning sciences, Takeuchi (2018) has forwarded diverse ways of knowing elevating mathematical practices in out-of-school practices in immigrant communities in Japan. Similarly, Hostetler, Sengupta, and Hollett (2018) extends critically anchored through Freire's work in human geographies through computational simulation based on agent-based modeling.

This piece proposes a deep engagement in critical literacy grounded in critical pedagogy. Munir Fasheh, an educator, researcher, and mathematician from Palestine, presents several concepts toward reclamation of situated knowledges and outlines a particular form of agency in learning. As he recounts his mathematical training within a British colonial education system, he names this mathematical knowledge as toxifying. Fasheh (2015) described toxification as knowledges advanced through a logic of universality, modernity, and bestowing high status while incongruent to everyday concerns and struggles. He was rewarded and regarded as intelligent and gifted through the mathematics knowledge and certifications he accumulated while his mother, a seamstress living in the West Bank, would be classified as illiterate and uneducated. Toxification served as a form of occupation, in the case of mathematics, an occupation of the mind. Fasheh's (2015) critical discovery emerged as he realized his mother's mathematical practices as a seamstress. Her mathematical practices were something that he was unable to understand despite his training and effectively confounding despite her apparent illiteracy/innumeracy with institutionalized knowledges. This compounded with a realization that the mathematics he was taught did not connect him to various forms of Palestinian resistance against Israeli occupation preceding and continuing during the first Intifada (Palestinian uprising) in 1987. Fasheh (2015) compared the mathematics he had been celebrated for as disconnected to the daily lived experience of people in the West Bank and Gaza. Fasheh reveals forms of agency through his conscientization around the politics of mathematical knowledge. Fasheh shares concepts/activities/processes through which critical literacy takes place. I bring forward one concept from Fasheh (2015) below:

> *Mujaawarah*: This term refers to a medium for learning and building in community. It is expressed through a group that freely decides to meet regularly in order to learn, understand, and act. It is about personal and communal freedom to learn and act. For example, neighborhood groups during first Intifada in 1987 were created when all schools and institutions were closed. These were the instantiations of mujaawarahs. Although critiques of military repression in academic settings were tolerated similar conversation through the medium of neighborhood groups, mujaawarahs, were seen as dangerous by those in power. These neighborhood committees consisted of people who thought and acted in freedom as part of a practice of critical literacy. (Fasheh, 2015, pp. 48–49)

This particular enactment is one that connects to critical numeracy in that a collaboratively created space is organized as part of transformative understanding and activity. Further, the concept of *mujaawarah* connects to a Crit-Trans heuristic emphasizing an on-going process of reflexivity and reflection.

Further, we engage critical pedagogy from the works of Freire:

> When people lack a critical understanding of their reality, apprehending it in fragments which they do not perceive as interacting constituent elements of the whole, they cannot truly know that reality. To truly know it, they would have to reverse their starting point: they would need to have a total vision of the context in order subsequently to separate and isolate its constituent elements and by means of this a clearer perception of the whole. (2000, p. 85)

In this vein, we assert that numbers are created, constructed, languaged, and communicated in a culture with a history of production. Numbers are not neutral. Following this, how do we make sense of people's lived experiences in relation to number?

Socio-political positioning to critical work. In many respects, this work emerges out of eruptions (social and political), deconstructions, and critical reflections (personal and collective); a cauldron of numerous experiences from living and being in a world of great injustice, schooling as a site of tracking and possibilities, active political consciousness and organizing, and an urgency to act in light of a legacy and continued violence of occupation in a settler colonial and anti-Black State. We can see this in the violent enactments organizing private property (stolen land, water, air, minerals, bodies and labor, and the excavating and mining, which is numerated as well as monetized) as it is necessary within capitalist development. Following this, we attempt to develop a theory and praxis that centralizes agency and social justice to a conception of number. In order to make sense of this proposed project, we must take a look at the terrain in which a historical, political look at numeracy and number has unfolded. We can look at how numbers are utilized and constructed and connect it to the realm of numeracy.

16.3 Theoretical Framework: Engaging with Numeracy and Number from a Crit-Trans Perspective

This following section focuses on emergent theoretical, onto-epistemological connections between numeracy, a globalized political economy, and knowing and learning. There are two main parts in this piece: One details the theoretical groundings and analysis around mathematics, numeracy, critical literacy, and critical pedagogy. The second serves to explicate these notions in practice. These two parts are linked as part of an on-going activity to make sense of a critical numeracy.

Historical and theoretical positioning of mathematics in society. We might have heard the common phrase that "numbers don't lie." From this phrase, numbers are defined as immutable facts as well as entities in and of themselves. Numbers have been used by humans in diverse ways to count, track, categorize, measure, and differentiate everything from humans or living things to our environment and even atomic particles.

However if we are to suspend an uninterrogated acceptance of the validity of number, we can then posit that numbers are descriptive, operationalized, and disseminated in a social, historical politicized world. Numbers and acts of numerating, in the form of data, are incredibly pervasive and increasingly influential to our lived experiences (i.e., points for college entry and credit scores).

"Western" mathematics positions math as the science of "real objects" and as a science of properties of measurable and calculable magnitude (Lasserre, 1964). Bertrand Russell extends this understanding of mathematics in the following statement, "Mathematics is the science concerned with the logical deduction of consequences from the general premise of all reasoning" (in Lasserre, 1964, p. 14). The implication from Russell's statement is that mathematics produces truth and therefore is an uncontestable reality.

Connected to this brief insight into mathematics, research in ethnomathematics has situated the use and purpose of mathematics in terms of a historical, cultural, social, and political relationship to production. Ethnomathematics is a study of how mathematical thinking and logic are tied to specific histories and cultures of people. Mathematics education researchers, Arthur Powell and Marilyn Frankenstein (1997), assert that Eurocentrism permeates mathematics education. Like other forms of math, Eurocentric or academic mathematics is tied to specific histories and cultures, in this case emerging from the project of modernity. Ethnomathematics grounds the emerging agenda of critical numeracy to situate the multiplicity of ways of number systems, classifications, logic systems and categorizations have been created within human relations. A historicized understanding of human relations is necessary to construct a critical consciousness of prevalent forms of abstraction, such as data or mathematics.

One example that highlights a historicity of mathematical practices/instruments is the Ishango bone, dated in some accounts to around 20,000 BCE. This particular bone was excavated in the 1950s by Belgian archeologist, Prof. de Heinzelin, nearby the Ishango fishing village on the shores of Lake Rutanzige (otherwise known as Lake Edward), which straddles the border of the Congo and Uganda in central Africa (Joseph, 2011). The asymmetrical markings on the bone point to mathematical practices such as the conceptualization of prime numbers and/or to the tracking of menstrual cycles. However, the Ishango bone is one of the oldest mathematical artifacts in human society. Even still, it is not widely acknowledged indicating that mathematical knowledges emerging from the African continent (Zaslavsky, 1999) stand in contradistinction to knowledge as produced through the project of Modernity (Wynter, 2003).

Ethnomathematics emphasizes and situates diverse mathematical practices across the world. The theories and assertions from this line of research connect to the area of numeracy to provide insight into the ways that Eurocentric and white supremacist ideologies are privileged and used as tools of domination over other systems of knowledge. Numbers, as particular forms of abstracting and making sense of our world, can be situated as part of broader, interconnected human activity as opposed to simply a "Western" phenomena.

Positioning number and numeracy. Mathematical practices as they emerged from Western Europe maintain a hegemonic relationship to positioning number and numeracy. I go behind/beyond the disciplinary silos from which a concept of number is put forth. Historically, numeracy is positioned as a policy and as an indicative measure for curriculum and instruction (i.e., numerate/innumerate). At the same time, various terms are used interchangeably with numeracy, such as quantitative literacy and quantitative reasoning, among many others. While there is an interrelationship between various terms from numeracy, quantitative literacy, critical mathematical literacy, mathemacy, matheracy, and statistical literacy (Niss & Jablonka, 2014), this piece will forefront the term numeracy. The term numeracy can be dated back to a British government report from the late 1950s, the Crowther Report. The Crowther Report forwarded the need in the emerging global order for an education of English citizens that included both literacy and numeracy. Additionally, the British Cockcroft Report of 1982 brought numeracy further into the policy lexicon. It emphasized and situated the need for numeracy as a tool for students to cope with adult life and work, forming the "good" citizenry. The need for numeracy was later popularized in the United States in the late 1990s through John Allen Paulos's 1988 book, *Innumeracy.*

Various definitions of numeracy include the following: "connoting mathematical topics woven into the context of work, community, and personal life" (Ginsburg, Manly, & Schmitt, 2006, p. 1); Johnston and Yasukawa state numeracy is, "the ability to situate, interpret, critique, use and perhaps create mathematics in context, taking into account all the mathematical as well as social and human complexities which come from that process unless it is political [it] cannot pretend to be objective and value free" (in Atweh, Forgasz, & Nebres, 2001, p. 279). While there is no consensus on its definition, this argument of numeracy concerns the utility and purpose to engage quantitative information as part of an on-going sense making and doing in the world. We share this to emphasize the particular connection of numeracy and mathematics to economic life across time. In other words, we make these connections to demonstrate the use of numeracy in reproducing particular relations in a capitalist society. The connection between numeracy and development of the citizen is thus necessary to underscore as we continue to develop this analysis.

Extending from Yasukawa and Johnston's stance above, critical numeracy is proposed as an on-going practice (praxis) as all humans engage in the production of life and making sense/struggle in the world. By praxis, we mean action that is formed by theory through ongoing reflexivity and reflection. In thinking through how critical numeracy is grounded in a social practice, this argument draws inspiration from various critically oriented scholars in mathematics education. Skovsmose (2012) offers,

> Mathematics simplifies the designating thing, reducing it to a single feature. It dismembers the thing, destroying its organic unity, treating its parts and properties as autonomous. It inserts the thing into a field of meaning which is ultimately external to it. (p. 120)

Number has overwhelmingly acquired an ontological status of truth. While we do not question the function of number in mathematics, we do interrogate the truth produced through mathematics on the presumption that number is always objective truth. Mathematics is seen as unveiling unquestionable truth in relationships. However, this assumption needs to be desettled in a world that is organized by global monopoly (and financialized) capital.

Numeracy has been defined in many ways. However, in many of those articulations, numeracy is currently theorized as divorced and uncontextualized from society. Number is theorized as static and not in movement. In contrast, this piece defines numeracy as a socially based activity embedded within historically derived social relations, integrating math, numbers/abstractions, and ways of being. In this conception, numeracy must be recognized as always in movement. Further, a critical transformative numeracy links an analysis of contradictions between labor and capital to political struggle. Crit-Trans also employs that numeracy as critical consciousness is functioning to (re)form what is real, what we know, and what is human. This engages more than just a critique of number, numerating, and the term, numeracy. Both mathematics education and numeracy must be situated, contextualized, historicized, politicized, and most importantly must connect to a creative life force that is rooted in all human activity.

16.4 An Illustrative Case: Examining a Praxis to Develop Critical Numeracy

I demonstrate an emerging and changing understanding of critical numeracy in one vignette in my on-going praxis of refining and defining a critical numeracy. A teacher organizing group called New York Collective of Radical Educators (NYCoRE) puts together Inquiry to Action Groups (ItAG) for teachers in New York City each year. One particular ItAG group met to critically examine STEM as well as their teaching practice in New York City schools. I facilitated a session using an emergent framework of critical numeracy. My goals for this session were to highlight what and how critical numeracy could be conceptualized and enacted. As a starting point, we watched and listened to a piece by renowned poet, Amiri Baraka, entitled *Why Is We Americans?* The poem powerfully emphasizes a long history of struggle and resistance of exploitation in the United States. Below is a short excerpt of this piece:

> What I want is me. For real. I want me and my self. And what that is is what I be and what I see and feel and who is me in the. What it is, is who it is, and when it me its what is be… .I'm gone be here, if I want, like I said, self determination, but I ain't come from a foolish tribe, we wants the mule the land, you can make it three hundred years of blue chip stock in the entire operation. We want to be paid, in a central bank the average worker farmer wage for all those years we gave it free. (Baraka, 2002)

Participants were prompted to reflect on and share major themes from his poem. With this introductory activity, educators engaged in analyzing social and political dynamics of a particular place—in this case the United States. Through engaging Baraka's compelling poetry, the teachers broadened their views, examining the legacy of racialized capitalist exploitation as well as social resistance. In essence, his piece served as a call to enlarge our political engagement with numeracy. This historicizing activity through Baraka's poem invites the possibility to link social-political understanding of history with numbers and human production relations (Fig. 16.1).

Following this introductory activity, participants engaged in a simulation as a means to examine the lived dynamics of exploitative relations, as well as latent unrehearsed activities of resistance. This teaching activity, called "The Organic Goodie Machine," was adapted from the work of Norm Diamond and Bill Bigelow in a curriculum called *The Power in Our Hands: A Curriculum on the History of Work and Workers in the United States* (1988).

In the simulation, participants were introduced to the conditions and structure of a "new" society: "everyone lives together in the same location and everyone needs food, clothing, and shelter to live. In this society, the organic goodie machine serves as the sole means to produce food." As the facilitator, I took on the role of owner of the organic goodie machine. From this basis, an entire economy including the wages for working the machine and taxes for maintaining a society was outlined (Fig. 16.2). Participants were either hired by the owner of the goodie machine (i.e., hired by me) as workers or unemployed. From this point onwards, we engaged in the simulation: enacting the activities of producing food in real time (for this version of simulation, assembling toy blocks together as "goodies"), earning wages, and using wages for food/resources, as well as fast forwarding through time to look across the immediate experience to assess broader social conditions/lived experiences. For example, in the dual role of facilitator and owner, I would at times apply pressure to working conditions by demanding more goodies or cut wages in shorter time frame. Periodically, I would also ask participants to share at particular points how they imagined life to be like (Fig. 16.2).

It is important to note that this is not a scripted or predetermined activity. Participants could choose to do a number of things over the course of the simulation. For example, some participants who were working in the factory would start sharing their earnings with those who were unemployed. Participants also organized to demand better wages and some even organized to wrest control over the organic

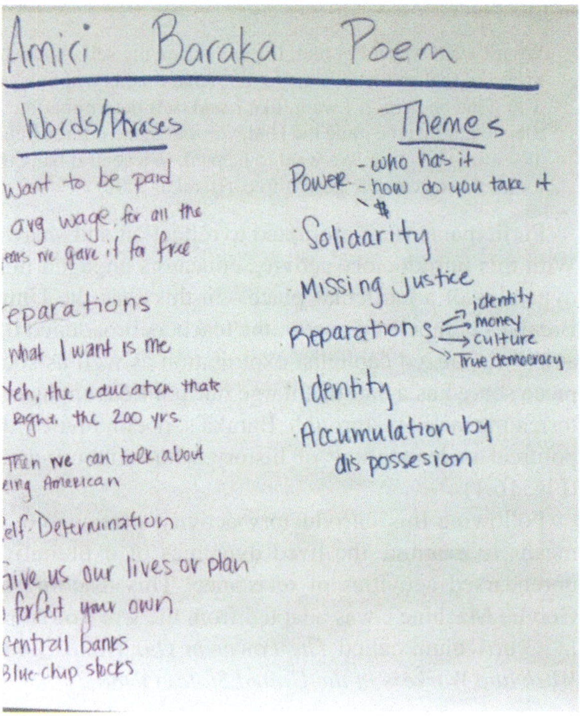

Fig. 16.1 Participants identified poignant words, phrases, and overarching themes from Amiri Baraka's poem

Organic Goodie Economy

Production = 11 x no. of workers

	Per day		
	Workers	*Unemployed*	*Owner*
Wages	$6 x no. workers	Nothing	Nothing
Taxes	-$1 x no. workers	+$2 x no. unemployed	-$1 x no. unemployed (see note)
Consumption	5 Organic Goodies x no. workers	2 x no. unemployed	6 Organic Goodies
Surplus	Nothing	Nothing	4 x no. workers - 6 for personal daily consumption

Example: If there were ten workers and ten unemployed, the owner would end up with 50 Goodies: 10 would go to unemployed, 6 would be consumed, leaving a remaining 34.

Note: Workers' and owner's tax needs to provide $2 to each unemployed person (taxes paid in Goodies).

Fig. 16.2 Outline of Organic Goodie Economy from Diamond & Bigelow (1988, p. 28)

goodie machine itself. As the owner and facilitator, I would respond to changes in accordance with the owner's central focus: accumulating profits. In one example, I (as owner) created a police force and a prison to protect the machine and to counter organized resistance. The simulation ends once participants make an organized effort to change not only their conditions but the basis of them as well. Engaging in such a simulation is always chaotic, confusing, revelatory, and inspiring as participants learn about power of collectivity and voice as a form of enacted numeracy, and as an ongoing praxis, I make connections between emerging political consciousness, learning contexts, and historically intersectional subjectivities.

Following the organic goodie simulation during the ItAG, participants and the facilitator reflected upon what they encountered. They recalled particular moments as well as overarching themes that resonated from this simulation. This experience was extended by describing and analyzing those very same relations through the use of number. Figures 16.3 and 16.4 show some of these processes. Figure 16.3 shows our calculation of how many goodies were produced by eight (8) workers (88 goodies) and the amount of goodies which were accumulated as part of surplus and profits at the end of 1 day (26 goodies). Figure 16.4 is a table that shows the amount of goodies that remained in the hands of workers (w), the unemployed (u), and the owner (o).

Through these activities, we repositioned the use and purpose of employing numbers. The starting point did not center the tool of number or use of arithmetic, but instead focused on the historical and unfolding relations between people. In the final portion of this session, we connected the simulation back to Amiri Baraka's poem (Fig. 16.5).

It was from this basis that we engaged in critical literacy, or reading the world, a practice of making meaning with number and numerating within a particular political and economic relationship. In this way, critical numeracy is a practice of critical literacy. Our discussions prompted a number of connections as can be seen in Fig. 16.5. In the process of making connections, as the facilitator, I asked: What

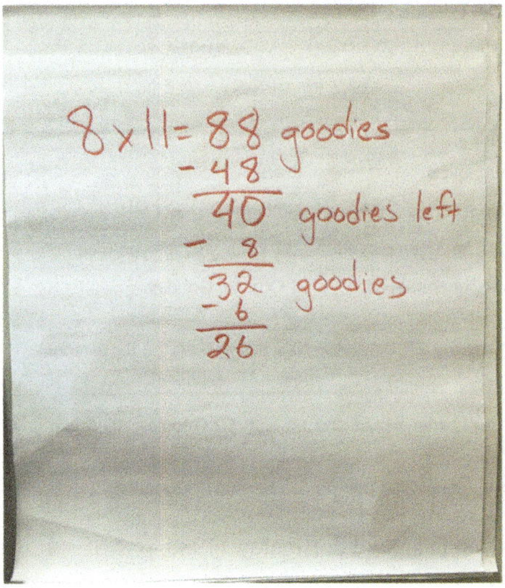

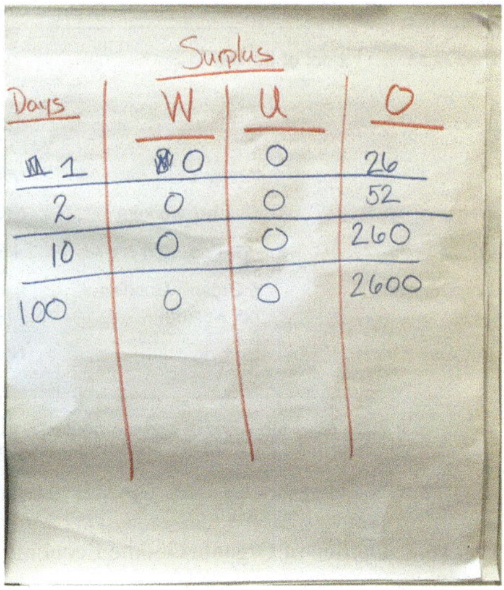

Figs. 16.3 and 16.4 (Left and right) Charts of numerically re-presenting the simulation during the debrief

relevance or use does number have to understand the world around us? How are particular historical conditions reified and transformed through the asserting of number? In what ways did Amiri Baraka's poem connect to other movements, such as in Black Lives Matter or No DAPL in Standing Rock? How do these unfolding/ political relations intersect with number?

16.5 Discussion

Theorizing a politicized understanding of learning and knowing. Zooming out from the account above, we forward the concept of situated knowledges as a means of knowing and learning, in this case about number and numeracy. Situated knowledges as oriented by Haraway (in Carpenter, 2012) relates that all knowledge is partial, located, grounded, and subjective and yet has a common materiality in its configuration. Number form is a partial representation of life processes enforcing rigidity and static-ness. The static-ness of number/numerating was challenged directly with a politicized engagement in the historical present. For example, as workers produced goodies, they eventually realized that the number of goodies they earned did not allow them to "survive." At a particular point in the simulation, workers and those who were unemployed would organize to challenge the number of goodies in their control. Situated knowledge foregrounds the possibility to challenge a normativity that is deeply embedded in the teaching and learning of number.

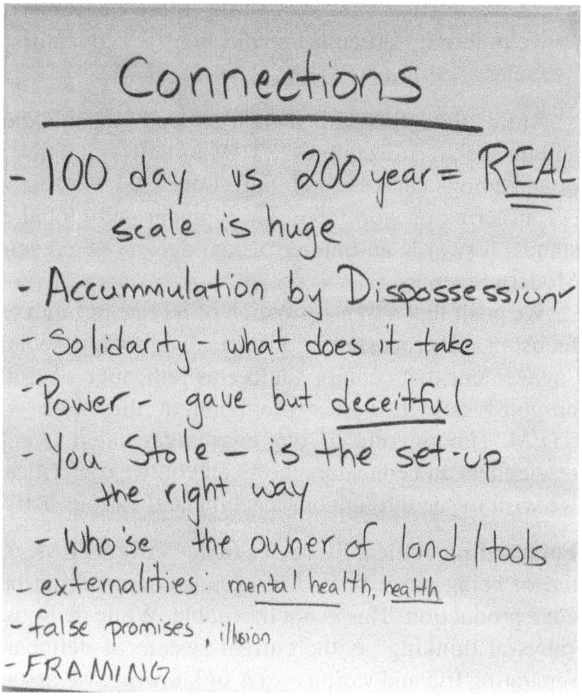

Fig. 16.5 Connection participants made between overarching themes from Amiri Baraka's poem and the Organic Goodie Machine simulation

Furthermore, Carpenter and Mojab (2017) assert, in "apprehending a phenomenon [with number or numeracy], these phenomena must be connected to *either part of* or as *the result of a historically developed social relations*" (p. 5, emphasis mine). Additionally, Carpenter's (2012) assertion of experience and learning extends beyond mere representation (static description of quantity or qualities) to reimagine a possibility of transforming experience. We must develop and connect our understanding of experience (which is simultaneously individual, social, and global) in order to change the conditions that we live in. In this way, number can be indicative of relationality in that it infuses difference, contradiction, and possibility. A transformative engagement with number can advance a more incisive tool to discern racialized production relations and class struggle.

A critical numeracy that centers production relations and human agency as the primary means of opposing all logics of exploitation and oppression can form the basis of a critical literacy of STEM. This criticality can be explored when thinking, operationalizing, and disseminating number.

Critical numeracy as on-going critical engagement with STEM. Why is this important for STEM? Critical numeracy must compel us to organize, teach, and practice toward human emancipation particularly as State violence continues to intensify forms of occupation in a settler colonial and anti-Black society (Hudson & McKittrick, 2014; Patel, 2015). Considering this, we want to come back to the following questions: Where do our ideas come from? Why do we educate? What is the world we wish to see? Answering these questions with STEM and the analysis above in mind, I articulate connections for educators and researchers using critical literacies.

At this time, STEM is representative of an amalgamation of historically distinct and interrelated disciplines. STEM initiatives are also backed by powerfully funded organizations such as Microsoft, Lockheed-Martin, Verizon, and IBM among others, in terms of workforce development and global competitiveness. STEM, as it stands, forwards an uninterrogated agenda of presumed objectiveness of Western Modern Science.

We posit that a transformation of STEM through critical literacies is possible by focusing on processes of historically moving social relations as opposed to an engagement that centers number as indicative of truth. This can work to interrupt on-going activities of exploitation at this time while broadening a scope of STEM. Having offered our own engagement, we further pose the question to researchers and educators: how can you/we use critical literacies to create the world we wish to see through our teaching and various activities?

Forwarding critical literacies to de-settle STEM. Critical literacy has the potential of being subordinated to agendas maintaining hegemonic power over knowledge production. This is not inevitable. While de Sousa Santos (2007) proposes that "abyssal thinking" is the current system of delineated thinking, dividing society separating life and various ways of knowing, we assert that this forced separation is possible to subvert. Not only is it unnecessary but transformative and unexpected ways of knowing emerge in moments of broad, embodied, and reflective dialectical explorations. Desettling literacy must question all taken-for-granted notions which force us to broaden our engagement. The critiques and connections that are emphasized by the Crit-Trans heuristic forefront the struggles and realities of all learners in a way that respects many *ways to* knowing.

Critical numeracy as articulated in this piece can be elucidated as an unrecognized and silenced aspect of knowing. Drawing from our earlier discussion of hierarchies of knowledge, we articulate this version of critical numeracy as a political tool of pedagogy that desettles relations of domination and control—both within STEM and beyond. This is one of many forms of critiquing and deconstructing that actively engages and transforms our very lives.

References

Adams, J., Luitel, B. C., Afonso, E., & Taylor, P. C. (2008). A cogenerative inquiry using postcolonial theory to envisage culturally inclusive science education. *Cultural Studies of Science Education, 3*(4), 999.

Amin, S. (1989). *Eurocentrism*. New York, NY: NYU Press.

Amin, S. (2014). *Capitalism in the Age of Globalization: The management of contemporary society*. London, UK: Zed Books Ltd.

Atweh, B., Forgasz, H., & Nebres, B. (2001). *Sociocultural research on mathematics education: An international perspective*. London, UK: Psychology Press.

Baraka, A. (2002). *Why is we Americans?* Newark, NJ: Razor.

Carpenter, S. (2012). Centering Marxist-feminist theory in adult learning. *Adult Education Quarterly, 62*(1), 19–35.

Carpenter, S., & Mojab, S. (2017). *Revolutionary learning: Marxism, feminism and knowledge*. London, UK: Pluto Press.

Crowther, G. L. (1959). *Crowther report 1959 Volume I - full text (central advisory council for Education)*. London: Ministry of Education. Retrieved from http://www.educationengland.org.uk/documents/crowther/crowther1959-1.html

de Sousa Santos, B. (2007). Beyond Abyssal thinking: From global lines to ecologies of knowledges. *Review (Fernand Braudel Center), 30*(1), 45–89.

DeJesus, I. (2016, May 6). After more than 100 years, American Indian children buried in Carlisle begin a journey home. *PennLive.com*. Retrieved from http://www.pennlive.com/news/2016/05/carlisle_indian_school_repatri.html

Dewey, J., & Small, A. W. (1897). *My pedagogic creed*. Battle Creek, MI: E.L. Kellogg & Company.

Diamond, N., & Bigelow, W. (1988). *Power in our hands* (Teacher ed.). New York: Monthly Review Press.

Fasheh, M. (2015). Over 68 years with mathematics: My story of healing with modern superstitions and reclaiming my sense of being and well-being. In *Eighth international Mathematics Education and Society Conference* (Vol. 1, pp. 33–60), Portland, OR.

Freire, P. (2000). *Pedagogy of the oppressed*. New York, NY: Continuum.

Ginsburg, L., Manly, M., & Schmitt, M. J. (2006). *The Components of Numeracy. NCSALL Occasional Paper*. National Center for the Study of Adult Learning and Literacy (NCSALL). Retrieved from http://eric.ed.gov/?id=ED495440

Giroux, H. A. (2004). Critical pedagogy and the postmodern/modern divide: Towards a pedagogy of democratization. *Teacher Education Quarterly, 31*, 1.

Hostetler, A., Sengupta, P., & Hollett, T. (2018). Unsilencing Critical Conversations in Social-Studies Teacher Education using Agent-based Modeling. *Cognition and Instruction, 36*(2), 139–170.

Hudson, P. J., & McKittrick, K. (2014). The geographies of blackness and anti-blackness: An interview with Katherine McKittrick. *The CLR James Journal, 20*(1/2), 233–240.

Joseph, G. G. (2011). *The crest of the peacock: Non-European roots of mathematics* (3rd ed.). Princeton, NJ: Princeton University Press.

Lasserre, F. (1964). *The birth of mathematics in the Age of Plato*. London: Hutchinson & Co.

Maldonado-Torres, N. (2007). On the coloniality of being. *Cultural Studies, 21*(2–3), 240–270.

McKittrick, K. (2015). Axis, bold as love: On Sylvia Wynter, Jimi Hendrix, and the Promise of Science. In K. McKittrick (Ed.), *Sylvia Wynter: On being human as praxis* Duke University Press. (pp. 142–163).

Niss, M., & Jablonka, E. (2014). Mathematical literacy. In S. Lerman (Ed.), *Encyclopedia of mathematics education* (pp. 391–396). Netherlands: Springer.

Oakes, J. (2005). *Keeping track: How schools structure inequality* (2nd ed.). New Haven, CT: Yale University Press.

Patel, L. (2015). *Decolonizing educational research: From ownership to answerability* (1st ed.). New York: Routledge.

Paulos, J. A. (1988). *Innumeracy: Mathematical illiteracy and its consequences* (1st ed.). New York: Holt McDougal.

Powell, A. B., & Frankenstein, M. (1997). *Ethnomathematics: Challenging eurocentrism in mathematics education*. Albany, NY: State University of New York Press.

Pratt-Clarke, M. (2010). *Critical race, feminism, and education: A social justice model*. Springer.

Quan, H. L. T. (2012). *Growth against democracy: Savage developmentalism in the modern world*. Lanham, MD: Lexington Books.

Robinson, C. J. (2005). *Black Marxism: The making of the black radical tradition*. Chapel Hill, NC: University of North Carolina Press.

Skovsmose, O. (2012). Symbolic power, robotting, and surveilling. *Educational Studies in Mathematics, 80*(1/2), 119–132.

Strong, L., Adams, J. D., Bellino, M. E., Pieroni, P., Stoops, J., & Das, A. (2016). Critical transdisciplinary STEM: An approach to science teaching and learning centered on the lived experiences of youth. *Mind, Culture, and Activity*, (Reimagining Science Education in Neoliberal Global Contexts: Sociocultural Accounts of Science Learning in Underserved Communities).

Takeuchi, M. A. (2018). Power and identity in immigrant parents' involvement in early years mathematics learning. *Educational Studies in Mathematics, 97*(1), 39–53.

Wynter, S. (2003). Unsettling the coloniality of being/power/truth/freedom: Towards the human, after man, its overrepresentation--An argument. *CR: The New Centennial Review, 3*(3), 257–337.

Zaslavsky, C. (1999). *Africa Counts: Number and Pattern in African Cultures* (3rd edition). Chicago, Ill: Chicago Review Press.

Chapter 17
Queering Virtual Reality: A Prolegomenon

Dylan Paré, Pratim Sengupta, Scout Windsor, John Craig, and Matthew Thompson

Abstract In this chapter, we investigate how innovations in STEM, such as Virtual Reality (VR) and 3D Sculpting, can support the development of critical literacies about gender and sexuality. Our work arises from the concern that the assumed "naturalness" of male/female binary categories in biology is often at the center of the queer, trans, and intersex panics in public education. Echoing sociologists and critical scholars of gender and sexuality, we posit that transgender and queer identities should be positioned as realms of playful, active inquiry. Further, we investigate how new forms of computational representational infrastructures can be leveraged to support productive and playful experiences of inquiry about gender and sexuality. We present a retrospective analysis of a design group meeting of a small group of friends in their early thirties with gender nonconforming and queer identities and life histories. The group interacted in VR-based environments, where they engaged in two different forms of constructionist learning experiences: creating 3D sculptures of personally meaningful objects, and re-creating their VR avatars in VR social media. Our analysis illustrates how such experiences can be productively analyzed using social constructivist perspectives that situate knowing as boundary play and figured worlds, and the roles that play and friendship have in supporting deep and critical engagement with complex narratives and marginalized experiences of gender and sexuality.

Keywords Virtual reality · Gender · Sexuality · LGBTQ · Figured worlds

---

D. Paré (✉) · J. Craig · P. Sengupta
Mind, Matter and Media Lab, Werklund School of Education, University of Calgary, Calgary, AB, Canada
e-mail: dylan.pare@ucalgary.ca

S. Windsor
Wakebold, Calgary, Canada

M. Thompson
Telus SPARK Science Centre, Calgary, Canada

P. Sengupta et al. (eds.), *Critical, Transdisciplinary and Embodied Approaches in STEM Education*, Advances in STEM Education,
https://doi.org/10.1007/978-3-030-29489-2_17

17.1 Introduction

In this chapter, we investigate how innovations in STEM, such as Virtual Reality (VR) and 3D Sculpting, can support the development of critical literacies about gender and sexuality. Our work is motivated by the concern that the assumed "naturalness" of male/female binary categories in biology (Bazzul & Sykes, 2011; Westbrook & Schilt, 2014) is often at the center of the queer, trans, and intersex panics in public education and sex-segregated spaces. This also undergirds oppressive legislation and cultural policing of trans people's access to sex-segregated bathrooms (Cavanagh, 2010), medical panics that create perceived "emergencies" around assigning a binary sex to intersex people (Davis, 2015; Davis & Murphy, 2013), and the panic in sports over whether and in which gender category intersex or trans athletes can compete (Karkazis, Jordan-Young, Davis, & Camporesi, 2012; Travers, 2017). To challenge this assumed naturalness of male/female binary categories especially in pedagogical contexts, echoing Thorne's (1993) call for reconsidering gender as social action and not merely as a category, Grace (2015) argued that we need to reconsider transgender and queer identities as "a realm of active inquiry, play and creative expression" (p. 49). Building on this work, our chapter investigates how new forms of computational representational infrastructures—e.g., 3D Sculpting in VR—can be leveraged to support productive and playful experiences of inquiry about gender and sexuality.

Empirical research has also shown that embodied interactions and projections in virtual worlds—the fundamental form of experience in VR environments—can help people take on perspectives that would be otherwise difficult for them to adopt, in complex topics ranging from physics and chemistry to animal rights, racism, and ageism (Abrahamson & Lindgren, 2014; Ahn, Le, & Bailenson, 2013; Groom, Bailenson, & Nass, 2009; Hostetler, Sengupta, & Hollett, 2018; Lindgren & Johnson-Glenberg, 2013; Oh, Bailenson, Weisz, & Zaki, 2016; Peck, Seinfeld, Aglioti, & Slater, 2013; Vea, 2019). Our chapter furthers this line of work by illustrating the roles of play, friendship, and embodiment in the context of creating and exploring complex representations of gender and sexuality in immersive VR. And, while sociologists and gender scholars have shown that informal gender play and conversations with friends shape and/or reify our understanding of gender and sexual identities (Risman & Banerjee, 2013; Thorne, 1993), there is little understanding of how such playful experiences with friends can be leveraged to support the development of a more socially just epistemology of queerness and transgender identities.

We present a retrospective analysis of a design group meeting involving all five authors, with a focus on the four authors who are friends in their early thirties with gender nonconforming and queer identities and life histories. These participants engaged in two different forms of constructionist learning experiences in VR-based environments: creating 3D sculptures of personally meaningful objects, and re-creating their VR avatars in VR social media. Our analysis illustrates how VR-based learning environments can be designed to support productive and playful learning experiences about gender and sexuality, and furthermore, how such experiences can be productively analyzed using social constructivist perspectives that situate knowing as boundary play and figured worlds (Holland, Lachicotte Jr., Skinner, & Cain, 1998; Sengupta & Shanahan, 2017).

17.2 Theoretical Background

17.2.1 Play, Friendships, Figured Worlds, and Pivots

Learning scientists have shown that playful learning environments can greatly facilitate learning of complex topics across a range of disciplines (Berland & Lee, 2011; Kim & Ho, 2018; Sengupta & Shanahan, 2017). Following Vygotsky (1980), Holland et al. (1998) argue, it is through play that imagination becomes embodied. In this view, play can be viewed as the *experience* of imagination, and Vygotsky also argues that it is deeply connected with desire, albeit in the context of children's development. Play thus creates contexts for us to experience virtuality, which as Leander and Boldt (2018) argue, is about "what could happen that is unplanned, undesigned" (p. 35). This is because in play, the meanings and imaginations of, rather than the mere physicality of, the objects shape the experience of the participants. The context of play thus brings into the picture learners' desires, which goes beyond both the notions of designing learning environments and the teachers' and designers' intentionalities (Leander & Boldt, 2018; Massumi, 2002). Leander and Boldt (2018) argue that this leads to *difference making*, where learners' experiences unfold in a non-deterministic manner, often differently from the curriculum—not necessarily within the curricular activities.

This is also resonant with the scholarship on playful learning, which shows that when learners are playfully engaged in a virtual environment, they can (re)shape the learning activities even within a structured setting, such that the activities are both personally meaningful and relevant to the disciplinary context of learning (Azevedo, 2018; Farris & Sengupta, 2016; Kim & Ho, 2018). These activities range from problem-solving to invention, modification, and transgression of the underlying rules that govern the virtual environment (Kim & Ho, 2018), and even bringing together, unexpectedly, different disciplines and realms of experience (Sengupta & Shanahan, 2017). In such contexts, learners are positioned in agentive roles, i.e., in ways that enable them to engage in a deeper exploration and inquiry of canonical practices and ideas through manipulating and even transgressing them, as well as through affective engagement (Kim & Ho, 2018).

One might then ask: What is the nature of imagination that can arise through such forms of play? The notion of *figured worlds* is useful in answering this question. Holland et al. (1998) argue that, "… [Play] allows for the emergence of new figured worlds, of refigured worlds that come eventually to reshape selves and lives in all seriousness" (p. 236). Figured worlds are "sociohistoric, contrived interpretations or imaginations that mediate behaviour and so, from the perspective of heuristic development, inform participants' outlooks. The ability to sense (see, hear, touch, taste, feel) the figured world becomes embodied over time, through continual participation" (Holland et al., 1998, pp. 52–53). In this view, figured worlds manifest themselves in the form of people's activities and practices.

Pivots play an important role in the experiences of our participants in the context of their representational work of building identity artifacts. Building on Vygotsky (1980), Holland et al. (1998) argue that pivots are culturally defined artifacts that shift the frame of an activity and evoke or "'open up' figured worlds" (p. 61). The elements of a figured world—its artifacts, storylines, characters and their concerns, and the activities—help in positioning oneself meaningfully in relation to the figured world and can also serve as pivots. Pivots enable participants to swivel across multiple figured worlds. This swiveling back and forth between multiple figured worlds can also take the form of boundary play (Sengupta & Shanahan, 2017), where the pivots act as boundary objects at the intersection of multiple social worlds. The primary feature of pivots is that the same objects are meaningful in multiple social worlds, even though those meanings may be different or even contradictory. Sengupta and Shanahan (2017) show that boundary play in computational worlds allows learners to meaningfully bring together multiple figured worlds in the context of interpreting and explaining complex phenomena. In the context of our work, we believe that the notions of figured worlds, boundary play, and pivots can be useful for understanding queer and trans experiences as they unfold in virtual reality environments.

Finally, we draw from a small but growing body of research on the relationship between friendship and learning. While some studies show that the dynamics within friendships may sometimes present challenges for learning (Esmonde, Brodie, Dookie, & Takeuchi, 2009; Mitchell, Reilly, Bramwell, Solnosky, & Lilly, 2004), Takeuchi (2016) found that group work with friends can also offer greater opportunities for access to a wider variety of complex, disciplinary work practices and positional identities, for example, by enabling students to take on roles of both experts and learners. Takeuchi's work further shows that learning with friends can be particularly helpful for marginalized learners, whose lived experiences outside the classroom often become the sources of social isolation and marginalization within the classroom. Furthermore, sociologists have also shown that informal interactions and conversations with friends play a formative role for the development of gender and sexual identities for children and youth. This is evident in several forms, such as informal gender play (Thorne, 1993), and informal discourse about gender, sex, and sexuality (Risman & Banerjee, 2013).

17.2.2 Emergent and Figured Worlds of Queerness, Body-Becoming, and Compulsory Heterosexuality

We draw from queer theory, specifically Butler's (2006) concept of the *heterosexual matrix*, Ahmed's (2006) *queer phenomenology*, as well as from Lane's (2009) call for trans and queer studies to engage with a fundamentally more complex, *new materialist* biology and the concept of the *body-becoming*. A proliferation of language to describe the assumed naturalness of the gender binary and how it is reinforced through heterosexuality arose out of lesbian and gay studies scholarship in the 1980s and 1990s, including Butler's (2006/1990) *heterosexual matrix*, Adrienne Rich's (1980) *compulsory heterosexuality,* Monique Wittig's (1980) *heterosexual contract*, and *heteronormativity* by Michael Warner in 1991 (Jeppesen, 2016). The heterosexual matrix highlights the social system of constraints shaping understandings of gender and sexuality, where bodies are expected "to cohere and make sense" by expressing a stable, binary sex and gender through compulsory heterosexuality (Butler, 2006, p. 208). Beyond Butler's (2006) notion of the heterosexual matrix, Ahmed (2006) further argued that bodies take shape as an effect of how they are continuously oriented within the world through the experience of compulsory heterosexuality. Ahmed wrote that bodies become "contorted" (Ahmed, 2006, p. 67), through repeating specific gestures (and not others), or through being orientated in specific directions (and not others). The metaphor of contortion indicates that our bodies get "twisted" into shapes that enable some action "only insofar as they restrict the capacity for other kinds of action. Compulsory heterosexuality diminishes the very capacity of bodies to reach what is off the straight line" (Ahmed, 2006, p. 67).

Ahmed's work offers a way to see how gender and sexuality are produced in everyday moments of interaction with objects, others, and the spaces we inhabit. In this light, our everyday experiences, including language, can be seen as oriented towards heteronormativity, which is reinforced socially, institutionally, politically, and culturally. However, as both Butler (2006/1990) and Ahmed (2006) pointed out, what makes this more complicated is that gender and sexuality are typically experienced as originating from within oneself, and the interactions which reproduce the gendered subject are displaced and hidden from view. Ahmed (2006) explains what it takes to go "off the straight line," using the example of being/becoming a lesbian: "It takes time and work to inhabit a lesbian body; the act of tending toward other women has to be repeated, often in the face of hostility and discrimination, to gather such tendencies into a sustainable form" (p. 78).

However, it has also been argued that by focusing primarily on homosexuality as a means to disrupt heteronormativity, earlier scholarship in lesbian, gay, and queer studies failed to recognize how transgender experiences could be antiheteronormative (Stryker, 2013). Thus, transgender studies emerged alongside—sometimes within, and sometimes in opposition to—lesbian and gay studies and queer studies, and from this history, developed its own language to highlight transgender experiences (Currah & Stryker, 2014). This language includes cisnormativity and cisheteronormativity, which indicate the co-occurrence and/or intertwinement of both cisnormativity and heteronormativity in our everyday experiences. In this context, it is important to note that Ahmed's approach to understanding queer experiences—i.e., non-dominant and marginalized experiences related to gender and sexuality—aligns with Lane's (2009) new materialist approach to queer and trans studies which argues that "feminist analysis needs to move from ideas of the body as constraining, fixed, and given toward ideas of the 'body *becoming*' as dynamic, transformative process" (p. 141). In this perspective, the body is both the place where gender is experienced, and a dynamic and public representation of one's gender and sexual identities. To actively take up the "body-becoming" approach to understanding gender and sexuality, we must recognize how gender and sexuality are dynamically shaped and experienced in the body within interactions. This is where Ahmed's work provides a productive analytic framework for understanding gender and sexuality phenomenologically by attending to how we turn towards or away from certain objects, others, and places as well as how objects, others, and places in turn might offer opportunities for the extension of our bodies.

The analytic framework we use combines the theoretical work in queer and trans studies with the literature on *figured worlds* (Holland et al., 1998). We posit that compulsory heterosexuality can be interpreted as a figured world because it comes to be inscribed upon the body, and embodied as part of one's identity, through everyday participation in reproducing the figured world. As will become evident in our analysis in the following sections, boundary play in VR involves the creation and use of pivots in the form of 3D sculptures, reflections and conversations around which can help us engage in both personally meaningful and socially disruptive discourse about heteronormativity.

17.3 Virtual Reality Environments Used in this Work

We used two applications during the project, Oculus Medium and Facebook Spaces. Oculus Medium is a VR digital sculpting application which allows the user to sculpt an object floating in space in front of them within the virtual environment. The user places digital "clay" in the environment and then can shape the clay into whatever object they desire using a variety of digital tools. The application also has a feature called Studio Share which allows a user to invite a friend (their Oculus accounts must be linked as friends) to sculpt in the same virtual space. The users can see and hear each other in real time and both can see what each person is sculpting. They can sculpt directly in the space that the other is sculpting, but they cannot alter the other person's sculpture. This application was chosen for two reasons. First, one of our participants, Scout, is a professional digital, VR sculptor. We designed the sculpting activity so that Scout could support other participants in learning to navigate the new application.

The second application used during the research was Facebook Spaces. We originally had not included Facebook Spaces in the design of the research, but during a break one participant, Matthew, suggested we try out the application. It is an explicitly social application that relies upon you and your friends all being available at the same time and Matthew saw the research event as an opportunity to try out the application since he had three of his friends available. In Facebook Spaces, one user invites up to three other people to join the same virtual environment. Since we only had three VR systems available, we could only have three participants in the virtual environment at one time. We used the version of Facebook Spaces that was available in late 2017. The application has since been updated and some of the functions described below have changed.

Within the Facebook Spaces virtual environment, the users embody avatars that are positioned around a circular table. Users can speak and gesture with their hands and head to friends around the table without requiring their physical presence. The application requires users to sign in with a Facebook social media account when launching the application. Once in the virtual environment, the user can play with virtual objects in the space such as dice, a drawing pen, a selfie stick, or virtual fishing game. The user can also connect with media they have shared on their Facebook account to display it in the virtual environment to their friends, or the user can access public media content available within the application, such as 360° photos and videos. Users can also modify their avatars while looking into a virtual mirror at their avatar, using preset options that are organized into the following categories: eyes, nose, head shape, skin color, hair style and colors, ears, mouth, eyebrows, facial hair, T-shirt color, and glasses. Color options for eyes include two shades of blue, four shades of green/hazel, and three shades of brown. Skin color includes 14 shades that run from pale beige to dark-medium brown. Seventy-two hairstyles are included and nine natural hair colors, including grey, orange, red, two shades of blond, three shades of brown, and black. Sixteen mouth shapes and 14 shades of lip colors are included, including one natural skin color that changes shades depending on the user's selected skin color, as well as non-natural lip colors, including blue, purple, turquoise, green, black, and various shades of pink. The user also has the option of having the application analyze a Facebook profile photo that the user selects from a list of polaroid-style photos on the right side of their mirror. The user can grab a polaroid photo and the application will generate five possible avatars from which the user can select one and further modify if they wish.

17.4 Method

17.4.1 Participants and Settings

We are presenting a retrospective analysis of conversations, interactions, and design artifacts from a design group meeting. As designers, we wanted to collaboratively explore the possibilities that existing virtual reality tools might have for representing gender and sexual identities. The design team is comprised of all five authors and all were present in the design meeting. Our initial idea was to bring together a design team of friends with interests in virtual reality development. The design team decided prior to the meeting to record their interactions in case they came up with ideas for new technology designs. The design team members agreed to video recordings because it was easier than making written notes which would have prevented us from participating in the virtual reality activities. As the conversations ensued and became more personal, the first and second author checked that the design team members still wished to share their experiences. Half-way through the meeting, there was a food break during which the authors discussed and agreed that there was value in collaborating and sharing the design conversations at an upcoming STEM symposium. All participants agreed to serve as co-authors and share their personal stories and pictures of their artifacts. The order of authors was decided based on the contributions to conceptualizing and writing the paper. Member-checking was done by the first and second author several times during the analysis and the writing of the paper in the form of informal meetings and phone and text conversations.

The analysis focuses on four of the design team members, who are all friends with each other, and in their early thirties. They all have gender nonconforming, queer identities and life histories. Some parts of their identities and life histories are shared and are presented in the analysis. The design team members who are the focus of this analysis are all white and were born and raised in Canada. The participants' pronouns are as follows: Scout (she/her), John (he/him), Matthew (he/him), and Dylan (they/them).

Scout is a professional 3D digital sculpture artist, who works in VR and non-VR applications. John is a computer programmer and works as an IT technician. Matthew is a game designer and worked as the manager of immersive technologies at the local science center. He has worked in the gaming industry in various roles for over a decade. Dylan is a PhD student in the Learning Sciences and has worked as an educator in gender and sexuality for post-secondary, community, and workplace settings. In addition, it is important to note that all participants had previous experiences with VR explorations. This enabled them to dive into the designed activities right away, instead of learning to use VR controllers and navigate VR worlds. This was important for the on-site time constraints that we were working within, although in future studies, our goal is to involve people who are new to VR.

We conducted the design meeting in the immersive technologies studio in a public museum in a Canadian metropolitan city, as part of a collaboration between the University researchers (authors of this paper) and the museum. The design meeting activities occurred over approximately 4 h. The first activity involved VR sculpting using the Oculus Medium VR sculpting application and Oculus Rift virtual reality systems (Fig. 17.1). The second activity used Facebook Spaces, a social VR space that includes forms of identity development and sharing, like avatar creation and "selfies," and involved undirected activity initiated by the one of the design team members that was not initially part of the design meeting planned activities. During the VR sculpting activity, the prompt provided to Scout and John asked them to think about their *experiences* of gender and sexuality, rather than normatively used identity labels such as "queer" or "trans." That is, we asked them to focus on

Fig. 17.1 Room setup with three oculus rift virtual reality systems

experiences that they felt were meaningful for their own coming into their gender and sexual identities. Specifically, the prompt to design a symbolic object was phrased as, "Try to sculpt something that represents your *experience* of gender and/or sexuality. For example, a time when your gender or sexuality was policed, challenged, or questioned." This shift away from a focus on normative gender and sexuality labels and towards complex, emergent expressions is grounded in queer theory.

The observed conversations and interactions between the participants were audio and video recorded. Recordings were transcribed and analyzed through a phenomenographic lens (Marton, 1981; Sengupta & Shanahan, 2017). Central to phenomenography is attending to people's experiences of a phenomenon and how these experiences shape their behaviors, conceptualizations, and interpretations related to the phenomenon. This is particularly important for our theoretical focus on figured worlds and boundary work, which involves not only how we act in the world, but also how we conceptualize and interpret our actions and the environment where we are situated. To this end, a phenomenographic approach attends to relationships between the participants and the world around them, including conceptual thought, immediate experience, and physical behavior. In addition, phenomenography is also based on the premise that there is a deeper structural dimension that shapes experience that underlies the diversity and variability of participants' experiences and sense-making (Marton, 1986, pp. 41–42). This is important for our work, given that our experiences of gender and sexuality are "oriented" (Ahmed, 2006) by the social and historical forces at and across multiple levels: personal, interactional, and institutional (Risman, 2004). Therefore, our analysis focuses how the participants' interactions, explanations, and responses reveal their ways of sense-making, and also some of the underlying structural forces that shaped their experiences, both historically and in-the-moment.

To this end, in our analysis, we focused on the relationship between participants' creative work in VR, their conversations with friends, and their gender and sexual life-histories and identities. A key characteristic of boundary objects is that they can be used by different people or groups in different ways—i.e., the same boundary object can take on different meanings in different social realms (Star, 1988; Shanahan, 2011). As Shanahan (2011) aptly put it, boundary objects have "interpretive flexibility," which is the form of data that we focus on in our analysis. Interpretive flexibility is deeply connected to the work of figuring, through which the participants constructed and represented their figured worlds. So, our analysis focuses on presenting relevant segments of the transcript and relevant images of the participants' work that make explicit both how they engaged in the work of *figuring*—the active construction of meaning—and their interpretive explanations of the meanings of their VR sculptures. Our analysis thus shows that the figured worlds of the participants are also emergent and dynamically constructed through their interactions with creative work in VR and with each other. We also identify how elements of the computational infrastructure served as *pivots*, enabling participants to shift across multiple social realms of their experiences. That is, the theoretical lenses of figured worlds and pivots offer us language to describe the *form* of participants' sense-making, while the "content" of their sense-making—a key commitment of phenomenographic approaches—is evident in the rich descriptions of the illustrative cases we present in the findings.

17.5 Findings

Our analysis, presented in the form of illustrative cases, shows how participants' figured worlds of gender and sexuality emerge through the creation and use of pivots (Finding 1), how engaging playfully with friends in VR enabled them to represent their appearance in VR and find affirmation of their identity (Finding 2), and how playful conversations and 3D sculpting together with friends enabled some of the participants to talk about their experiences of hurt pertaining to their gender and queerness (Finding 3).

17.5.1 Finding 1: Pivots and Figured Worlds of Gender and Sexuality

When prompted to create an object in VR that represented a personally meaningful aspect of her gender and/or sexual identities, Scout, a professional VR artist and sculptor, initially sculpted a three-dimensional "gender key." She explained her work as follows:

> So when I was sculpting, it was really interesting because you asked us to sculpt something that had to do with a time that our gender, specifically, a time that our, like, people felt that we were too much one way or another, and so, I was trying to think, like, 'Okay, what can I sculpt?' And so I started sculpting a key without thinking because it was the first thing that came to mind. And I was like, 'Oh, it's like the key of… the gender key!' But then as I was sculpting, another thing came to mind, and I was like, oh, I remember when I was younger, I didn't get my ears pierced until I was seventeen and I only got one ear pierced. And that's, like, really unusual for women, um, to only have one ear pierced.

In this excerpt, Scout explains that, for her, sculpting and imagining are deeply intertwined, in a manner that is analogous to Pickering's (2010) description of scientific practice as a "mangle"—an inescapable intertwinement of conceptual and representational work. Rather than first imagining what the sculpted object should be, Scout started with a key because that is what came to her mind. The key, or what became the "gender key," is a pivot between figured worlds of heteronormativity and queerness. Pivots serve to shift us between different figured worlds, moving us from one representation and figured world to another. The pierced ear is a representation of the figured world of queerness, whereas the gender key serves as a meta-representation by serving as the pivot between two figured worlds—a figured world of heteronormativity and a figured world of queerness. For Scout, the gender key represented the struggle of being between conflicting figured worlds. In order to reflect upon and represent her experiences of gender and sexuality, Scout first began with representing this struggle of being between conflicting figured worlds. The meta-representation of the "gender key" signified the "in-betweenness" of two figured worlds. As a pivot, the "gender key" symbolically unlocked the movement from heteronormativity to queerness that arose in Scout's process of sculpting her life experience. The action of rendering shape to the initial sketch of the key reminded Scout of how she shaped her own body by piercing her one ear in order to represent her queer identity when she was a teenager. Sculpting the key became a pivot for Scout into a figured world of queerness. The gender key served as a meta-representational pivot—a key that would unlock the worlds of queerness for her by allowing her to belong to that world through re-making her body, thus shifting her from a figured world of heteronormativity into a figured world of queerness. The pierced ear then became a *pivot* for Scout to launch into deeper reflections about her queer past. She further explained:

> So yeah, I only got one ear pierced and that was really unusual and I remember thinking at the time, 'Uhh, I don't know which ear is the gay ear.' But I wasn't out yet. […] And a lot of gay men used to use that, right? But I got one ear pierced. I believe it was my right ear and uhh, I left it like that for two years and so that was a little, that was something that was, like, not quite correct for my gender, to only have one ear pierced, and so that was a memory that came up while I was sculpting that wasn't something I was thinking about. I actually hadn't thought about it at all until.. […] I was already, like, a little bit queer compared to some of the, you know, girls in my school. I'd already been told I was too masculine by some of the boys.

Sculpting the gender key led Scout to remember a moment in her past when she had been reflecting on queer identity markers and ways of signaling queerness through re-making the body—an example of *body-becoming* (Lane, 2009), which emerged in Scout's reflections as a central part of her figured world of gender and sexual identity. She mentioned that her memory occurs prior to her coming out—meaning prior to publicly claiming a label outside of cisheteronormative expectations of gender and sexual identities. This suggests that her figured world of gender and sexuality was shaped through her own sense of incongruity with social expectations, as expressed by her peers who thought she was too masculine. She expresses her emerging awareness of this incongruity with gender and sexual norms in terms of being "a little bit queer compared to some of the, you know, girls in my school" and being told she was "too masculine by some of the boys."

Her story highlights a significant experience of body-becoming—i.e., her experience of inscribing gender and sexuality upon her own body—and it also presents a complex and nuanced image of her figured worlds of gender and sexuality as a teenager. Her figured world was at once improvisational and rooted in her implicit desire to be recognized as queer, while at the same time being grounded in her own interpretation of the heterosexual matrix.

17.5.2 Finding 2: Play, Intimacy, and Desire for Recognition in VR

The second activity in the design meeting (the Facebook Spaces application) was initiated by Matthew, whose life history of play shaped his desire to direct his friends into playful activities. He explains his ongoing commitment to play in his work:

> I often will try to encourage, like people [at work], to play more. And very often in a work environment, you feel this automatic pushback, like, "Yeah, we value play," but like, the play has to have a purpose or we have to like be doing it, you know, to try to get at some learning. It has to be for "team-building" or something like that. And I try to push back against that has hard as I can to say that play has value for play on its own. And I know that there are benefits to development and benefits to socialization, but I feel like we shouldn't have to defend play. It should just be allowed to be play. Like, yes, it has benefits, but even if it doesn't, it has intrinsic value, I think.

In the context of our design meeting as well, Matthew explained that he was not even necessarily trying to consciously make everyone play. Matthew's own figured world has been shaped by and is centered around play and his desire to encourage play among others. This influenced him to invite all of us to play both in the sense of inviting us to use his work space to meet and in the sense of inviting us to play in the Facebook Spaces application.

Play among friends in Facebook Spaces became influential to the exploration of gender and sexual representation by developing intimacy and encouraging exploration of bodily representation and modification. One way in which play and body modification occurred was through drawing modifications. Scout and Matthew had learned that it was possible to draw items using the VR pencil that could then be attached to a player's avatar when Scout had drawn a pair of glasses which Matthew had been able to attach to his avatar's face. Scout went on to draw another object which Matthew then picked up, duplicated, and put on his avatar's ears to turn the objects into earrings. When Dylan entered into the application, they also attached the earrings to their avatar. In the following exchange, Scout, John, and Matthew discuss the significance of these objects:

Scout: I just love that you're both wearing those ridiculous earrings. They actually look fun. It's kinda cool. Do you guys want to draw new earrings for yourselves?
John: They want to wear "Scout."
Scout: They want to wear..? Oh yeah! Earrings by Scout.
Matthew: A Scout original.

Scout suggests to Matthew and Dylan that they could create their own objects to wear, but John and Matthew both affirm that the objects created by Scout hold more meaning, signifying the importance of friendship that is attributed to the objects created by Scout. Earrings had been an important part of gender and sexual representation for Scout as demonstrated in her initial sculpt in the first activity. Earrings became a significant object in Facebook Spaces as well. They signified intimacy and friendship as objects created by Scout and worn by Dylan and Matthew, and they signified gendered objects that were playfully worn in ways that challenged their gendered meaning.

Another way that body modification became important to exploring gender and sexuality was through avatar modification. Within Facebook Spaces, the player can modify their avatar as they face a virtual mirror of their avatar's body, including head, torso, arms, and hands. Their avatar projection in the mirror moves with the player's movement, increasing the sense of embodying the avatar. Facebook Spaces requires the player to login to their Facebook social media account and in the avatar modification menu, the application will load profile pictures from the player's Facebook account. The player can grab a profile picture and the application will offer suggestions for avatars based upon the picture the player chooses. When Dylan entered the avatar modification menu, they noted that the avatar choices were not divided into binary genders where the player first chooses a gender and then subsequent avatar modification options are limited by the initial choice of gender. Instead, the application did not ask for a gender and offered a variety of modification choices as detailed in the above section on virtual reality environments used. However, when Dylan picked up a profile picture to see the suggested avatars the application would generate based on the picture, the suggested avatars all had long blond hair. Dylan tried choosing other profile pictures, but continued to receive the same suggestions despite having neither long nor blond hair.

Dylan went on to modify their avatar without using the suggested avatars to approximate a likeness to their real self. They faced another limitation in hair color. For Dylan, who often colors their hair non-natural colors, the available hair colors were a significant factor in limiting their ability to recognize themselves in their avatar. As Dylan encountered the limitations of the application, they voiced their frustration to their friends. Upon exiting the avatar modification menu, Dylan joined Matthew and Scout around the virtual table. Scout suggested that she might be able to modify Dylan's avatar to look more like them by drawing purple into their hair. Scout stretched her virtual arm out towards Dylan's head and Dylan exclaimed surprise at how real it felt to have Scout move into their personal (virtual) space. Once Scout colored Dylan's hair with purple, Dylan used the virtual selfie stick to see how they looked and posed for a picture (Fig. 17.2).

Dylan shared this selfie to their Facebook social media from within the application. Upon having their hair modified by Scout, Dylan gained a sense of recognition with their avatar and thus shared this representation of themselves. Body modification in real life is a part of Dylan's life history of *body-becoming* that could not be achieved in Facebook Spaces' avatar creation tool. Dylan's desire for self-recognition and identity affirmation in VR could be achieved through intimacy and informal gender play with friends. Later, Matthew came back to the selfie of Dylan and remarked on how he felt about it:

> This is my favourite picture of Dylan so I'm going to make it really big. [Puts picture on VR "wall". Matthew and Dylan laughing.]

Although the Facebook Spaces application imposed limitations upon Dylan that impeded their ability to recognize themselves—a process of virtual body-becoming—Matthew and Scout found ways to support Dylan in modifying and being recognized in their avatar. Dylan explained after the experience what they had been thinking when Scout colored their hair:

> It made me remember when I was in high school with Matthew. We used to get to school early and go to the men's washroom to style each other's hair. It was such a heteronormative, strict school, and I could have gotten into trouble for going into the men's washroom.

Fig. 17.2 Dylan's avatar posing for a selfie within Facebook Spaces

> We got into trouble with teachers for even saying the word 'gay' at that time. I remember how Matthew and I would do things like style each other's hair in the men's washroom and it was kind of a way to challenge that.

The gesture of Scout standing close to Dylan in VR and reaching into Dylan's personal (virtual) space to color Dylan's hair was an intimate one, not merely because of the physical proximity, but on a more profound note, because it helped Dylan recognize themselves in their avatar. Recognition of oneself in virtual reality is also an affirmation of Dylan's queerness through their appearance in virtual space—a form of body-becoming, essential for being queer. Similar to Ahmed's (2006) reminder of the work that it takes to be lesbian (as mentioned earlier in our theoretical discussion of compulsory heterosexuality), this is an example of the work that it takes to see oneself as queer in a virtual environment. It took a significant amount of work and collaboration—alongside intimacy—in order for the Dylan to be able to recognize themselves in VR, and this example also illustrates how intimacy with close friends can help significantly in this process. The selfie of Dylan's avatar, once Dylan felt comfortable with their recognition of themselves in the virtual world, became another object attributed with the meaning of supporting recognition—of virtual body-becoming, achieved through intimate, informal gender play. Matthew's expression of appreciation for the selfie and gesture of support by putting the picture up on the VR "wall" further reinforced this meaning.

Overall, these intimate gestures of friends with each other around appropriating virtual artifacts to express their queerness created a shared figured world among the participants that was centered around supporting the desires of friends to recognize themselves and to be recognized by others in their genders and sexualities. This began with Matthew's desire to encourage play which created the context for intimacy and allowed for the support from friends that was necessary to modify avatars. At the same time, Scout's intimate gesture and interaction with Dylan's avatar acted as a pivot for Dylan into a past memory of being with Matthew and challenging the heteronormative and homophobic context of high school.

In these examples, we can see how play is an important context for expressing and affirming gender and sexual identities. As Ahmed (2006) demonstrates, a queer phenomenology orients us towards questions of *who* and *what* we are oriented towards as well as *how* we are oriented by others, which we can extend to our consideration of how the context of friendship and play orients us. In the context of our design meeting, figured worlds of queerness could be explored because of the factors of *who, what,* and *how.* The participants were oriented towards each other (*who*), because of their investment in their friendships, and towards each other's queer artifacts (*what*), because the artifacts (such as Scout's earrings in Oculus Medium and Facebook Spaces, or Dylan's avatar) became an object that a friend could interact with that could be played with in order to affirm (*how*) their meaning as queer objects. The participants were also oriented by the activities to explore their *experiences* of coming to understand their gender and sexuality and this orientation queered the context of learning and creating in virtual reality. The participants further engaged in this activity of queering virtual reality by affirming each other's queer identities and experiences.

17.5.3 Finding 3: Sculpting "Hurt" Playfully: Journeys in Virtual Body-Becoming

For Scout and John, gender and sexuality were shaped through their experiences of exclusion and incongruity in an existing cisheteronormative context among peers. As they proceeded further, their sculptures began to embody their experienced incongruities more viscerally and explicitly in the form of *hurt*. Interestingly, despite the gravity of the stories and emotions represented in their sculptures, play was an important element of this experience. In this section, we describe the final versions of the sculptures, and explain how the sense of playful exploration can create a context for *virtual* explorations in body-becoming, which in turn can facilitate deep explorations of issues pertaining to gender and sexual identities.

When John and Scout returned to the Oculus Medium application and finished their sculptures after playing in Facebook Spaces, their sculptures looked very different from what they had completed earlier. John continued with his sculpture of hangers and clothing, but now he used stamps from within the application of gendered male body parts (an arm and a torso) and hung the body parts from the hangers instead of using clothing as he had previously (Fig. 17.3).

John explains why his sculpture took this turn from clothing to body parts as directly linked to the play experienced through the Facebook Spaces application:

> In play it was all about being able to try something new and different, uh, colouring on each other, creating jewelry out of nothing, that, "hey, this is cool! do-do-do-do-do. Hey, look, I put it on my ear!" Little bits of creativity like that to almost exercise and get the mind going. Especially in that area of body exploration, which, as I say it, definitely, it brought forth ideas of my own body image and how it refers to gender, uh, and such. [….] Umm, in play it felt like I was being someone different, uh, being someone else, being playful, uh, being able to try different things as a different person. Here [in the second iteration of sculpting] was completely about how I perceived myself as I am, not really about what I want to be, or.. something I want to try. So I guess in that sense, being able to play and experience the different sides of myself gave me more perspective as to how I view myself and who I think I really am. If that makes sense?

John created a figured world where he could explore different sides of himself by creating and playing with gendered objects that sparked thinking and conversations about body exploration. This became the context for virtual exploration of John's body-becoming, as gendered objects (such as jewelry) became pivots for John into the figured worlds of the gendered body. When he returned to sculpting, John was able to reflect not only on how he challenged gendered, heteronormative ideas about clothing as a youth, but also his ongoing relationship with his body within the context of a cisheteronormative society. John reflected upon his K-12 experiences of enforced masculinity and cisheteronormativity; for example, he reflected upon

Fig. 17.3 John's part of the sculpture showing a dress, two arms, and a torso (gendered male) hanging from hangers

when he wore a dress to play a female character in a school play. John explained that wearing a dress was treated as a joke by his peers, even though he experienced it as a form of playful gender expression, a form of body-becoming, that allowed him to explore aspects of his femininity.

Upon returning to sculpting, Scout's sculpture also took a stronger turn towards the body and gendered assumptions. Scout sculpted a mannequin torso and head with mixed gendered characteristics (Fig. 17.4).

On one side of the mannequin, she sculpted short hair and long eyelashes, and on the other side she sculpted the reverse and added a beard. On the torso she painted two checkboxes with the letter F, representing female, and the letter M, representing male. Stakes were driven through each box and into the torso. Scout explained that the sculpture represents how "if you express a mix of masculine and feminine characteristics, you are likely to experience double the injuries." Scout's addition to her sculpture further expresses her struggle of being between figured worlds of cisheteronormativity and queerness. She captures in her final sculpture the very painful and visceral limits one faces when trying to express one's gendered and sexual identity through the body when we cannot fully escape the cisheteronormative discourse of society which imposes the interpretive lens of "male" and "female," regardless of how we see and experience ourselves.

Fig. 17.4 Scout's part of the sculpture showing a pierced ear on the side of a window frame with curtains and a mannequin torso

17.6 Conclusions and Discussion

17.6.1 Analytic Summary: Intimacy, Pivots, and Virtual Becoming

Overall, our analysis reveals how by creating boundary objects in the form of 3D sculptures and engaging in playful, creative work and discussions with friends, Scout and John were able to represent and voice how gender and sexuality come to be inscribed upon the body both through their own improvisational, playful acts of body-becoming and through the violence of cisheteronormativity imposed upon them. Play and intimacy were key elements of the participants' experiences. Gender-becoming is also body-becoming, and this was also true in VR. Scout's initial sculpture (a pierced ear) represented her own experience and desire to express her gender and sexuality through body modification, a playful act of body-becoming through re-making her body. This improvisation in re-making her body demonstrates an exploration of her own figured world of gender as dynamic and playful, and in the process, she shared a story that she had never shared before with anyone besides her partner. Play and intimacy also created an environment that made the participants comfortable enough to share their hurt around their gender and sexual identities. For example, both of their final sculptures contained expressions of gendered violence inflicted upon bodies that do not conform to societal expectations of gender and sexuality.

In the cases we have presented, virtual objects serve as pivots in two senses. First, it is noteworthy that in the participants' improvisations, virtual body-becoming through 3D sculpting and avatar creation served as pivotal experiences for the participants in the sense that it enabled them to pivot between their lived histories and imagined selves. Once again, we see here the value of intimacy among friends, which created a context in which such experiences could be easily discussed without fear. Gender and sexuality becoming is not merely an individual act. As Butler (2006) and Lane (2009) poignantly argue, we experience gender and sexuality through our bodies as socially and culturally sanctioned ideologies of cisheteronormativity that we are expected to conform to with our bodies, and which is often imposed through gender violence. At the same time, virtual body-becoming through avatar creation applications can pose limitations for self-recognition, particularly in this case for those who use body modification or who are otherwise nonconforming in their self-expression, as was evident in the case of Dylan, who was simply assigned a virtual avatar algorithmically by Facebook Spaces, which was inconsistent with their gender identity and expression. Play and intimacy between close friends, combined with tools for open modifications, like drawing and attaching virtual objects to one's avatar, can offer potential opportunities in such cases, especially for those with queer and trans identities. Here, virtuality serves as a pivot between an unwanted, algorithmically and socially ascribed gender expression and a more desirable one, as well as a shift away from gender expression being a solitary experience to being a *place* for being queer *along with* close friends. This turn towards a more social and more intimate experience of gender-becoming, along with friends, created opportunities for the participants to interweave their histories of being and becoming queer with playful, intimate interactions with their friends.

Sociologists have long reported that informal conversations and learning about sex, gender, and sexuality play an important role in shaping our gender and sexual identities (Risman & Banerjee, 2013; Thorne, 1993). However, those studies also highlight how heteronormativity is reinforced through these informal conversations and interactions. Our work, in contrast, provides an illustrative example where innovations in STEM—for example, Virtual Reality and 3D Sculpting—can be *queered* to create a positive space for engaging in conversations and computational creative work about queer and trans identities by leveraging informality and intimacy between close friends.

17.6.2 *Queer Phenomenology, STEM, and the Learning Sciences*

Our work challenges the learning sciences to take up questions of not only *what* we are oriented towards (i.e., what is worth learning), but also *who* we are oriented towards, and *how* we are oriented in our learning. On one hand, these questions are deeply phenomenological in nature (e.g., see Sengupta, Dickes, & Farris, 2018), as they are invitations for us to peel off the sphere of givenness—the world *as-is*—i.e., as is given to us both in the form of canonical disciplinary knowledge in the classroom, and in the form of socially and culturally mandated ways of performing cisheteronormativity. The phenomenological agenda must then lead us to a place of *originary* sense-making (Merleau-Ponty, 1962), which, following Ahmed (2006), can directly challenge our entrenchment in cisheteronormativity. What is being learnt here, we believe, is not simply a matter of knowing in the disciplines; in contrast, the experiences reported here offer a fundamental shifting away from the "what" to the "who" as the topic of learning. That is, such queer and trans encounters with technology can move us away from reifying domains of knowledge, such as STEM, towards questions of *who we were, who we are,* and *who we can become*.

Such experiences can redefine our relationships with STEM innovations, such as VR and 3D sculpting, which can then become spaces and tools that serve the purpose of re-orienting us towards dynamic, emergent experiences of gender-becoming, while at the same time, de-orienting us away from our ubiquitous immersion in cisheteronormativity. Furthermore, in creating such spaces, we can also potentially reframe the purpose of learning sciences research as not merely phenomenological inquiry of learning, but also queering the phenomenology of learning (Ahmed, 2006). It enables us to imagine new ways of learning where we can engage in creative and incisive re-constructions of our individual histories and narratives, for example, in the contexts of gender and sexuality—that often carry a great deal of hurt and are usually silenced in public education—through the collaborative creation of imagined futures with close friends.

Acknowledgments Funding from NSF Award # 1842358 and Imperial Oil Foundation are gratefully acknowledged. All opinions are the authors' and not endorsed by the funding agencies.

References

Abrahamson, D., & Lindgren, R. (2014). Embodiment and embodied design. In R. K. Sawyer (Ed.), *The Cambridge handbook of the learning sciences 2* (pp. 358–376). New York, NY: Cambridge University Press.

Ahmed, S. (2006). *Queer phenomenology: Orientations, objects, others*. Durham, NC: Duke University Press.

Ahn, S. J. G., Le, A. M. T., & Bailenson, J. (2013). The effect of embodied experiences on self-other merging, attitude, and helping behaviour. *Media Psychology, 16*, 7–38.

Azevedo, F. S. (2018). An inquiry into the structure of situational interests. *Science Education, 102*(1), 108–127.

Bazzul, J., & Sykes, H. (2011). The secret identity of a biology textbook: Straight and naturally sexed. *Cultural Studies of Science Education, 6*, 265–286.

Berland, M., & Lee, V. R. (2011). Collaborative strategic board games as a site for distributed computational thinking. *International Journal of Game-Based Learning (IJGBL), 1*(2), 65–81.

Butler, J. (2006). *Gender trouble: Feminism and the subversion of identity*. New York, NY: Routledge. (Original work published 1990.).

Cavanagh, S. (2010). *Queering bathrooms: Gender, sexuality, and the hygienic imagination*. Toronto, ON: University of Toronto Press.

Currah, P., & Stryker, S. (2014). Postposttranssexual: Key concepts for a twenty-first-century transgender studies. *TSQ: Transgender Studies Quarterly, 1*, 1.

Davis, G. (2015). *Contesting intersex: The dubious diagnosis*. New York, NY: New York University Press.

Davis, G., & Murphy, E. L. (2013). Intersex bodies as states of exception: An empirical explanation for unnecessary surgical modification. *Feminist Foundations, 25*(2), 129–152.

Esmonde, I., Brodie, K., Dookie, L., & Takeuchi, M. (2009). Social identities and opportunities to learn: Student perspectives on group work in an urban mathematics classroom. *Journal of Urban Mathematics Education, 2*(2), 18–45.

Farris, A. V., & Sengupta, P. (2016). Democratizing children's computation: Learning computational science as aesthetic experience. *Educational Theory, 66*(1–2), 279–296.

Grace, A. P. (2015). *Growing into resilience: Sexual and gender minority youth in Canada*. Toronto, ON: University of Toronto Press.

Groom, V., Bailenson, J. N., & Nass, C. (2009). The influence of racial embodiment on racial bias in immersive virtual environments. *Social Influence, 4*(3), 231–248.

Holland, D., Lachicotte, W., Jr., Skinner, D., & Cain, C. (1998). *Identity and agency in cultural worlds*. Cambridge, MA: Harvard University Press.

Hostetler, A., Sengupta, P., & Hollett, T. (2018). Unsilencing critical conversations in social-studies teacher education using agent-based modeling. *Cognition and Instruction, 36*(2), 139–170.

Jeppesen, S. (2016). Heteronormativity. In A. Goldberg (Ed.), *The SAGE encyclopedia of LGBTQ studies* (pp. 493–496). Thousand Oaks,, CA: SAGE Publications, Inc.

Karkazis, K., Jordan-Young, R., Davis, G., & Camporesi, S. (2012). Out of bounds? A critique of the new policies on hyperandrogenism in elite female athletes. *The American Journal of Bioethics, 12*(7), 3–16.

Kim, B., & Ho, W. (2018). Emergent social practices of Singapore students: The role of laughter and humour in educational gameplay. *International Journal of Child-Computer Interaction, 16*, 85–99.

Lane, R. (2009). Trans as bodily becoming: Rethinking the biological as diversity, not dichotomy. *Hypatia, 24*(3), 136–157.

Leander, K. M., & Boldt, G. (2018). Design, desire, and difference. *Theory Into Practice, 57*(1), 29–37.

Lindgren, R., & Johnson-Glenberg, M. (2013). Emboldened by embodiment: Six precepts for research on embodied learning and mixed reality. *Educational Researcher, 42*(8), 445–452.

Marton, F. (1981). Phenomenography — Describing conceptions of the world around us. *Instructional Science, 10*(2), 177–200.

Marton, F. (1986). Phenomenography—A research approach to investigating different understandings of reality. *Journal of Thought, 21*(3), 28–49.

Massumi, B. (2002). *Parables for the virtual*. Durham, NC: Duke University Press.

Merleau-Ponty, M. (1962). *The phenomenology of perception*. London: Routledge and Kegan Paul. (Original work published in 1945.).

Mitchell, S. N., Reilly, R., Bramwell, F. G., Solnosky, A., & Lilly, F. (2004). Friendship and choosing groupmates: Preferences for teacher-selected vs. student-selected groupings in high school science classes. *Journal of Instructional Psychology, 31*(1), 20–32.

Oh, S. Y., Bailenson, J., Weisz, E., & Zaki, J. (2016). Virtually old: Embodied perspective taking and the reduction of ageism under threat. *Computers in Human Behavior, 60*, 398–410.

Peck, T. C., Seinfeld, S., Aglioti, A. M., & Slater, M. (2013). Putting yourself in the skin of a black avatar reduces implicit racial bias. *Consciousness and Cognition, 22*(3), 779–787.

Pickering, A. (2010). *The mangle of practice: Time, agency, and science*. Chicago, IL: University of Chicago Press.

Rich, A. (1980). Compulsory heterosexuality and lesbian existence. *Signs: Journal of Women in Culture and Society, 5*(4), 631–660.

Risman, B. J. (2004). Gender as a social structure: Theory wrestling with activism. *Gender & Society, 18*(4), 429–450.

Risman, B. J., & Banerjee, P. (2013). Kids talking about race: Tween-agers in a post-civil rights era. *Sociological Forum, 28*(2), 213–235.

Sengupta, P., Dickes, A., & Farris, A. (2018). Toward a phenomenology of computational thinking in STEM education. In M. Khine (Ed.), *Computational thinking in the STEM disciplines* (pp. 49–72). New York, NY: Springer.

Sengupta, P., & Shanahan, M.-C. (2017). Boundary play and pivots in public computation: New directions in STEM education. *International Journal of Engineering Education, 33*(3), 1124–1134.

Shanahan, M.-C. (2011). Science blogs as boundary layers: Creating and understanding new writer and reader interactions through science blogging. *Journalism, 12*(7), 903–919.

Star, S. L. (1988). The structure of ill-structured solutions: Boundary objects and heterogeneous distributed problem solving. In A. H. Bond & L. Gasser (Eds.), *Readings in distributed artificial intelligence*. San Mateo, CA: Morgan Kaufman.

Stryker, S. (2013). (De) subjugated knowledges: An introduction to transgender studies. In S. Stryker & S. Whittle (Eds.), *The transgender studies reader* (Vol. 1, pp. 1–17). New York, NY: Taylor & Francis.

Takeuchi, M. A. (2016). Friendships and group work in linguistically diverse mathematics classrooms: Opportunities to learn for English language learners. *Journal of the Learning Sciences, 25*(3), 411–437.

Thorne, B. (1993). *Gender play: Girls and boys in school*. New Brunswick, NJ: Rutgers University Press.

Travers, A. (2017). Transgender issues in sport and leisure). In L. Mansfield, J. Caudwell, B. Wheaton, & B. Watson (Eds.), *The palgrave handbook of feminism and sport, leisure and physical education* (pp. 649–665). London, England: Palgrave Macmillan.

Vea, T. (2019). The ethical sensations of im-mediacy: Embodiment and multiple literacies in animal rights activists' learning with media technologies. British Journal of Educational Technology, 50: 1589–1602.

Vygotsky, L. S. (1980). *Mind in society: The development of higher psychological processes*. Cambridge, MA: Harvard University Press.

Westbrook, L., & Schilt, K. (2014). Doing gender, determining gender: Transgender people, gender panics, and the maintenance of the sex/gender/sexuality system. *Gender & Society, 28*(1), 32–57.

Wittig, M. (1980). The straight mind. *Gender Issues, 1*(1), 103–111.

Chapter 18
Decolonizing Complexity Education: A Mayan Perspective

Marilú Lam-Herrera, Ixkoj Ajkem Council, and Pratim Sengupta

Abstract Our goal is to advance scholarship in STEM education by illustrating how computational modeling of complex, emergent phenomena can be re-imagined from an Indigenous perspective. Several Indigenous and Western scholars have argued that focusing on complexity and interdisciplinary forms of engagement with science can be synergistic with Indigenous ways and forms of knowing (Aikenhead & Mitchel, 2011; Bang et al., 2018). Building on their work, we offer a novel re-imagination of computational modeling for teaching and learning about complex systems that is grounded in Mayan traditions of garment weaving, an approach we co-designed in partnership with the Ixkoj Ajkem Council in Xenacoj, Guatemala. We will first present a theoretical review of synergistic frameworks from both Indigenous and Western perspectives for understanding and experiencing complex systems. We then introduce *Grafemos*, a learning environment that we have been designing in partnership with the Ixkoj Ajkem Community Council, in order to understand emergent phenomena, a key characteristic of complex systems. Finally, we present an example of the modeling experience in Grafemos, that integrates both Indigenous practices of storytelling in the context of fabric design and representational practices that are also valued by Western scholarship on complex systems education.

Keywords Indigenous perspectives · Complexity · Decolonization · STEM · Agent-based modeling

M. Lam-Herrera (✉) · P. Sengupta
Mind, Matter and Media Lab, Werklund School of Education, University of Calgary, Calgary, AB, Canada
e-mail: marilu.lamherrera@ucalgary.ca

Ixkoj Ajkem Council
Consejo Comunitario de Mujeres Tejedoras, Santo Domingo Xenacoj, Guatemala

P. Sengupta et al. (eds.), *Critical, Transdisciplinary and Embodied Approaches in STEM Education*, Advances in STEM Education,
https://doi.org/10.1007/978-3-030-29489-2_18

18.1 Introduction

Research on complex systems in K-12 education has been typically grounded in Western perspectives, emphasizing the importance of understanding emergent phenomena through building progressively complex relationships among individuals, and between individuals and their environments (Danish, 2014; Dickes, Sengupta, Farris, & Basu, 2016; Levy & Wilensky, 2008; Wilensky & Resnick, 1999). Emergent phenomena are complex, non-linear processes (e.g., ecological interdependence), which arise from simple, linear interactions between many individual agents (e.g., interactions between predators and prey). In a deeply synergistic manner, the sense of relationship and reciprocity with the greater-than-human world is foundational to Indigenous cultures and research (Bastien, 2003; Cajete, 1994). Indigenous scholars have therefore noted that complex emergent systems, as a domain of inquiry, offer a great opportunity for us to integrate Western and Indigenous perspectives and epistemologies in science and science education (Bang, Marin, & Medin, 2018).

However, integrating Indigenous traditions and practices in the context of teaching and learning about complex systems using computational modeling necessitates designing new forms of modeling in partnership with Indigenous communities, rather than relying solely on commonly used platforms for modeling or programming complex systems designed by Western scholars. To this end, this chapter offers a novel re-imagination of computational modeling grounded in Mayan traditions of textile weaving, co-designed in partnership with the Ixkoj Ajkem Group in Xenacoj, Guatemala. Theoretically, our work builds on prior scholarship by Indigenous scholars in STEM education, as well as synergistic scholarship on complexity by Western scholars. From a technology design perspective, our work illustrates how the experience of computational modeling of complex systems, when grounded in Mayan traditions of weaving and storytelling, can be re-imagined without the use of a computer. We also discuss the implications of our work in terms of challenging technocentrism (Papert, 1987; Sengupta, Dickes & Farris, 2018) from Indigenous perspectives.

18.2 Theoretical Background

18.2.1 Complex Systems and Computational Modeling in K-12 Education

Over the past two decades, complex systems have become an important area of focus in K-12 science and STEM education (Chi, 2005; Danish, 2014; Dickes et al., 2016; Sengupta & Wilensky, 2011; Wilensky & Resnick, 1999; Wilkerson-Jerde & Wilensky, 2015). Emergence, as a key characteristic of complex systems, gained educational researchers' attention in designing learning environments for complex systems (Chi, 2005; Danish, Peppler, Phelps, & Washington, 2011; Holland, 2000;Mitchell, 2009; Wilensky & Reisman, 2006). Emergence can be understood as aggregations of simple, local interactions between many individual actors which give rise to complex and often counterintuitive global pattern (e.g., traffic jam moving backward while cars move forward) (Mitchell, 2009; Wilensky & Reisman, 2006). While students at all levels find understanding emergent processes challenging (Chi, 2005; Resnick & Wilensky, 1998), agent-based computational models (ABMs) have been shown to be successful in helping novices understand complex ecological systems (Danish, 2014; Klopfer, Yoon, & Perry, 2005; Klopfer, Yoon, & Um, 2005; Resnick, 1994; Wilensky & Resnick, 1999).

A computational agent is an individual object or actor in an ABM, and the behaviors and interactions between computational agents and their environment give rise to emergent, system-level behavior (Wilensky & Resnick, 1999). Studies show that curricula which utilize ABMs can help students understand complex systems and emergence by grounding emergent phenomena in terms of their embodied, agent-level intuitions (Danish, 2014; Klopfer, Yoon, & Um, 2005; Resnick, 1994; Wilensky & Resnick, 1999). This body of work shows that learners can build upon their agent-level (individual level) intuitions in order to develop deep, multi-level understandings of emergent phenomena through creating and modifying embodied and computational simulations (Dickes et al., 2016; Levy & Wilensky, 2008; Wilensky & Resnick, 1999). This can be greatly facilitated by helping learners represent individual-level interactions and mid-level "events," and it is by summarizing and understanding the relationships between these events over time that even elementary grade children can develop understandings of complex phenomena such as disease spread (Levy & Wilensky, 2008), ecological interdependence (Dickes et al., 2016), and electrical conduction (Sengupta & Wilensky, 2009). Dickes et al. (2016) argued that three main insights have emerged that can guide the design of pedagogy: (a) agent-level (individual level) provides an intuitive foundation to start thinking about complexity, (b) embodied cognition can be very helpful for understanding both individual-level behaviors as well as interactions with the environment, (c) a progression from embodied activities and agent-level interactions to mid-level and aggregate level interactions through both iteratively building and refining embodied, physical, mathematical and computational models can be effective in deepening students' understanding of complexity using ABMs.

The history of agent-based modeling in education is inseparably tied to constructionism (Papert, 1980; Papert & Harel, 1991). Constructionism is both an epistemological perspective and a pedagogical approach. As Berland, Baker, and Blikstein (2014) pointed out, in the constructionist paradigm, learning involves building and designing creative artifacts that require complex forms of intellectual engagement, alongside high levels of affective involvement (Harel & Papert, 1990; Kafai, 2006; Papert, 1980). Constructionist learning is typically grounded in learners' use of generative computational technologies in disciplinarily as well as personally meaningful forms. The illustrations of constructionist learning often involve the creation of public artifacts and working in social spaces such that learners can collaborate to deepen their learning experiences (Harel, 1991; Kafai & Burke, 2013; Sengupta, Krishnan, Wright, & Ghassoul, 2015). A constructionist approach to learning supports "detached kinds of knowing" (Papert & Harel, 1991, p. 11) where different worlds, identities, languages, and understandings can co-exist and intermingle in a particular space or moment.

The history of constructionism is deeply connected with Seymour Papert's invention of the LOGO turtle and programming language. Working with the LOGO turtle recast learning formal mathematics and science as an intuitive, embodied and deeply qualitative experience. As Sengupta, Dickes, and Farris (2018) remind us, Papert (1980) argued that working with the LOGO turtle involves getting to know the turtle, through exploring what it can or cannot do. Sengupta, Dickes, and Farris (2018) further remind us that Papert's vision was not to reduce all ideas to computational terms. Instead, he argued that the early experience with turtles is a good model of learning and could provide foundations for deeper disciplinary inquiry. That is, "… it is a good way to 'get to know' a subject by 'getting to know' its powerful ideas" (Papert, 1980, p. 138), in the same way that we get to know a person. Wilensky (1991) termed this form of knowing "*concretion,*" i.e., an experience through which an idea *becomes* concrete as learners engage in building multiple and complex relationships with it.

Several LOGO-like programming languages and modeling environments such as NetLogo (Wilensky, 1999), Scratch (Resnick et al., 2009), AgentSheets (Repenning, 1993), CTSiM (Basu & Biswas, 2016; Sengupta, Kinnebrew, Basu, Biswas, & Clark, 2013), and ViMAP (Sengupta et al., 2015) have since been developed to support children, and K-12 students engage in computational modeling and programming. Computational models developed in such languages are more generally known as agent-based models (ABMs). A key characteristic of ABMs is that the agent-level interactions, attributes, and behaviors are often body-syntonic (Papert, 1980)—i.e., they can be explained and understood through simple embodied actions such as moving and turning. This, in turn, supports even young children to model mathematical processes and complex scientific phenomena using such forms of computing (Danish, 2014; Dickes et al., 2016; Levy & Wilensky, 2008; Papert, 1980). In such environments, students learn by taking on the perspectives of the agents and different elements of the phenomena being modeled. This form of thinking has also been shown to be effective even for professional and established scientists. For example, Evelyn Fox Keller's biography of the biologist Barbara McClintock argues that thinking like the agent (e.g., a chromosome) enabled McClintock to make significant advances in her research on human genetic structures (Keller, 1983). Similarly, Ochs, Gonzales, and Jacoby (1996) identified that scientists' sensemaking in the domain of physical sciences involves such mental projections of the self into the phenomenon of inquiry.

18.2.2 *Indigenous Perspectives in Science Education: Re-Imagining Place and Discourse*

Indigenous scholars in science education have argued that central to decolonizing science education is the work of re-positioning human–environment relationships from Indigenous perspectives, with a particular focus on our relationship with land (Bang et al., 2014; Bang & Marin, 2015; Lowan-Trudeau, 2018; Montejo, 2010). In Indigenous perspectives, *place* is not merely a location where people live; people are spiritually connected to the place, and it represents their point of reference in their lives. In many Indigenous cultures, place is experienced spiritually as the Earth is seen as our Mother (Hatcher, Bartlett, Marshall, & Marshall, 2009). However, the systematic erasure of Indigenous histories and ways of knowing from settler-colonial societies through both genocide and symbolic violence (e.g., renaming Shikaakwa to Chicago) has resulted in the dominance of *settled expectations* of nature–culture relations (Bang & Marin, 2015; Bang, Warren, Rosebery, & Medin, 2012). Settled expectations are the set of assumptions, privileges, and benefits that accompany the status of being white (González Ponciano, 2013; Harris, 1995, p. 277). Science education is no exception. Such expectations in science learning "tends to structure learning in ways that restrict experienced and possible forms of agency, identities, and relations" (Bang & Marin, 2015, p. 531). Bang and Marin noted that the key manifestation of settler colonialism is in the form of settled expectations of how humans position themselves in relationship with land. They argued:

> The fundamental tenet of settler-colonial societies is the acquisition of land as property, followed by the establishment of settler lifeways as the normative benchmark from which to measure development. These are accomplished through: (1) erasure of Indigenous presence, (2) staged inheritance of indigeneity by Whites (Reardon & Tallbear, 2012) and (3) erasure of African descendants humanity through the structuration of slavery and resultant reduction to and control of black bodies (Wolfe, 2006; Bang & Marin, 2015, pp. 532)

Bang and Marin (2015) have noted that the separation of humans from the environment is a foundational tenet of Western scientific epistemology, and this leads to ontological assumptions of non-human actors as objects without agency, whereas human beings are viewed as intentional agents. They argued that both of these perspectives are inconsistent with many Indigenous ways of knowing and learning, where both human and non-human actors viewed as intentional and agentic actors in the world, and neither do humans occupy a "privileged" status that divests us of responsibility, humility, and reciprocity in terms of our relationships to the world (Cajete, 2000; Kawagley, 1993, 2006).

Snively and Corsiglia (2001) positioned the overlooking of Indigenous perspectives in Western sciences and science education as "systematic racism" (p. 28). They argued that the "spiritual base" of Indigenous perspectives coupled with their "universalist and relativist position towards nature and natural sciences" (p. 28) is largely regarded as dissonant from Western scientific perspectives. In Indigenous perspectives, the *spirit* holds a unifying ontological presence. It is the *spirit*—which

Kawagley, Norris-Tull, and Norris-Tull (1998) referred to as "the fifth element" (p. 79)—that connects our actions with our environment and our cultural and familial histories. Caring for the spirit is caring for sustainability, conservation, and important strategies to protect biodiversity and ecology. Rather than viewing the spirit as a mystical, mythical, or superstitious representation in folk culture, the spirit therefore signifies ecological and historical interconnectivity and links well-being with ecological sustainability. It makes explicit that well-being—which has recently received significant attention in the Western academy as fundamental to education and societal growth (e.g., Russell-Mayhew, Arthur, & Ewashen, 2007)—is deeply intertwined with ecological sustainability and caring for the environment. This in turn can be conceptualized as an explicit acknowledgment that the ecological system, as experienced by human beings, "exists within an entirely different cultural experience and set of values" (Johnson, 1992, p. 13).

In Indigenous perspectives, space and time are often viewed as both historical and cyclical (Snively & Corsiglia, 2001). The community's ongoing relationship with nature is historically grounded, because "cycles" of observation are made over "several lifetimes…in all seasons" in one place (Snively & Corsiglia, 2001, p. 18). In contrast, Bang and Marin (2015) argued that settled expectations of nature–culture relations result in structurally shaping time-space presence in ways that are reflected in forms of revised "memory traces," which are typically devoid of Indigenous ways of knowing. For example, Indigenous histories associated with different locations along the Shikaakwa river have been erased from public presence through place and land renaming, as well as through erasure of these histories from public K-12 science education (Bang et al., 2014). Bang's work provides a rich imagination of pedagogical approaches that can "desettle" such erasures by actively engaging Indigenous youth in "counter mapping" activities (Taylor & Hall, 2013) through nature walks in conversations with family members and elders. Central to this notion of counter mapping is questioning settler-colonial practices of naming places through Indigenous erasure, and explicitly focusing attention to Indigenous histories of places, which in turn involve engaging in both intentional acts of noticing and conversations about histories and imagined futures grounded in local Indigenous histories and narratives. Lowan-Trudeau (2018) also presented a similar perspective for re-imagining environmental education, by highlighting the importance of working in partnership with Indigenous elders and communities and incorporating local, historical (Indigenous) narratives that can help us reposition our relationships with land and place.

In re-positioning curricula from Indigenous perspectives, Hermes (2000) poignantly argued that curricula that foreground immersion in Indigenous cultures need to be based on practices and theories "that assume 'cultures' are living, that is, cultures are able to influence and be influenced without losing their substance, cohesion, and distinctiveness of being a 'culture'" (p. 389). Even in Native American schools, Hermes (2000) reported divisions between "academic" and "cultural" work, where academic classes are those derived from a Western European discipline and "cultural" referred to classes that focus on Indigenous topics. Keeping Indigenous cultural practices and ways of knowing separate and distinct from academic work reifies dichotomies between Western and Indigenous ways of knowing; furthermore, as Hermes, Bang, and Marin (2012) noted, it keeps discourse attached to academic disciplines "often disconnected from place, shared meaningful localities and the everyday lives of children and families (Gruenewald, 2003; Hawkins, 2004; Schleppegrell, 2004)." (Hermes et al., 2012; p. 389). Along similar lines, Takeuchi (2018) showed that even when informally learnt Japanese finger computing practices can help young children learn about academic mathematics, such practices remain hidden and disconnected from the visible classroom discourse. These historically grounded practices are not typically taught in the classroom, but learnt by the children through informal engagement with family elders at home.

18.3 Toward Some Potential Synergies

Several authors have highlighted that modeling complex systems, and computational modeling and programming, share epistemic and representational practices such as abstraction, simulation, problem-solving, define values, and others (DiSessa, 2000; Harel & Papert, 1990; Papert, 1980; Papert & Harel, 1991; Perkins & Simmons, 1988; Sengupta et al., 2013; Sengupta, Dickes, & Farris, 2018; Wing, 2006). Recently, Sengupta, Dickes, and Farris (2018) went further, arguing that "computing and computational thinking, should be viewed as discursive, perspectival, material and embodied experiences" (p. 3), involving both "epistemic and representational work" (p. 6). These authors repositioned computational thinking in STEM education using a set of *phenomenological* (Merleau-Ponty, 1962) lenses that foreground the qualitative experiences of learners through which code and computational abstractions become meaningful to them. In this perspective, learning computational abstractions is neither the goal nor the signature of such learning experiences; rather the goal is to provide a greater account of the richness of the learning experiences as fundamentally heterogeneous, extending far beyond code (Sengupta, Dickes, & Farris, 2018).

We also believe that constructionist learning environments can offer learners opportunities to experience complex disciplinary forms of knowing as *place-making*. That is, learners can position themselves as creators of disciplinary knowledge, and they do so by projecting themselves or their loved ones within the phenomena being modeled (e.g., see Farris & Sengupta, 2016). Learning with computational agents can also become a rich context for incorporating personal histories and desires, without necessarily displacing disciplinary learning goals (Resnick, Berg, & Eisenberg, 2000). Through such learning experiences, disciplines can be transformed into *lived-in* spaces (Nemirovsky, Tierney, & Wright, 1998). It is thus no surprise that constructionist learning using agent-based computational modeling foregrounds personal excursions, narratives, and aesthetic experiences, where learners' personal lives become deeply intertwined with computational abstractions (Farris & Sengupta, 2016; Resnick et al., 2000). It has also been shown to be effective in supporting collaboration, leveraging (rather than ignoring) intersubjectivity as key elements of the learning experience (Sengupta, Krishnan, et al., 2015).

We therefore believe that focusing on (a) complex systems as the *domain* of computational modeling, (b) agent-based modeling as the *form* of modeling, and (c) constructionism as the pedagogical approach, can offer us a potential synergistic milieu for Western and Indigenous perspectives. Our position builds on previous scholarship by Indigenous scholars such as Cajete (2000), Bang et al. (2018), Aikenhead and Michell (2011). For example, Cajete (2000) has poignantly positioned Indigenous science as being fundamentally relational in nature: "everything is related, that is, connected in dynamic, interactive, and mutually reciprocal relationships" (p. 75). The central objective of complex systems educators, even when grounded in Western perspectives, has been to support students develop a deep understanding of interrelationships between organisms and their environments through computational modeling (Danish, 2014; Wilensky & Resnick, 1999; and others). This is also consistent with Bang et al. (2018), who argued that complex systems theory is also well aligned with relational epistemologies in Indigenous science.

It is noteworthy that Indigenous scholars such as Snively and Corsiglia (2001) emphasized that children should learn "how science relates to action" (p. 42). Multimodality emphasizes action in various forms: perspectival noticing, embodied movements and modeling, creating physical and computational representations (Rosebery, Ogonowski, DiSchino, & Warren, 2010; Sengupta, Dickes, & Farris, 2018). Furthermore, computational modeling of complex systems involves creating and analyzing different forms of spatio-temporal patterns through spatial simulations and graphs (Dickes et al., 2016; Wilkerson-Jerde & Wilensky, 2015); and along similar lines, the Indigenous scholars, Hatcher and colleagues mentioned that some of the strengths of Indigenous perspectives are the focus on the observation of "spatio-temporal patterns" (Hatcher et al., 2009; p. 147). To this end, Aikenhead and Michell (2011) offered several recommendations for integrating Western and Indigenous perspectives in science education, including: a) valuing students' prior knowledge and experiences and building upon them, and b) using Indigenous practices such as demonstrations, Sharing Circles, and storytelling as integral parts of the pedagogy. Aikenhead and Michell (2011) also argued for integrating science with other topics such as social studies and Indigenous studies.

The goal of this chapter is to advance this line of work by illustrating how Eurocentric and Indigenous perspectives in science and science education can be brought together through re-imagining what agent-based computational modeling activities might look like for modeling complex systems, when grounded in a particular Indigenous perspective. In the next section, we introduce Grafemos, an activity system for modeling emergence that we have been co-designing alongside the Ixkoj Ajkem Council in Guatemala, and we also discuss how it has been shaped by Mayan fabric design and storytelling practices. We will then illustrate how working with Grafemos can become a context for putting some of the recommendations mentioned above into practice.

18.4 The Grafemos Activity System

Grafemos is a collaborative modeling activity that involves working with 4–6 people. The activity begins with the group selecting a specific scenario or a problem that is both relevant to their communities or lived experiences, that concerns both their individual experiences and a social issue (e.g., creosote contamination in Calgary; plastic contamination of drinking water in Xenacoj, Guatemala). Specifying the problem involves small group conversations about their concerns for, possible explanations of and emotions related to the topic, eventually establishing a common narrative within the small group that then becomes the phenomenon to be modeled. The next stage of the discussion involves identifying potential "characters"—i.e., agents—in the narrative, as well as their emotional states, relationships among them, and the geographical location and constraints. In partnership with the Ixkoj Ajkem Council, we have identified a few key categories that these "agents" belong to: humans; non-human animals and plants; seasons; time; duration; natural elements (e.g., water, wind, etc.); and emotions of agents. Simply put, each small group should have identified agents from each of these categories in their narratives and descriptions of the phenomena.

In the next stage, students design iconic representations of the different agents using a Grafemos meaning card, as shown in Fig. 18.1, and the Grafemos interaction card, as shown in Fig. 18.1b. This is a multi-step process. In order to create these cards, participants first need to use the Grafemos physical printing device, as well as consult and draw upon from the Grafemos library of motifs that has been co-designed with the Ixkoj Ajkem Council. Each motif has been designed based on traditional and historical motifs used by Mayan women weavers. Once the participants select, modify, and appropriate motifs for each agent, they are given physical shape using thick threads that are pasted on small circular pieces of fabric (see Fig. 18.2). These pieces are then wrapped, with a rubber band, around small cylinders with sponge, to form stamps. These imprints are then placed in the meaning card (as shown in Fig. 18.1), where the participants also explain their rationale for selecting each motif to represent specific agents. Participants then begin to fill out

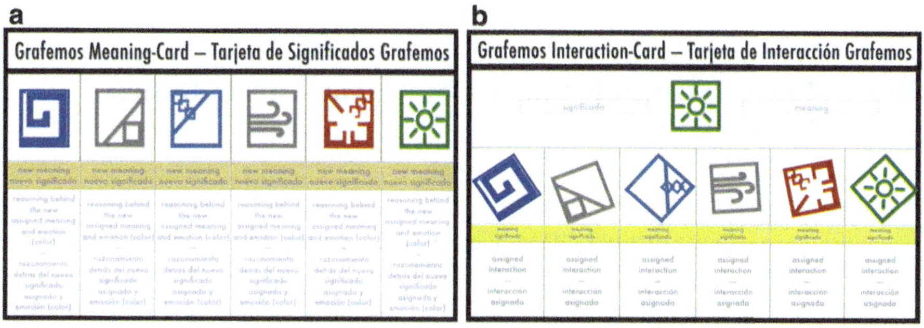

Fig. 18.1 (**a**) Grafemos meaning card. (**b**) Grafemos interaction card

Fig. 18.2 Grafemos motif-seals inked, and the Grafemos printing device

the interaction card, one for each agent, which explains what happens when the agent interacts with other agents in the system. These interactions could include, for example, change in states of motion (slowing down or speeding up or stopping), change in emotional states, change in quantity, etc.

Once the motif stamps, meaning and interaction cards are designed, the motif stamps can be inked and placed in slots of a dodecahedron (Fig. 18.2), which is the physical printing device used in the Grafemos modeling activity. Participants then roll the dodecahedron on a large paper on the floor, and begin to "print" their model, as the inked motifs get imprinted on the paper (Fig. 18.3). As the model is being "printed," the participants continue to collectively observe the emergent patterns in the form of the emerging trajectories. At the same time, they also engage in collectively discussing possible interpretations of the emergent patterns using the meaning and interaction cards.

18.5 How Grafemos Is Shaped by Mayan Fabric Design Practices

The conceptual design of this modeling environment was principally shaped by the weaving and design practices of Mayan textiles of Guatemala, supported by the Ixkoj Ajkem Community Council of Santo Domingo Xenacoj. The Mayan Textile Art is more than a canvas formed by means of a textile technology. There are also brocade symbols that carry oral narratives and legends, details that connect it with a

Fig. 18.3 Grafemos dodecahedron along with motifs stamps

place, born from the heart and history of its inhabitants, their ancestors, of love, and the need for conservation and sustainability of their land, their traditions, and their daily life. Each textile is also unique, represents a different point of view and meanings, tells a different story, and represents interconnectedness through its symbolic patterns (Lam-Herrera, 2011; Otzoy, 1992).

The design, making, and teaching of textiles are mostly a profession for women in the Mayan communities in Guatemala. According to the Ixkoj Ajkem Community Council of Santo Domingo Xenacoj, it is through this practice that the *souls of their grandmothers are kept alive by being transmitted from mothers to daughters in their textiles* (Ixkoj Ajkem Community Council of Santo Domingo Xenacoj, personal communication, 2018). These involve transmission of emotions, thus the perception of color as *bright* or *dull* represents the designer's personal feeling of happiness or sadness (conversation with García-García, Community Council member, 2016). Grafemos follows this color scheme, so participants use *bright* and *saturated* colors to express that a symbol is positive or happy; *dull* and *lusterless* colors to show a symbol's sadness or negative emotions.

Weaving is not just the interaction between the weaver, the materials, and the backstrap loom; for members of the Ixkoj Ajkem Community Council of Santo Domingo Xenacoj, it represents the thinking, psychology, creativity, and the cultural values behind it. It is an integral activity. The smallest unit of embroidering to form symbolic patterns is called the ***grain***. Each grain follows a designed complex sequence—including color, intention, narratives, locality, and perpetuation—to form images rich in history, tradition, and identity. Metaphorically, Grafemos' structure provides a learning space through group work, using small symbolic units that can form complex patterns to observe, in which the participants' identities play an essential role. Hence its name Grafemos, a combination of ***grapheme***, that is, the smallest unit of a writing system; and the Spanish suffix ***~mos***, that implies a conjoint action, an action that includes all of ***Us***.

In computational terms, each motif represents an agent (human, animal, plant, or object), seasons, time, natural elements, and such; further, Grafemos allows to add emotions to each agent. These categories follow the practices used by Mayan women weavers to represent their cultural historical narratives through the images embroidered in their textiles. It is also important to note that Mayan weaving practices also involve representation of narratives from multiple perspectives. This is also a point of alignment between Western and Indigenous epistemologies, because negotiating multiple perspectives (points of views) have been shown to be important for understanding complex systems (Danish, 2014; Dickes et al., 2016). In Grafemos, participants are encouraged to represent the same narrative from different perspectives, using motifs that also represent their emotional states. For example, a scenario of cats hunting squirrels, if the topic centers on cats, the cat will be happy each time he catches a squirrel because he gets more food for its kittens; however, if the model is observed from the squirrels' point of view, the cat is a negative or sad agent because each time a cat catches a squirrel it limits their survival, valuing emotional states.

The final stage in each Grafemos iteration is a Sharing Circle. Participants gather in a circle to observe the printed system and share their interpretations of the overall emergent picture, relevant experiences, discoveries, and insights. The practice of Sharing Circles (Hatcher et al., 2009) creates the ideal space for telling stories to examine and exercise problem-solving, provide even opportunities to everybody to talk and participate, as well as an opportunity to share multiple perspectives between members of the community (intergenerational groups). We believe that the Indigenous practices of storytelling and learning circles can serve a deeply reflexive means to learn about complex systems, especially creating contexts for reflective conversations about the *emergent* themes in the model.

18.6 Designing Together: Beginnings of a Journey

We have begun a partnership with the Mujeres Tejedoras Ixkoj Ajkem Council, who belong to the Maya Kaqchikel community of Santo Domingo Xenacoj. The first author (Marilú), currently a PhD student in Canada, was born and raised in Guatemala and has worked with the women weaver-designers in the Maya Kaqchikel community for over 10 years. Marilú visited the community for a few days in early 2018, and upon invitation of the elders of the community, worked with some elders and children for 5 days to discuss how they wanted to use and change Grafemos to model stories from their lives, as well as complex issues that they are faced with.

In partnering with the community at this stage of the project, our goal is to follow the criteria for *methodological Métissage* (Lowan-Trudeau, 2012). Methodological Métissage involves recognizing Indigenous practices and traditions such as storytelling and symbolizing from Indigenous perspectives as critical for teaching, research and the researchers, as well as working with the elders and members of the community during all stages of the research: conceptualizing the research questions and study design, conducting the study, and interpreting and analyzing data. In this chapter, we only present highlights from the initial discussions around Grafemos, including two Grafemos models that emerged from the discussions.

As mentioned earlier, working with Grafemos involves modeling the target phenomenon (or a story) by first selecting six individual elements or agents that are relevant to the target phenomenon or story, assigning an image to provide meaning to each element, and then creating different combinations of the elements to model interactions between the agents. Elders and children discussed the importance of plastic contamination as a key issue in their community. Further conversations with elders revealed that the deterioration of the quality of drinking water, a known effect of chlorinated plastic contamination even in Western science (Steffan et al., 1999), to be an important issue for the community.

Through intergenerational conversations, the community members identified particular objects (e.g., specific forms of plastic used in everyday life) and actions (e.g., throwing away plastic objects without care) as the "agents" and "actions" that lead to contamination. The community members also spatially represented a map of their community on the chart paper, where they used Grafemos to imprint their model selection (See Fig. 18.4). This map included relative locations of drainage tubes, the public school, the village market, and an overall geographical boundary of their community. As the group began to imprint their model, they explained that the emergence of contamination resulted from the use and discarding of the different elements of plastic which would accumulate in the drainage system. The group noted to Marilú that being able to work with elders and younger members on these issues would give them opportunities to think about both what each individual can do in their daily lives to help reduce contamination, as well as how they could come together as a community to address this issue. They reflected that working with Grafemos could help them visualize the overall (i.e., *emergent*) effect of their indi-

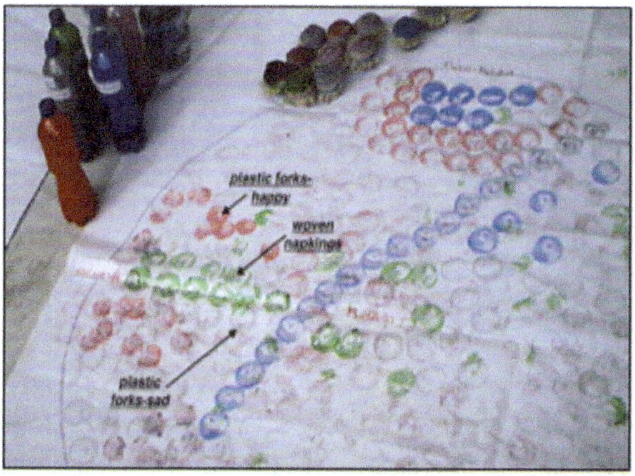

Fig. 18.4 Grafemos model of plastic contamination

vidual actions in terms of reducing contaminants in the community. We believe that this is similar to the notion of multi-level reasoning used by Western scholars of complex systems education, including some of our own previous work (e.g., Wilensky & Resnick, 1999; Dickes et al., 2016). The intergenerational group further argued that the introduction of handwoven fabric and napkins instead of plastic bags and containers could potentially reduce the contamination.

The group also chose a well-known tragic legend of Guatemala called "La Llorona" [The Weeping Woman] as a traditional narrative that they wanted to model using Grafemos (see Fig. 18.5). This is a story about a woman who regretted killing her children and cries for them eternally, showing her pain and sorrow while wandering on the streets of Guatemala. This legend is taught in schools in Guatemala. It is also part of the oral tradition between parents and children, often used by parents as a warning for children when they are misbehaving. Children in the community typically interpret the story as a horror story, as mentioned to us by the group. The initial Grafemos elements that the group selected were water, sons or daughters, a woman, her husband, and the action of hanging a person. The individual-level feeling of sadness was also included as the sixth element, represented by choice of colors. As the group began to model began empathizing with both the sadness of La Llorona (her helplessness) and the sadness of the families whose children she was killing. The group was taken aback by how "sad and empty" their pattern began to look as the modeling progressed. Sadness was thus represented as an individual-level feeling, as well as a *mid-level* (Levy & Wilensky, 2008) interaction, because it also emerged as La Llorona would interact with others, as was also discussed by the group. They even decided to stop representing the progressive elimination of the

Fig. 18.5 Grafemos model of "La Llorona" [the Weeping Woman]

different elements—a consequence of the interactions between La Llorona with others—because they were also feeling sad and uncomfortable with the overwhelming emergence of sadness in their model, evident very visibly by the decreasing number of agents and the colors in their model. The elders also observed that the younger members of their group had begun seeing the story as one of "deep sadness," as opposed to a funny horror story, as was their previous experience. Thinking *with* the agents, in this case, created an empathetic bond between the modelers and the actors in the story, which in turn resulted in changing the overall, emergent meaning of the story from horror to sadness.

In future visits, we hope to engage in deeper conversations with the community members to co-design ways of engaging in modeling with Grafemos that further integrate the traditional craft of fabric design (as noted by some community members), as well as co-designing analytic frameworks for looking at the discourse during modeling. In particular, we noted that elders in the community framed the modeling activities as an opportunity to engage in teaching and practicing the traditional, Indigenous language—Kaqchikel—with the younger members in group, suggesting that discursive practices that would help the group delve deeper into scientific discourse may also be shaped by the deepening of the groups' understanding and practice of Kaqchikel. This partnership can also help us identify more clearly synergies and relationships between the noticings of both individual-level actions and emergent outcomes by the community members, and commonly used frameworks for analyzing multi-level explanations in Western scholarship of complex systems (e.g., Dickes et al., 2016).

18.7 Conclusion and Discussions

Our work illustrates that Indigenous motifs and traditional forms of representation can serve as productive anchors for designing toolkits and languages for modeling complex systems. It also suggests that by positioning Indigenous communities as authoring partners, we can co-create opportunities for identifying critical socio-ecological issues within the community, as well as grounds for working together toward potential solutions. The work presented here is thus an important example of voicing *with*, rather than speaking for, Indigenous communities. We hope that this can provide a useful starting point for understanding the affordances of a different pedagogical approach for modeling complex systems, one that is grounded in both Indigenous and Western ways of knowing and representing complexity.

Our work, as we have mentioned earlier, is an attempt to put into practice important recommendations that have already been made by science education scholars. For example, Bang and colleagues have argued that complexity could be a place for unifying both Western and Indigenous perspectives (Bang et al., 2018), and Aikenhead and Michell (2011) argued for a more interdisciplinary approach for science education. They emphasize building on students' personal and cultural repertoires that they bring with them to the institutional settings, but are almost always left out of the curricular activities. To this end, co-designing *Grafemos*—both its material forms and modeling activities—in partnership with the Ixkoj Ajkem Council aligns the symbolic elements as well as the representational and discursive practices in computing with weaving practices in the Mayan community. It also attends to the critical problem of language practice and revitalization in Indigenous communities as noted by Hermes et al. (2012): As noted earlier, elders in the community framed the modeling activities as an opportunity to engage in teaching and practicing the traditional, Indigenous language—Kaqchikel with the younger members in group.

Finally, our work is an invitation to re-think the design of computational modeling languages and activities from a decolonized perspective. Even some Western scholars have noted problems inherent in much of educational computing that arise from an overtly *technocentric* approach (Papert, 1987). Technocentrism refers all questions about technology to the technology itself (Papert, 1987), rather than focusing on phenomenological perspectives that can help us understand learners' and teachers' experiences and enframings of the technology (Sengupta, Dickes & Farris, 2018). In a technocentric approach, learning computational modeling has become synonymous with regurgitation of a narrowly defined set of computational abstractions (Sengupta, Dickes, & Farris, 2018). Our work further extends this critique by challenging the very notion that computing, especially for K-12 students, should be viewed only through the Western lenses of computational thinking and abstractions (Wing, 2006) that is also largely individualistic (Ames, 2018). In contrast, we present an illustrative example where the traditional weaving practices of Mayan women can offer a community-based, decolonized approach for modeling various forms of emergent patterns, also inviting us to fundamentally reimagine what computing can look like beyond a technocentric image of digital technologies.

References

Aikenhead, G., & Michell, H. (2011). *Bridging cultures: Indigenous and scientific ways of knowing nature*. Don Mills, ON: Pearson Education.

Ames, M. G. (2018). Hackers, Computers, and Cooperation: A Critical History of Logo and Constructionist Learning. Proceedings of the ACM on Human-Computer Interaction, 2(CSCW), 18.

Bang, M., Curley, L., Kessel, A., Marin, A., Suzukovich, E. S., III, & Strack, G. (2014). Muskrat theories, tobacco in the streets, and living Chicago as indigenous land. *Environmental Education Research, 20*(1), 37–55.

Bang, M., & Marin, A. (2015). Nature–culture constructs in science learning: Human/non-human agency and intentionality. *Journal of Research in Science Teaching, 52*(4), 530–544.

Bang, M., Marin, A., & Medin, D. (2018). If indigenous peoples stand with the sciences, will scientists stand with us? *Daedalus, 147*(2), 148–159.

Bang, M., Warren, B., Rosebery, A. S., & Medin, D. (2012). Desettling expectations in science education. *Human Development, 55*(5–6), 302–318.

Bastien, B. (2003). The cultural practice of participatory transpersonal visions: An Indigenous perspective. *ReVision, 26*(2), 41–49.
Basu, S., & Biswas, G. (2016). *Providing adaptive scaffolds and measuring their effectiveness in open ended learning environments*. Singapore: International Society of the Learning Sciences.
Berland, M., Baker, R. S., & Blikstein, P. (2014). Educational data mining and learning analytics: Applications to constructionist research. *Technology, Knowledge and Learning, 19*(1–2), 205–220.
Cajete, G. (1994). *Look to the mountain: An ecology of indigenous education*. (p. 81301). Durango, CO: Kivaki Press.
Cajete, G. (2000). *Native science: Natural laws of interdependence*. Clear Light Pub.
Chi, M. T. (2005). Commonsense conceptions of emergent processes: Why some misconceptions are robust. *The Journal of the Learning Sciences, 14*(2), 161–199.
Danish, J. A. (2014). Applying an activity theory lens to designing instruction for learning about the structure, behavior, and function of a honeybee system. *Journal of the Learning Sciences, 23*(2), 100–148.
Danish, J. A., Peppler, K., Phelps, D., & Washington, D. (2011). Life in the hive: Supporting inquiry into complexity within the zone of proximal development. *Journal of Science Education and Technology, 20*(5), 454–467.
Dickes, A. C., Sengupta, P., Farris, A. V., & Basu, S. (2016). Development of Mechanistic Reasoning and Multilevel Explanations of Ecology in Third Grade Using Agent-Based Models. *Science Education, 100*(4), 734–776.
DiSessa, A. A. (2000). *Changing minds: Computers, learning, and literacy*. Cambridge, MA: MIT Press.
Farris, A. V., & Sengupta, P. (2016). Democratizing children's computation: Learning computational science as aesthetic experience. *Educational Theory, 66*(1–2), 279–296.
González Ponciano, J. R. (2013). The shumo challenge. White class privilege and the postrace, post-genocide alliances of cosmopolitanism from below. *War by other means: Aftermath in post-genocide Guatemala*, pp. 307–329.
Gruenewald, D. A. (2003). Foundations of place: A multidisciplinary framework for place-conscious education. *American Educational Research Journal, 40*(3), 619–654.
Harel, I. (1991). *Children designers: Interdisciplinary constructions for learning and knowing mathematics in a computer-rich school*. Westport, CT: Ablex Publishing.
Harel, I., & Papert, S. (1990). Software design as a learning environment. *Interactive Learning Environments, 1*, 1–32.
Harris, T. (Ed.). (1995). *Popular culture in England, c-1500–1850*. London, England: Macmillan International Higher Education.
Hatcher, A., Bartlett, C., Marshall, A., & Marshall, M. (2009). Two-eyed seeing in the classroom environment: Concepts, approaches, and challenges. *Canadian Journal of Science, Mathematics and Technology Education, 9*(3), 141–153.
Hawkins, M. R. (2004). *Language learning and teacher education: A sociocultural approach* (Vol. 48). Bristol, England: Multilingual Matters Ltd..
Hermes, M. (2000). The scientific method, Nintendo, and eagle feathers: Rethinking the meaning of "culture-based" curriculum at an Ojibwe tribal school. *International Journal of Qualitative Studies in Education, 13*(4), 387–400.
Hermes, M., Bang, M., & Marin, A. (2012). Designing indigenous language revitalization. *Harvard Educational Review, 82*(3), 381–402.
Holland, J. H. (2000). *Emergence: From chaos to order*. Oxford: OUP.
Johnson, M. (Ed.). (1992). *Lore: Capturing traditional environmental knowledge*. Ottawa, ON: IDRC.
Kafai, Y. B. (2006). Constructionism. In R. K. Sawyer (Ed.), *The Cambridge handbook of the learning sciences* (1st ed., pp. 35–46). New York, NY: Cambridge University Press.

Kafai, Y. B., & Burke, Q. (2013, March). The social turn in K-12 programming: moving from computational thinking to computational participation. In *Proceeding of the 44th ACM technical symposium on computer science education* (pp. 603–608). ACM.
Kawagley, A. O. (1993). *A Yupiaq world view: Implications for cultural, educational, and technological adaptation in a contemporary world* (Doctoral dissertation, University of British Columbia).
Kawagley, A. O. (2006). *A Yupiaq worldview: A pathway to ecology and spirit*. Long Grove, IL: Waveland Press.
Kawagley, A. O., Norris-Tull, D., & Norris-Tull, R. A. (1998). The indigenous worldview of Yupiaq culture: Its scientific nature and relevance to the practice and teaching of science. *Journal of Research in Science Teaching, 35*(2), 133–144.
Keller, E. F. (1983). *A feeling for the organism* (pp. 45–46). San Francisco, CA: Freeman.
Klopfer, E., Yoon, S., & Perry, J. (2005). Using palm technology in participatory simulations of complex systems: A new take on ubiquitous and accessible mobile computing. *Journal of Science Education and Technology, 14*(3), 285–298.
Klopfer, E., Yoon, S., & Um, T. (2005). Teaching complex dynamic systems to young students with StarLogo. *Journal of Computers in Mathematics and Science Teaching, 24*(2), 157–178.
Lam-Herrera, M. (2011). *Organización para la Puesta en Marcha de una Microempresa de Confección, Aplicada a la Asociación Ixil Para el Desarrollo Económico Y Social (AIDES), en Nebaj, Quiche* [Organization, Processes and Methods, for the implementation of a micro clothing, addressed to the Ixil Development Association (AIDES) in Acul, Nebaj, Quiché, Guatemala]. Retrieved February 12, 2011, from glifos.unis.edu.gt, http://glifos.unis.edu.gt/digital/tesis/2011/26958.pdf
Levy, S. T., & Wilensky, U. (2008). Inventing a "mid level" to make ends meet: Reasoning between the levels of complexity. *Cognition and Instruction, 26*(1), 1–47.
Lowan-Trudeau, G. (2012). Methodological métissage: An interpretive indigenous approach to environmental education research. *Canadian Journal of Environmental Education (CJEE), 17*, 113–130.
Lowan-Trudeau, G. (2018). From reticence to resistance: Understanding educators' engagement with indigenous environmental issues in Canada. *Environmental Education Research*, 1–13.
Merleau-Ponty, M. (1962). *Phenomenology of perception* (translated by C. Smith). London, England: Routledge & Kegan Paul.
Mitchell, M. (2009). *Complexity: A guided tour*. Oxford, England: Oxford University Press.
Montejo, V. D. (2010). *Maya intellectual renaissance: Identity, representation, and leadership*. Austin, TX: University of Texas Press.
Nemirovsky, R., Tierney, C., & Wright, T. (1998). Body motion and graphing. *Cognition and Instruction, 16*(2), 119–172.
Ochs, E., Gonzales, P., & Jacoby, S. (1996). "When I come down I'm in the domain state": Grammar and graphic representation in the interpretive activity of physicists. *Studies in Interactional SocioLinguistics, 13*, 328–369.
Otzoy, I. (1992). Identidad y trajes mayas [Maya Identity and Costumes]. *Mesoamérica, 13*(23), 95–112.
Papert, S. (1980). *Mindstorms: Children, computers, and powerful ideas*. New York, NY: Basic Books.
Papert, S. (1987). Information technology and education: Computer criticism vs. technocentric thinking. *Educational researcher, 16*(1), 22–30.
Papert, S., & Harel, I. (1991). Situating constructionism. *Construction, 36*(2), 1–11.
Perkins, D. N., & Simmons, R. (1988). Patterns of misunderstanding: An integrative model for science, math, and programming. *Review of Educational Research, 58*(3), 303–326.
Reardon, J., & TallBear, K. (2012). "Your DNA is our history" genomics, anthropology, and the construction of whiteness as property. *Current Anthropology, 53*(S5), S233–S245.

Repenning, A. (1993, May). Agentsheets: a tool for building domain-oriented visual programming environments. In *Proceedings of the INTERACT'93 and CHI'93 Conference on Human Factors in Computing Systems* (pp. 142–143). ACM.

Resnick, M. (1994). *Turtles, termites and traffic jams: Explorations in massively parallel microworlds*. Cambridge, MA: MIT Press.

Resnick, M., Berg, R., & Eisenberg, M. (2000). Beyond black boxes: Bringing transparency and aesthetics back to scientific investigation. *The Journal of the Learning Sciences, 9*(1), 7–30.

Resnick, M., Maloney, J., Monroy-Hernández, A., Rusk, N., Eastmond, E., Brennan, K., … Kafai, Y. (2009). Scratch: Programming for all. *Communications of the ACM, 52*(11), 60–67.

Resnick, M., & Wilensky, U. (1998). Diving into complexity: Developing probabilistic decentralized thinking through role-playing activities. *The Journal of the Learning Sciences, 7*(2), 153–172.

Rosebery, A. S., Ogonowski, M., DiSchino, M., & Warren, B. (2010). "The coat traps all your body heat": Heterogeneity as fundamental to learning. *The Journal of the Learning Sciences, 19*(3), 322–357.

Russell-Mayhew, S., Arthur, N., & Ewashen, C. (2007). Targeting students, teachers and parents in a wellness-based prevention program in schools. *Eating Disorders, 15*(2), 159–181.

Schleppegrell, M. J. (2004). *The language of schooling: A functional linguistics perspective*. Mahwah, NJ: Lawrence Erlbaum.

Sengupta, P., Brown, B., Rushton, K., & Shanahan, M. C. (2018). Reframing coding as "Mathematization" in the K–12 classroom: Views from teacher professional learning. *Alberta Science Education Journal, 45*(2), 28–36.

Sengupta, P., Dickes, A., & Farris, A. V. (2018). Toward a phenomenology of computational thinking in K-12 STEM. In M. S. Khine (Ed.), *Computational thinking in STEM discipline: Foundations and research highlights*. Cham, Switzerland: Springer.

Sengupta, P., Dickes, A., Farris, A. V., Karan, A., Martin, D., & Wright, M. (2015). Programming in K-12 science classrooms. *Communications of the ACM, 58*(11), 33–35.

Sengupta, P., Kinnebrew, J. S., Basu, S., Biswas, G., & Clark, D. (2013). Integrating computational thinking with K-12 science education using agent-based computation: A theoretical framework. *Education and Information Technologies, 18*(2), 351–380.

Sengupta, P., Krishnan, G., Wright, M., & Ghassoul, C. (2015). Mathematical machines & integrated STEM: An intersubjective constructionist approach. *Communications in Computer and Information Science, 510*, 272–288.

Sengupta, P., & Wilensky, U. (2009). Learning electricity with NIELS: Thinking with electrons and thinking in levels. *International Journal of Computers for Mathematical Learning, 14*(1), 21–50.

Sengupta, P., & Wilensky, U. (2011). Lowering the learning threshold: Multi-agent-based models and learning electricity. In *Models and Modeling* (pp. 141–171). Netherlands: Springer.

Snively, G., & Corsiglia, J. (2001). Discovering indigenous science: Implications for science education. *Science Education, 85*(1), 6–34.

Steffan, R. J., Sperry, K. L., Walsh, M. T., Vainberg, S., & Condee, C. W. (1999). Field-scale evaluation of in situ bioaugmentation for remediation of chlorinated solvents in groundwater. *Environmental Science & Technology*, 33(16), 2771–2781.

Takeuchi, M. A. (2018). Power and identity in immigrant parents' involvement in early years mathematics learning. *Educational Studies in Mathematics, 97*(1), 39–53. https://doi.org/10.1007/s10649-017-9781-4

Taylor, K. H., & Hall, R. (2013). Counter-mapping the neighborhood on bicycles: Mobilizing youth to reimagine the city. *Technology, Knowledge and Learning, 18*(1–2), 65–93.

Wilensky, U. (1991). *Abstract meditations on the concrete and concrete implications for mathematics education*. Cambridge, MA: Epistemology and Learning Group, MIT Media Laboratory.

Wilensky, U. (1999). Center for connected learning and computer-based modeling. In *Netlogo*. Evanston, IL: Northwestern University.

Wilensky, U., & Reisman, K. (2006). Thinking like a wolf, a sheep, or a firefly: Learning biology through constructing and testing computational theories—An embodied modeling approach. *Cognition and Instruction, 24*(2), 171–209.
Wilensky, U., & Resnick, M. (1999). Thinking in levels: A dynamic systems approach to making sense of the world. *Journal of Science Education and Technology, 8*(1), 3–19.
Wilkerson-Jerde, M. H., & Wilensky, U. J. (2015). Patterns, probabilities, and people: Making sense of quantitative change in complex systems. *Journal of the Learning Sciences, 24*(2), 204–251.
Wing, J. M. (2006). Computational thinking. *Communications of the ACM, 49*(3), 33–35. Chicago.
Wolfe, P. (2006). Settler colonialism and the elimination of the native. *Journal of Genocide Research, 8*(4), 387–409.

Part IV
Reflections

Chapter 19
Engaging with Complexities and Imaging Possibilities Across the Boundaries of STEM

Shakhnoza Kayumova

Abstract The emphasis on STEM education in public debates and policies often centers on arguments for improving academic proficiency and workforce readiness. Going beyond such instrumentalist calls, the chapters in this edited collection present critical perspectives on design and epistemology in STEM education research and practice by highlighting the inherent ambiguities, tensions, and challenges. This essay explores several common threads woven throughout the book, which include the complexities and opportunities associated with research, teaching, studying, and designing STEM learning environments, as well as the productive possibilities, from which learning scientists and the educational community can (re)imagine and (re)configure the emerging field of STEM education.

Keywords STEM education · Power · Equity · Justice · Epistemological pluralism · Complexity

I am grateful for this opportunity to participate in a discussion related to STEM education and learning. We live in an era where the emphasis on improving academic proficiency and broadening participation in STEM, the acronym for "science, technology, engineering, and mathematics," has become one of the "hot" topics that occupy the public debate, as well as the discourses of educational policy. In spite of what seems like an "obsession" to dramatically increase STEM proficient citizens, the papers in this edited collection present critical perspectives about some of the ambiguities, tensions, and challenges that come with the current STEM initiatives. In doing so, the studies also provide a vision for possibilities from which learning scientists and the educational community can (re)imagine and (re)configure this

S. Kayumova (✉)
Department of STEM Education and Teacher Development, University of Massachusetts, Dartmouth, MA, USA
e-mail: skayumova@umassd.edu

P. Sengupta et al. (eds.), *Critical, Transdisciplinary and Embodied Approaches in STEM Education*, Advances in STEM Education,
https://doi.org/10.1007/978-3-030-29489-2_19

emerging field. I noticed several common threads woven throughout the book, which include the complexities and opportunities associated with research, teaching, studying, and designing STEM learning environments. My goal in writing this chapter is to engage with these ideas, and also provide my views and braid my own "critical and contextual struggle[s]" back into this text (Fine, 1994 p. 71).

What does an integrated STEM approach entail, and how do we teach it? Davis, Chandra, and Bellochi's (Chap. 1) paper raises important issues related to complexities inherited within the integrated STEM approach to learning, and the kinds of pedagogical challenges it poses on teachers and teacher candidates. As it stands, STEM is an interdisciplinary as well as a transdisciplinary field, constituted of unique epistemic domains, practices, values, and norms (Shanahan, Burke & Francis, 2016). Situating their research within initial teacher education programs, Davis and colleagues propose the notion of epistemic fluency to address the pedagogical demands put on teacher candidates as it relates to the integrated STEM approach to learning. Much of my own work with teachers and teacher candidates resonates with the pedagogical challenges raised by Davis, Chandra, and Bellochi. However, in the context of the United States where I am situated, I also cannot help to notice how, both in research and public discourse, teachers are yet again positioned in deficit terms, in which their knowledge, practices, and beliefs need to be developed in congruence with the new STEM initiatives and reforms (Kayumova & Tippins, 2018). In my experience, this has been especially the case for the early childhood, elementary, and middle grade teachers who are constantly disempowered by deficit discourses. As described by some other contributors to this edited book, calls for urgency in STEM education are premised on instrumentalist and neoliberal notions of autonomous and economically driven subjects (see Sengupta et al. (Chap. 1) for a review). STEM integration in public education is thus situated within the discourse of national global competitiveness. This reminds me to be cognizant of what Britzman (2012) called "a contest between what the technological society demands of its teachers and the personal visions of those teachers," and how "often repressed notions of what is desirable and what ought to be" makes teachers feel as though they are "being watched, and viewing the self through the eyes of others" (p. ix).

As I engage with these ideas, I have to keep reminding myself of my own power and privilege when I work with teachers and teachers' candidates, and make sure that I do not contribute to epistemic (knowing) and ontic (being) injustices. As Bobis (Chap. 12) argued, the integrated STEM is not only a new way of thinking and knowing (epistemic) that teachers are expected to grapple with, but it is also a new way of teaching (being). Bobis' research emphasizes the importance of providing teachers and teachers' candidates with ample opportunities to enact their knowledge and ways of knowing to better prepare them to respond to all the complexities of STEM teaching. There are a number of papers in this volume that attempt to provide a variety of theoretical positions and empirical evidence on how to support teachers in integrating STEM disciplinary practices and also advance new models of integrated STEM pedagogy. For instance, Kim, Rasporich, and Gupta's (Chap. 3) research demonstrates how a teacher, Stefan, enacted STEM integration with eighth-grade students by supporting students in planning and building a sustainable village in a virtual fantasy world. The authors discuss how Stefan's teaching practices helped students create sustainable communities in the Minecraft environment as they interacted with their imagined land and characters. The findings demonstrate how teaching practices situated in humanistic and *aesthetic* approaches (Dewey, 1934; Farris & Sengupta, 2016) can help students and teachers not only to engage in sophisticated STEM learning, but also to engage in topics regarding equity, societal norms, values, and practices. I find that Kim, Rasporich, and Gupta's research provides an excellent example of how STEM learning cannot be reduced to only disciplinary knowledge and skills, but must also include axiological orientations for students and teachers. To this end, Alonzo-Yanez and colleagues (Chap. 2), also provide valuable insights about why axiology is as crucial as epistemology and ontology by highlighting the importance of focusing on critical sustainability as a context for STEM integration (I will return to the relationship between axiology, epistemology and ontology later).

Another good example is presented by Paré, Sengupta, and colleagues (Chap. 16) who present a case how STEM could be also an environment for supporting deeply personal and yet playful experiences of inquiry about gender and sexuality. The authors present a re-imagination of how gender and sexuality can be experienced using innovations in STEM such as 3D sculpting and Virtual Reality through engaging in deep conversations and joint artifact creation in VR with close friends. An implication of this work is that the authors showcase how STEM can also potentially become a safe pedagogical space for students who are marginalized due to their gender and sexual identities, particularly in pedagogical contexts.

The chapters reviewed so far show that teaching practices do not happen in a vacuum; they are dynamically constituted with students and learning – a worlds in themselves that are quite heterogeneous. This implies that research and evidence related to student learning has to be an integral part of teacher preparation. For that purpose, I find studies by Lee (Chap. 11), Rahm (Chap. 14), Takeuchi and Dadkhahfard (Chap. 10)—to have important insights about the nature of STEM learning. These studies demonstrate how STEM practices require the sophisticated deployment of multiple communicative modalities and systems of representation (e.g., see Lee, Chap. 11). As a collective, these studies underline the importance of supporting students' learning by making their learning visible in ways that go beyond the use of privileged linguistic forms and promote the use of students' bodies for both expressivity and meaning-making. For instance, Lee's (Chap. 11) study underlines the importance of providing students with direct bodily learning opportunities to better support student's STEM participation with "complex forms of expression." Lee critiques the current overemphasis and privilege given to language as one of the primary forms of communication, as evidence for learning or knowing.

A similar critique is provided by Takeuchi and Dadkhahfard (Chap. 10), and their critical approach allows researchers to examine how students' bodies become implicated in relations of power. As Foucault (1982) said, the body becomes "directly involved in a political field: power relations have an immediate hold upon it; they invest it; mark it; train it; torture it; force it to carry out tasks; to perform ceremonies; to emit signs" (p. 25). In Takeuchi and Dadkhahfard's study, researchers attend to power and privilege issues in learning, which allows them to account for how minority students' informal knowledge of mathematics is embodied in their bodies, despite being subjugated within the formal spaces of learning, which also carry cultural and historical potentialities that can help desettle privileged ways of knowing. These expressive and productive dimensions of body in learning processes are also discussed by Rahm (Chap. 14). Rahm's analysis of learning within informal spaces goes beyond multimodal learning to show how body and affect imbued within learning spaces allow students to mobilize their agencies as they connect their learning to their everyday lives and geographies. The authors of both studies, Rahm, Takeuchi, and Dadkhahfard, effectively raise concerns about how the current STEM initiatives driven by functionalist and instrumentalist discourses reduce bodies and the value of learning to economic rationalities.

Similar critiques are provided by O'Neill (Chap. 6) and Dickes and Farris (Chap. 7) in relation to the current computational thinking movement. O'Neill (Chap. 6) argues that driven by functionalist rationality and corporate interests, the current hype with computational thinking, spearheaded by Kids Code Movement, lacks an intellectual foundation and evidence in relation to its claims about long-term benefits. This reminds me of Stephen Ball's (1990) argument that the field of education is not only subject to social, political, and economic discourses, but also part of "the selective dissemination" and "social appropriation of discourse" (p. 3). Ball's research showed how in the 1970s, scholarship sociology of education "informed and reinforced the 'problem of working-class underachievement'" (p. 4). Ball states that the educational studies and research findings of the time constructed working-class families as underachieving and represented them as being deprived of social, cultural, and epistemic capital. O'Neill's historical analysis also showcases how arguments about the importance of teaching children coding and programming are being constructed by social, political, and economic discourses and corporate values and visions of the present age. One such structuring has to do with instrumentalization of coding and programming and the need for STEM workers with these skills. After all, STEM is "a very human activity whose focus of interest and theoretical dispositions in any historical period were, and are, very much a part of and not apart from the dominant cultural and political issues of the day" (Lemke, 2001, p. 298). Gupta, Turpen, Philip, and Elby (Chap. 13) provide further context to consider how entangled the social, technical, and political are. By drawing on *social constructivist* epistemology, the authors showcase how society and technology are inextricably intertwined. The authors call attention to "considering the short- and long-term, intended and unintended effects, direct stakeholders, and those affected indirectly."

Based on these readings, it seems to me that for an emerging field, we might be acting too quick and overly enthusiastic to adopt, develop, and design new ecologies and approaches for learning by approximating epistemic practices of S-T-E-M disciplines. Partly, these approximations are made possible due to the representational affordances of computational tools and current technologies, which are transforming not only how knowledge can be constructed and expressed, but also understood within and across fields. However, despite the advances in technologies and representational capabilities of our current tools, there seems to be uncertainty about how to characterize and account for diverse ways of knowing and learning when engaging with representational affordances of computational tools. Moreover, "neglecting the political dimension[s]" of STEM learning and assuming neutrality behind practices, materialities, representations, and tools in the construction of students' knowledge(s), skills, and their relationships to these disciplines, means presupposing that a unanimous and ahistorical reality is "out there" to be represented; and such assumptions also dismiss the interpretive, sociocultural, and political nature of learning and practice. This reminds me of ethnographies of science (see Knorr-Cetina, 1983, 2013) and feminist studies in sciences (Haraway, 1988). These scholars systematically examined scientific practices, thereby exposing the socially constructed nature of science (Knorr-Cetina, 1983, 2013; Haraway, 1988) and also called for the recognition of gendered, raced, and classed history of STEM, emphasizing that knowledge-making is an inherently value-laden practice (Haraway, 1988). In relation to the history of science, as Ruth Hubbard (1990) cogently described,

> Science is both a social activity and the kind of knowledge about nature that this activity produces. The people who "do" science and so produce that knowledge are called scientists. Because of the time and place in which modern Western science—and that is what people nowadays mean by science—developed, scientists predominantly have been men. As this kind of knowledge and its applications have expanded into a dominant world view over the past two or three centuries, men have become so identified with all phases of it—the activity, product, application, and world outlook—that science as both product and activity has come to be thought of as essentially masculine. (p. 41)

The history of STEM is not so much different from science. Given the historical legacy of present inequities in STEM (see Das and Adams' Chap. 15), as well as the overtly colonized form of scholarship in STEM education (see Lam-Herrera, Ixoqui Ajkem Council and Sengupta's chapter (Chap. 17)), I wonder if we are ready to really and genuinely attend to the epistemic heterogeneity in the sense that Farris and Sengupta (2016) positioned Aesthetic Experience (Dewey, 1934) as fundamental to learning STEM – i.e., "seeing more" rather than merely "seeing as" – and thus reifying epistemic, systemic and historical injustices in STEM education. For instance, decades of research showed that the issue of underrepresentation in STEM fields is directly connected with the ways in which culture of science (which includes tools and practices) has maintained two forms of injustices – epistemic (systems of knowing) and ontic (categorizations of being) (Haraway, 1988) – often by failing to value, recognize, and legitimize socially and culturally diverse ways of being, doing, knowing, and experiencing the world, and by conceiving this heterogeneity in deficit terms. Several chapters in this book extend this body of work and provide more in-depth critiques and empirical analysis (see Das & Adams, Chap. 15; Gupta et al., Chap. 13, and Rahm Chap. 14). The argument is that many of the conventional and taken-for-granted knowledge-making practices in STEM education/research are rooted in histories of dominance, which continue to manifest in contemporary forms of representational and epistemic hierarchies (for more see Kayumova, Zhang, & Scantlebury, 2018). For many decades, various groups, such as people of color, women, and indigenous communities, have questioned the presumed neutrality of tools and practices used in STEM, and whose ways of being and knowing are legitimized in understanding and defining the world (see Harding, 1990). Black, Chicana, Indigenous, and Aboriginal feminist scholars, for example, have emphasized the entangled nature of being, doing, knowing, and living in the world, highlighting the inseparability of education, learning, and pedagogy from a researcher's own epistemological, ontological, methodological, and ethical assumptions (see Tuck, McKenzie, & McCoy, 2014).

These concerns also resonate with increasing calls for epistemologically pluralistic views of STEM education and research practices: ones that go beyond, what Bang and Marin (2015) call, "settled expectations" of Western and Eurocentric dichotomous/binary thinking of the nature-culture, subject-object, self-other divide, to those that value subject–subject relations, partnering, with equal right to exist and respond (Barad, 2003; Bang & Marin, 2015; Kayumova, McGuire, & Cardello, 2019; see also Lam-Herrera, Ixoqui Ajkem Council and Sengupta's chapter (Chap. 17)). This also speaks to previous research by Black, Chicana, Indigenous, and Aboriginal feminist scholars who have called attention to the entangled nature of being, doing, knowing, and living in the world, and have also documented how education and pedagogy do not stand apart from epistemological, ontological, and methodological assumptions and visions of researchers (see for example Delgado-Bernal & Elenes, 2011). As Ladson-Billings (2000) argues, although "there are well-developed systems of knowledge, or epistemologies, that stands in contrast to the dominant Euro-American epistemology" (p. 258), they are rarely, if at all, recognized, validated, or carefully attended within conventional knowledge-making practices of research on learning or curriculum development, resulting in inequitable educational practices and experiences for children from historically minoritized communities.

The collection of papers in this edited volume prompts us to examine the ways in which STEM practices, culture, tools, and materialities (such representations) get entangled with students' learning, identities, and subjectivities, provoking us to rethink taken-for-granted assumptions in our knowledge-making practices. They urge us to tread carefully and humbly on the entangled and evolving tapestry of human understanding that promotes epistemic and ontic pluralism and heterogeneity. Maxine Greene (1997) said that imagination and possibility is "a matter of awakening and empowering … to name, reflect, to imagine, and to act … matter of enabling … to remain in touch with dread and desire" (p. 2). It is imagination, Green argued, that allows us to "think of things as if they could be otherwise; it is the capacity that allows a looking through the windows of the actual towards alternative realities" (p. 2). The papers in this edited volume allowed me to see alternative realities and imagine possibilities, within the complexities of STEM field we all are in.

References

Bang, M., & Marin, A. (2015). Nature–culture constructs in science learning: Human/non-human agency and intentionality. *Journal of Research in Science Teaching, 52*(4), 530–544.

Barad, K. (2003). Posthumanist performativity: Toward an understanding of how matter comes to matter. *Signs: Journal of Women in Culture and Society, 28*(3), 801–831.

Bernal, D. D., & Elenes, C. A. (2011). 4 Chicana Feminist Theorizing. *Chicano school failure and success: Past, present, and future*, 99.

Britzman, D. P. (2012). *Practice makes practice: A critical study of learning to teach*. Suny Press.

Dewey, J. (1934). *Art As Experience*. New York: Minton, Balch, and Company.

Farris, A. V., & Sengupta, P. (2016). Democratizing children's computation: Learning computational science as aesthetic experience. *Educational Theory, 66*(1–2), 279–296.

Fine, M. (1994). *Working the hyphens. Handbook of qualitative research*. Thousand Oaks: Sage.

Foucault, M. (1982). The subject and power. *Critical Inquiry, 8*(4), 777–795.

Greene, M. (1997). Teaching as possibility: A light in dark times. *Journal of Pedagogy, Pluralism and Practice, 1*(1), 1–10. https://doi.org/10.1057/9780230100893_8

Haraway, D. (1988). Situated knowledges: The science question in feminism and the privilege of partial perspective. *Feminist Studies, 14*(3), 575–599.

Harding, S. (1990). Starting thought from women's lives: Eight resources for maximizing objectivity. *Journal of Social Philosophy, 21*(2–3), 140–149.

Hubbard, R. (1990). *The politics of women's biology*. Rutgers University Press.

Kayumova, S., & Tippins, D. (2018). Obsessed with Accountability? Science Teachers Under the Microscope. In Lynn Bryan & Kenneth Tobin (Eds). Thirteen Questions in Science Education. Peter Lang: New York, NY. https://doi.org/10.3726/b11305

Kayumova, S., McGuire, C. J., & Cardello, S. (2019). From empowerment to response-ability: rethinking socio-spatial, environmental justice, and nature-culture binaries in the context of STEM education. *Cultural Studies of Science Education, 14*(3), 257–270.

Kayumova, S., Zhang, W., & Scantlebury, K. (2018). Displacing and disrupting colonizing knowledge-making-practices in science education: Power of graphic-textual illustrations. *Canadian Journal of Science, Mathematics and Technology Education, 18*(3), 257–270.

Knorr-Cetina, K. D. (1983). *The ethnographic study of scientific work: Towards a constructivist interpretation of science* (pp. 115–140).

Knorr-Cetina, K. D. (2013). The manufacture of knowledge: An essay on the constructivist and contextual nature of science. Elsevier.

Ladson-Billings, G. (2000). Fighting for our lives: Preparing teachers to teach African American students. *Journal of Teacher Education, 51*(3), 206–214.

Lemke, J. L. (2001). Articulating communities: Sociocultural perspectives on science education. *Journal of Research in Science Teaching, 38*(3), 296–316.

Shanahan, M. C., Carol-Ann Burke, L. E., & Francis, K. (2016). Using a Boundary Object Perspective to Reconsider the Meaning of STEM in a Canadian Context. *Canadian Journal of Science Mathematics and Technology Education, 16*, 129–139.

Tuck, E., McKenzie, M., & McCoy, K. (2014). Land education: Indigenous, post-colonial, and decolonizing perspectives on place and environmental education research. *Environmental Education Research, 20*(1), 1–23. https://doi.org/10.1080/13504622.2013.877708

References

Index

P. Sengupta et al. (eds.), *Critical, Transdisciplinary and Embodied Approaches in STEM Education*, Advances in STEM Education,
https://doi.org/10.1007/978-3-030-29489-2

The manufacturer's authorised representative in the EU is Springer Nature Customer Service Centre GmbH, Europaplatz 3, 69115 Heidelberg, Germany. If you have any concerns regarding our products, please contact ProductSafety@springernature.com

Printed and bound by CPI Group (UK) Ltd, Croydon, CR0 4YY
22/09/2026
02221048-0001